CONTINUUM
THEORY

PURE AND APPLIED MATHEMATICS

A Program of Monographs, Textbooks, and Lecture Notes

EXECUTIVE EDITORS

Earl J. Taft
Rutgers University
New Brunswick, New Jersey

Zuhair Nashed
University of Delaware
Newark, Delaware

CHAIRMEN OF THE EDITORIAL BOARD

S. Kobayashi
University of California, Berkeley
Berkeley, California

Edwin Hewitt
University of Washington
Seattle, Washington

EDITORIAL BOARD

M. S. Baouendi
University of California, San Diego

Donald Passman
University of Wisconsin–Madison

Jack K. Hale
Georgia Institute of Technology

Fred S. Roberts
Rutgers University

Marvin Marcus
University of California,
Santa Barbara

Gian-Carlo Rota
Massachusetts Institute of
Technology

W. S. Massey
Yale University

David L. Russell
Virginia Polytechnic Institute
and State University

Leopoldo Nachbin
Centro Brasileiro de Pesquisas Físicas
and University of Rochester

Jane Cronin Scanlon
Rutgers University

Anil Nerode
Cornell University

Walter Schempp
Universität Siegen

Mark Teply
University of Wisconsin–Milwaukee

MONOGRAPHS AND TEXTBOOKS IN
PURE AND APPLIED MATHEMATICS

Additional Volumes in Preparation

CONTINUUM THEORY

An Introduction

Sam B. Nadler, Jr.

West Virginia University
Morgantown, West Virginia

Marcel Dekker, Inc.　　　　New York • Basel • Hong Kong

Library of Congress Cataloging-in-Publication Data

Nadler, Jr., Sam B.
    Continuum theory : an introduction / Sam B. Nadler, Jr.
        p.   cm.
    Includes bibliographical references and index.
    ISBN 0-8247-8659-9
    1. Continuum (Mathematics)   I. Title.
QA611.28.N33   1992
514'.32--dc20   92-1338
                                                                            CIP

This book is printed on acid-free paper.

MARCEL DEKKER, INC.
270 Madison Avenue, New York, New York 10016

Current printing (last digit):
10 9 8 7 6 5 4 3 2 1

PRINTED IN THE UNITED STATES OF AMERICA

To my wife, Elsa, my daughter, Annette,
and my two sons, Bill and David

Preface

Most of the material in this book is in the metric setting. Though many notions are defined for general topological spaces, almost all results are stated and proved for the metric case.

The book consists of thirteen chapters and is divided into two parts. Chapters I–VII comprise the first part and deal with the general structure of continua. Chapters I–III give some general construction techniques; Chapters IV–VI are concerned with the global analysis of continua; Chapter VII gives a general method for constructing continuous functions—this method is used a number of times. Chapters VIII–XIII comprise the second part and are concerned with specific types of continua and maps. Chapters VIII–X are about Peano continua; Chapters XI and XII examine two other important special classes of continua; Chapter XIII investigates some special types of maps. There are several threads that unify various aspects of the material, but the most pervasive is the nested intersection technique introduced in Chapter I. It is used throughout the book to construct continua and maps and to prove theorems. Indecomposable continua and hereditarily indecomposable continua are constructed early (Chapter I) and are investigated and discussed in various places.

In view of the vast literature on continuum theory, I could cover only a limited number of topics, and I could not give comprehensive treatment of any of them. Thus, for example, I do not discuss cyclic element theory, flows, periodic points of maps, shape theory (except briefly), or a number of special types of maps and particular continua of interest. I do not give a systematic treatment of plane continua, homogeneous continua, the fixed point property, curve classification, hyperspaces of sets, or homotopy techniques (though all these topics are examined in various places as a natural consequence of other developments).

The book presents a mixture of classical and modern ideas and techniques. I hope that readers will find it an appropriate introduction that gives the uninitiated a foundation on which to build an understanding of continuum theory.

Only a minimal knowledge of basic concepts from topology is assumed. The reader should be familiar with the elementary properties of compactness, connectedness, continuous functions, and the topology of metric spaces. More specifically, the reader should know that, in metric spaces, compactness (meaning every open cover has a finite subcover) is equivalent to every infinite set (or every sequence) having a limit point (or a convergent subsequence, respectively). The reader should know basic facts about unions and closures of connected sets (i.e., the results on p. 132 of Kuratowski's *Topology*, Vol. II) and should be familiar with the fundamental properties of continuous functions, including the Tietze extension theorem and its application to neighborhood extensions of maps into spheres. The reader should know that countable products of metric spaces are metrizable and that such products preserve compactness and connectedness. Finally, the reader should know the Baire theorem (p. 9 of Kuratowski's *Topology*, Vol. II).

Therefore, those who have had the equivalent of a standard one-semester topology course should be able to read and understand the book. Appropriate references are provided whenever any background material is used.

There are many exercises at the end of each chapter; they are an integral part of the book. They are not designed to test the memory of specific results—indeed, most exercises contain hints referencing the particular results to be used. A number of the hints are informal sketches of proofs or solutions. Some exercises are mainly a discussion of ideas in the chapter and the relationship of these ideas to material in other chapters or in the literature. Other exercises clarify minor points or give motivation that would be a distraction if included in the main body of the chapter. All exercises can be done with the material given—in particular, no solution depends on ideas or techniques from other areas of mathematics.

Instructors who wish to use this book in a one-semester course should cover the first eight chapters (moving fairly quickly through Chapters I and III) and some of each of Chapters IX and X. When the book is used in a one-year course, the entire book can be covered fairly easily. However, the instructor may want to skip sections 4 and 5 of Chapter XI and thereby make Sorgenfrey's theorem "user friendly" for Chapter XII.

I express my gratitude to Dr. James H. Lightbourne III, Chairman of the Department of Mathematics at West Virginia University, for providing some departmental resources and, most importantly, for his repeated encouragement during the writing of the book. I also thank Henry W. Gould, Professor of Mathematics at West Virginia University, for helping me to persevere during the writing of the manuscript. I express my gratitude to W. J. Charatonik of the Department of Mathematics at the University of Wrocław (Wrocław, Poland) with whom I had numerous stimulating discussions regarding some of the material while he was a visiting professor at West Virginia University. I am also most grateful to Joann Mayhew for her splendid typing of the original manuscript and for proofreading the final typescript. Special thanks go to the students in my class (Clyde Campbell, C. Brad Davis, Yongping Luo, Gary A. Seldomridge, Timothy Swyter, and Cheng Zhao) who allowed themselves to be test subjects as the first students to study continuum theory from this book. I thank my wife Elsa G. Nadler for her help with the final manuscript, including her drawing of most of the figures. Finally I thank the people at Marcel Dekker, Inc., for their cooperation and assistance.

Sam B. Nadler, Jr.

Contents

Contents

CONTINUUM
THEORY

Part One
GENERAL ANALYSIS

I

Examples of Continua and Nested Intersections

A *continuum* is a nonempty, compact, connected metric space. A *subcontinuum* is a continuum which is a subset of a space. We shall on occasion have reason to consider nonempty, compact, connected T_1-spaces (or Hausdorff spaces); we shall refer to such spaces as T_1-*continua* (or *Hausdorff continua*). However, we emphasize that the term continuum by itself always means a metric continuum.

When referring to a space, the term *nondegenerate* means the space consists of more than one point.

1. SOME SIMPLE EXAMPLES

We begin our study of continua by giving some simple examples in 1.1–1.6. Simultaneously, we shall introduce some basic definitions and notation.

1.1 Arcs. An *arc* is any space which is homeomorphic to the closed interval $[0,1]$. Since $[0,1]$ is a continuum, an arc is a continuum. Let A be an arc, let h be a homeomorphism from $[0,1]$ onto A, and let $p = h(0)$ and $q = h(1)$. It is easy to prove that if h' is any homeomorphism of $[0,1]$ onto A, then

$$\{h'(0),h'(1)\} = \{p,q\}.$$

3

It is therefore the case that p and q are special points of A; they are called the *end points of A* (comp., 6.25). When we say *A is an arc from p to q* we mean A is an arc with end points p and q (we do not mean to imply there is an ordering $<$ of A with $p < q$).

1.2 n-Cells. We let R^n denote Euclidean n-space and, for each point $x = (x_i)_{i=1}^n \in R^n$, we let

$$\|x\| = \left(\sum_{i=1}^n x_i^2\right)^{1/2}.$$

An *n-cell* is a space which is homeomorphic to the n-dimensional closed ball B^n in R^n, where

$$B^n = \{x \in R^n: \|x\| \le 1\}, \quad \text{each } n = 1, 2, \ldots.$$

Clearly, an n-cell is a continuum since B^n is a continuum.

1.3 n-Spheres. An *n-sphere* is a space which is homeomorphic to the n-dimensional sphere S^n in R^{n+1}, where

$$S^n = \{x \in R^{n+1}: \|x\| = 1\}, \quad \text{each } n = 1, 2, \ldots.$$

An n-sphere is a continuum. A 1-sphere is called a *simple closed curve*.

All of our continua so far have been finite-dimensional. The following one is not.

1.4 Hilbert Cube. A *Hilbert cube* is a space which is homeomorphic to the countable cartesian product

$$\prod_{i=1}^\infty I_i, \quad \text{each } I_i = [0,1]$$

with the product topology [12, p. 147]. A Hilbert cube is a continuum by Theorem 4 of [13, p. 17], Theorem 11 of [13, p. 137], and [12, pp. 212–213]. Let us also note that every continuum can be topologically embedded in a Hilbert cube by Theorem 1 of [12, p. 241]. We take this opportunity to state the following elegant topological characterization of the Hilbert cube due to Toruńczyk [19]: *A compact, metric, absolute retract X is a Hilbert cube if and only if the identity map $i: X \to X$ is a uniform limit of Z-maps* (a Z-set in X is a closed set A in X such that there are continuous functions of X into $X - A$ arbitrarily close to the identity map, and a Z-map is a continuous function whose range is a Z-set in X; see [6]).

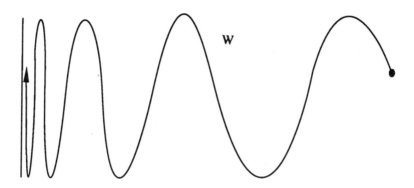

Figure 1.5

As stated, this result is a special case of Theorem 1 of [19] as can be seen using (*E*) of [19]. There are many applications of this result, one of which is in [19, p. 39].

The continua in 1.1–1.4 all have the property that they are *arcwise connected* (meaning, any two points can be joined by an arc). The following continuum is not arcwise connected.

1.5   The sin(1/*x*)-Continuum.   The *sin(1/x)-continuum* is the closure $\overline{W}$ of *W* where

$$W = \{(x,\sin(1/x)) \in R^2: 0 < x \le 1\}$$

(see Figure 1.5). We remark that the sin(1/*x*)-continuum is an example of an arc-like continuum (see 2.23). We leave as an exercise that it is not arcwise connected (1.12).

1.6   The Warsaw Circle.   This is the name given to any continuum homeomorphic to *Y* ∪ *Z* where *Y* is the sin(1/*x*)-continuum (1.5) and *Z* is the union of three convex arcs in R^2, one from $(0, -1)$ to $(0, -2)$, one from $(0, -2)$ to $(1, -2)$, and one from $(1, -2)$ to $(1,\sin(1))$. We note that the Warsaw circle is arcwise connected but that, like the sin(1/*x*)-continuum, it has points which do not have small connected neighborhoods (e.g., $(0,0)$). Continua, each point of which has a base of connected neighborhoods, are called Peano continua (8.1). The Warsaw circle is a circle-like continuum (2.24).

2. NESTED INTERSECTIONS

One of the most important techniques for obtaining interesting examples
of continua is the use of nested intersections. We shall illustrate this in
1.10, 1.11, 1.23, Chapter II, and in many other places in the book. In
fact, it may be said that the nested intersection technique is central to
continuum theory. It is not only used to construct examples, but it is the
key idea for the proofs of many theorems. It is even used in constructing
continuous functions (Chapter VII). The following two results lay the
foundation for using this technique.

1.7 Proposition. Let $\{X_i\}_{i=1}^{\infty}$ be a sequence of compact metric
spaces such that $X_i \supset X_{i+1}$ for each $i = 1, 2, \ldots$, and let

$$X = \bigcap_{i=1}^{\infty} X_i.$$

If U is an open subset of X_1 such that $U \supset X$, then there exists N such that

$$U \supset X_i \qquad \text{for all } i \geq N.$$

In particular: If each $X_i \neq \varnothing$, then $X = \varnothing$ (and, clearly, compact
metric).

Proof. Suppose that for each $i = 1, 2, \ldots$, there exists $x_i \in X_i - U$.
Since $X_1 - U$ is a compact metric space, we may assume that the se-
quence $\{x_i\}_{i=1}^{\infty}$ converges to some point $p \in X_1 - U$. For each k, $x_i \in X_k$
for all $i \geq k$. Hence $p \in X_k$ for each k. Thus, $p \in X$ which, since $p \notin U$,
contradicts our assumption that $U \supset X$. Hence, there exists N such that
$U \supset X_i$ for all $i \geq N$. This proves the first part of the proposition. The
fact that $X \neq \varnothing$ if each $X_i \neq \varnothing$ follows from the first part by assuming
$X = \varnothing$ and taking $U = \varnothing$ to see that $X_N = \varnothing$. This completes the
proof of 1.7. $\blacksquare$

1.8 Theorem. Let $\{X_i\}_{i=1}^{\infty}$ be a sequence of continua such that $X_i \supset$
X_{i+1} for each $i = 1, 2, \ldots$, and let

$$X = \bigcap_{i=1}^{\infty} X_i.$$

Then, X is a continuum.

Proof. By 1.7, X is a nonempty, compact metric space. Now, suppose that X is not connected. Then, $X = A \cup B$ where A and B are

disjoint, nonempty, closed (hence, compact) sets. Since X_1 is a normal space, there are disjoint open subsets V and W of X_1 such that $A \subset V$ and $B \subset W$. Let $U = V \subset W$. Then, by 1.7, $U \supset X_n$ for some n. Hence,

$$X_n = (X_n \cap V) \cup (X_n \cap W).$$

Since $X_n \supset X = A \cup B$ and since $A \neq \varnothing$ and $B \neq \varnothing$, we see that $X_n \cap V \neq \varnothing$ and $X_n \cap W \neq \varnothing$. It now follows easily that X_n is not connected, a contradiction. Therefore, X is connected. This completes the proof of 1.8.

3. EXAMPLES USING NESTED INTERSECTIONS

We give two examples of continua which are constructed using nested intersections.

Let us observe that all the continua in 1.1–1.6 have the property that they can be written as the union of two proper subcontinua. Note the following definition.

1.9   Definition of Decomposable, Indecomposable Continuum.   A continuum X is said to be *decomposable* provided that X can be written as the union of two proper subcontinua. A continuum which is not decomposable is said to be *indecomposable*.

For persons uninitiated in continua theory, it may seem that all continua are decomposable (except a one-point continuum). However, as the following example shows, this is not the case. In fact, "most" continua are indecomposable (1.17; also, see 13.57). We shall show even more in 1.23.

1.10   A Nondegenerate Indecomposable Continuum.   We construct a nondegenerate indecomposable continuum X in the plane R^2 as follows. Call a finite indexed collection $\mathfrak{C} = \{A_1, ..., A_n\}$ of subsets of a set S a *simple chain from x to z through y* provided that $A_i \cap A_j \neq \varnothing$ if and only if $|i - j| \leq 1$, $x \in A_i$ if and only if $i = 1$, $z \in A_i$ if and only if $i = n$, and $y \in A_i$ for some i. The members A_i of a simple chain are called *links*. Now, let a, b, and c be distinct points in R^2. It is easy to see that there are simple chains $\mathfrak{C}_n$ in R^2, $n = 1, 2, ...$, whose links are 2-cells (disks) of diameter $< 2^{-n}$ satisfying (1) and (2) below (see Figure 1.10):

(1)   For each $n = 0, 1, 2, ...$, $\mathfrak{C}_{3n+1}$ goes from a to c through b, $\mathfrak{C}_{3n+2}$ goes from b to c through a, and $\mathfrak{C}_{3n+3}$ goes from a to b through c;

(2)   For each $n = 1, 2, ...$, $\cup \mathfrak{C}_n \supset \cup \mathfrak{C}_{n+1}$.

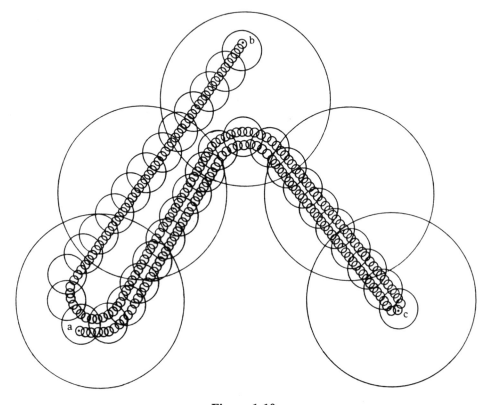

Figure 1.10

The desired continuum X is the nested intersection of the unions of these chains, i.e.,

$$X = \bigcap_{n=1}^{\infty} (\cup \ \mathbb{C}_n).$$

By 1.8, X is indeed a continuum. By using the fact that

$$X = \bigcap_{n=0}^{\infty} (\cup \ \mathbb{C}_{3n+1})$$

it can be verified that no proper subcontinuum Y of X contains $\{a,c\}$ (if such a Y existed, then let $p \in X - Y$ and fix k large enough so that the links of $\mathbb{C}_{3k+1}$ containing p miss Y; then, use the links of $\mathbb{C}_{3k+1}$ to contradict the connectedness of Y—you are asked for the details in 1.15). Similarly, no

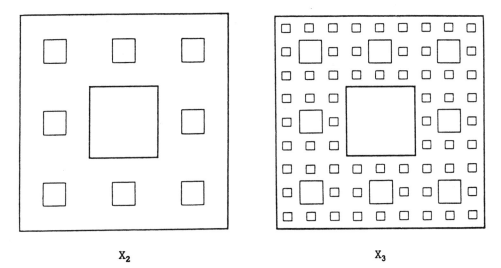

$$X_2 \qquad\qquad\qquad\qquad X_3$$

Figure 1.11 (From Ref. 13, with permission of Academic Press and Panstwowe Wydawnictwo Naukowe, Warsaw.)

proper subcontinuum of X contains any two of the three points a, b, c. It follows easily from this that X is indecomposable (use the pigeonhole principle). We remark that like 1.5, X is an arc-like continuum (comp., 12.11).

In relation to 1.10, we strongly encourage the reader to work through the exercise in 1.23, especially parts (b) and (c).

As another example of the nested intersection technique, we construct the following continuum.

1.11   The Sierpiński Universal Curve.   Partition the solid square $S = [0,1] \times [0,1]$ into nine congruent squares, and let

$$X_1 = S - (1/3, 2/3) \times (1/3, 2/3).$$

Similarly, partition each of the remaining eight squares into nine congruent squares, and let X_2 be the continuum obtained by removing the interiors of each of the resulting central eight squares (see Figure 1.11). Continue in this fashion to define X_3, X_4, ... Let

$$X = \bigcap_{i=1}^{\infty} X_i.$$

Then, X is a continuum by 1.8 and is called the *Sierpiński Universal*

Curve. The term universal refers to the fact that this one-dimensional continuum in the plane contains a topological copy of every one-dimensional continuum in the plane (the word universal is not always used in this way). The proof of the universality of X is accomplished by showing that every one-dimensional continuum in the plane is contained in a nested intersection of topological copies Y_i of the sets X_i, the sets Y_i being chosen in the plane in such a way that their total intersection is homeomorphic to X (1.16).

In connection with the Sierpiński Universal Curve, we mention the Menger Universal Curve which is a one-dimensional Peano continuum containing a topological copy of every one-dimensional separable metric space. For this and other results about the Menger curve, see [16]. We shall construct some other universal continua in later chapters (10.37, 10.49, and 12.22).

In the next two chapters, we discuss some other important general methods of constructing continua and give a number of examples.

EXERCISES

1.12 Exercise. Prove that there is no continuous function f from $[0,1]$ into the $\sin(1/x)$-continuum X (1.5) such that $f(0) = (1, \sin(1))$ and $f(1) = (0,0)$. Thus, clearly, X is not arcwise connected.

1.13 Exercise. Prove that there is no continuous function from $[0,1]$ onto the Warsaw circle (1.6).

1.14 Exercise. In relation to 1.8, give an example of a nested intersection of connected metric spaces which is not connected. [Hint: Consider certain subsets of the continuum in 1.5.]

1.15 Exercise. Verify the details that no proper subcontinuum of X in 1.10 contains any two of the three points a, b, c and, thus, X is indecomposable. In connection with these ideas, see 11.20 and the discussion in the paragraph preceding it.

1.16 Exercise. Using the idea mentioned in 1.11, prove that the Sierpiński curve X is universal. [Hint: For each i, select Y_i carefully enough so that it is easy to define a homeomorphism h_i from X_i onto Y_i as a pair of linear homeomorphisms f_i, g_i and so that the sequence $\{h_i\}_{i=1}^{\infty}$ converges uniformly to a homeomorphism of X onto $\cap_{i=1}^{\infty} Y_i$. The only fact about the condition of one-dimensionality you need to use is that a one-dimensional subset of the plane contains no nonempty open subset of the plane.]

1.17 Exercise. Let M denote an n-cell (1.2), $n \geq 2$, or the Hilbert cube (1.4), and let d denote a metric for M. For this exercise, we use the notions in 4.1 and the fact that $C(M)$ with the metric H_d (4.1) is a compact metric space (4.17). Let

$$\mathfrak{L} = \{X \in C(M): X \text{ is a nondegenerate indecomposable continuum}\}.$$

Give meaning to the statement that "most" continua are indecomposable (preceding 1.10) by proving that $\mathfrak{L}$ is a dense G_δ in $C(M)$ (G_δ means a countable intersection of open sets). [Hint: Use the sets $\mathfrak{C}_n (n = 1, 2, \ldots)$, defined by

$$\mathfrak{C}_n = \{X \in C(M): X = A \cup B \text{ where } A \text{ and } B \text{ are subcontinua of } X,$$
$$A \not\subset N_d(1/n,B), \text{ and } B \not\subset N_d(1/n,A)\},$$

to show that $\mathfrak{L}$ is a G_δ in $C(M)$; use the example in 1.10 to show that $\mathfrak{L}$ is dense in $C(M)$.] See (d) of 1.23.

1.18 Exercise. Let S be a connected topological space, and let $A_1, \ldots, A_n, n < \infty$, be closed, connected subsets of S such that $\cup_{i=1}^n A_i = S$. Prove that there is an ℓ, $1 \leq \ell \leq n$, such that $\cup_{i \neq \ell} A_i$ is connected.

1.19 Exercise. A *fixed point* of a function f is a point p such that $f(p) = p$. A topological space X is said to have the *fixed point property* provided that every continuous function from X into X has a fixed point. Prove (a)–(g) below.
- (a) The fixed point property is a topological invariant (i.e.: If X has the property, then any space homeomorphic to X has the property).
- (b) If X has the fixed point property, then X is connected.
- (c) An arc has the fixed point property. [Hint: For any continuous function $f: [0,1] \to [0,1]$, consider $\{t: t \leq f(t)\}$.] See (d).
- (d) If f is any continuous function from any connected space S onto $[0,1]$, then, for any continuous function g from S into $[0,1]$, $f(s) = g(s)$ for some $s \in S$. Note that this is a generalization of (c). See 12.23.
- (e) The continua X in 1.5 and 1.6 have the fixed point property. [Hint: Make the following idea into a rigorous proof—X is a road along which a dog is chasing a rabbit where, for each $x \in X$, x and $f(x)$ represent the positions of the dog and the rabbit, respectively; start the chase at an appropriate point, and eventually use (c).]

(f) Let W be as in 1.5; then, $W \cup \{(0,1)\}$ has the fixed point property. Thus, non-compact spaces may have the fixed point property.

(g) If X is a continuum and there is a sequence $\{f_i\}_{i=1}^{\infty}$ of continuous functions of X into X converging uniformly to the identity map on X such that $f_i(X)$ has the fixed point property for each i, then X has the fixed point property. Two specific applications are in 1.20.

We note that (c) is a special case of the Brouwer Fixed Point Theorem [12, p. 313] which we shall prove for 2-cells in 12.39. It is also a special case of 12.30, as is the part of (e) concerning 1.5, and of 10.31. The descriptive hint for (e) comes from [4] which is an excellent, readable, expository article on the fixed point property (Question 2 in [4] has been answered negatively [1]). With respect to (f), the cartesian product of $W \cup \{(0,1)\}$ with itself does not have the fixed point property [7, p. 977]; in connection with this, see [11] and [14]. With respect to (g), see [8].

1.20 Exercise. Show how to use (c) and (g) of 1.19 to prove the continuum in 1.5 has the fixed point property. The *Warsaw disk* is the Warsaw circle (1.6) together with the bounded component of its complement in R^2. Using 12.39 and (g) of 1.19, prove that the Warsaw disk has the fixed point property.

1.21 Exercise. Let X be a topological space. A continuous function $r: X \to X$ is called a *retraction* provided that $r \circ r = r$, i.e., r is the identity on its range. A subset A of X for which there is a retraction of X onto A is called a *retract of* X. Prove that if X has the fixed point property, then every retract of X has the fixed point property. Use this and 12.39 to show that the continuum $Y = Y_1 \cup A \cup Y_2$, where

$Y_1 = \{(x,y) \in R^2 : (x + 1/2)^2 + y^2 \leq 1/16\}$

$A = \{(x,0) \in R^2 : -1/4 \leq x \leq 1/4\}$

$Y_2 = \{(x,y) \in R^2 : (x - 1/2)^2 + y^2 \leq 1/16\},$

has the fixed point property.

For the theory of retracts, see [5]. We also remark that retractions are important in the study of semigroups of selfmaps since the retractions are the idempotents of these semigroups—see [15].

1.22 Exercise. A continuum X is said to be *aposyndetic at p with respect to q*, where $p, q \in X$, provided that there is a subcontinuum A of $X - \{q\}$ such that p is a point of the interior (in X) of A. A continuum X is said to be *aposyndetic at p* provided that X is aposyndetic at p with respect to each point $q \in X - \{p\}$. A continuum is said to be *aposyndetic* provided that it is aposyndetic at each of its points. At which points p are the continua in 1.1–1.6, 1.10, and 1.11 aposyndetic? Let X be the continuum in 1.5, and let

$$Y = X \cup \{(x,1) \in R^2 \colon 0 \le x \le 1\}.$$

Show that Y is aposyndetic at $(0,1)$, aposyndetic at $(0,0)$ with respect to $(0,-1)$, but not aposyndetic at $(0,0)$ with respect to $(0,1)$.

The notion of aposyndesis is due to Jones [9] (a survey of early results is in [10]). See (e) of 8.44.

1.23 Exercise. We have constructed a nondegenerate indecomposable continuum (1.10), and we have seen that most continua are indecomposable (1.17). A continuum is said to be *hereditarily indecomposable* provided that each of its subcontinua is indecomposable. In (b) and (c) below, we construct a specific nondegenerate hereditarily indecomposable continuum and, in (d), we strengthen 1.17. First, we need some terminology and notation involving simple chains (defined in 1.10).

Let $\mathcal{U} = \{U_1, \ldots, U_m\}$ and $\mathcal{V} = \{V_1, \ldots, V_n\}$ be simple chains whose links are open subsets of a space S. We say that $\mathcal{V}$ is a *subchain* of $\mathcal{U}$ provided that each link of $\mathcal{V}$ is a link of $\mathcal{U}$ (i.e., $\mathcal{V} \subset \mathcal{U}$); if $\mathcal{V}$ is a subchain of $\mathcal{U}$ and $\{V_1, V_n\} = \{U_i, U_j\}$, we often denote $\mathcal{V}$ by $\mathcal{U}(i,j)$. We say that $\mathcal{V}$ *refines* (*strongly refines*) $\mathcal{U}$ provided that each link (the closure of each link) of $\mathcal{V}$ is contained in a link of $\mathcal{U}$. We say that $\mathcal{V}$ is *crooked* in $\mathcal{U}$ provided that $\mathcal{V}$ refines $\mathcal{U}$ and (#) below holds:

(#) Whenever $\mathcal{V}(i,j)$ is a subchain of $\mathcal{V}$ with $V_i \cap U_h \ne \emptyset$ and $V_j \cap U_k \ne \emptyset$ where $|h - k| \ge 3$, then there are r and s such that

$$\mathcal{V}(i,j) = \mathcal{V}(i,r) \cup \mathcal{V}(r,s) \cup \mathcal{V}(s,j)$$

where $(s - r)(j - i) > 0$ and V_r and V_s are subsets of links of $\mathcal{U}(h,k)$ which intersect U_k and U_h respectively.

To improve your understanding of the condition in (#), do (a) below.

(a) Let $\mathcal{U} = \{U_1, \ldots, U_5\}$ be the simple chain from p to q drawn below whose links are interiors of round disks in R^2. Draw a simple chain $\mathcal{V}$ from p to q which is crooked in $\mathcal{U}$ such that the links of $\mathcal{V}$ are connected open sets (in $\cup \mathcal{U}$) none of which is contained in the intersection of any two links of $\mathcal{U}$. [Hint: $\mathcal{V}$ can be drawn with six changes in direction, but not with five or less such changes.]

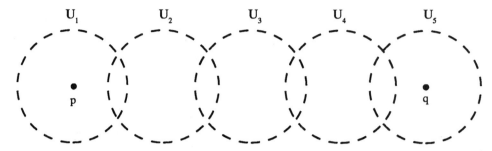

Prove (b)–(d) below.

(b) Fix $p, q \in R^2$ with $p \neq q$. Then, there is a sequence $\{\mathcal{U}_\ell\}_{\ell=1}^{\infty}$ of simple chains from p to q whose links are connected, open subsets of R^2 and such that, for each $\ell = 1, 2, ...,$ $\mathcal{U}_{\ell+1}$ strongly refines $\mathcal{U}_\ell$, $\mathcal{U}_{\ell+1}$ is crooked in $\mathcal{U}_\ell$, and each link of $\mathcal{U}_\ell$ has diameter < $1/\ell$.

(c) Let $\{\mathcal{U}_\ell\}_{\ell=1}^{\infty}$ be as in (b), and let $X = \bigcap_{\ell=1}^{\infty} (\bigcup \mathcal{U}_\ell)$. Then, X is a nondegenerate hereditarily indecomposable continuum. [Hint: Use 1.8 to prove X is a continuum. Suppose Y is a decomposable subcontinuum of: X, $Y = A \cup B$ where A and B are proper subcontinua of Y. Find ℓ large enough so that there are links U_r, U_s, and U_t of $\mathcal{U}_\ell$, with s being between r and t, such that either (1) $U_r \cap X$ and $U_t \cap X$ are contained in $A - B$ and $U_s \cap X$ is contained in $B - A$ or (2) $U_r \cap X$ and $U_t \cap X$ are contained in $B - A$ and $U_s \cap X$ is contained in $A - B$. Then, using ideas we used in 1.10, (1) (respectively, (2)) can be used to contradict the connectedness of A (respectively, B).

(d) Most continua are hereditarily indecomposable in the following sense: If M, d, $C(M)$, and H_d are as in 1.17 and $\mathcal{H} = \{Z \in C(M): Z$ is a nondegenerate hereditarily indecomposable continuum$\}$, then $\mathcal{H}$ is a dense C_δ in $C(M)$. [Hint: Modify the sets $\mathcal{C}_n$ in 1.17 in a straightforward way to obtain that $\mathcal{H}$ is a G_δ in $C(M)$. Use the construction in (b) and the result in (c) to show that $\mathcal{H}$ is dense in $C(M)$.]

The hereditarily indecomposable continuum X in (c) is arc-like by 12.11 (to apply 12.11, intersect each link of $\mathcal{U}_\ell$ in (b) with X, for each ℓ, to produce a $(1/\ell)$-chain in X (12.8)). Up to homeomorphism, there is only one hereditarily indecomposable arc-like continuum (Theorem 1 of [3] and 12.11), and it is called the *pseudo-arc*. Thus, in particular, any continuum constructed as in (b) and (c) is a pseudo-arc (comp., Theorem 5 of [18]). The pseudo-arc has many fascinating properties, a few of which are mentioned along with some historical comments in the second paragraph of Chapter XII. For now, we note only that we have constructed our first example of a nondegenerate continuum which contains no arc! An hereditarily decomposable continuum which contains no arc is constructed in 2.27, and an even "stronger" example is in 2.32. Regarding the statement that the pseudo-arc is our *first* example of a nondegenerate arcless continuum, we note that if the simple chains used to construct the continuum in 1.10 are assumed to be "as straight as possible in each other for the construction" (e.g., as in Figure 1.10), then it can be shown that every nondegenerate proper subcontinuum of the continuum in 1.10 is an arc. Thus, assuming the construction in 1.10 is as you thought it was when you read it, the continuum in 1.10 contains lots of arcs.

The notion of crookedness we used above is from [2, p. 730]; it is a modification of a prior notion in [18, p. 583]. These notions are closely related to the type of "crooked arc" constructed in Lemma 6 of [17, p. 152]. Part (c) is Theorem 10 of [2]; comp., Theorem 6 of [18]. Part (d) for $M = B^2$ is the theorem in [17]. In relation to (d), we note that most continua are pseudo-arcs (Theorem 2 of [3])—see the comments near the end of 12.70.

REFERENCES

1. David P. Bellamy, A tree-like continuum without the fixed-point property, Houston J. Math., 6(1980), 1–13.

2. R. H. Bing, A homogeneous indecomposable plane continuum, Duke Math. J., 15(1948), 729–742.

3. R. H. Bing, Concerning hereditarily indecomposable continua, Pacific J. Math., 1(1951), 43–51.

4. R. H. Bing, The elusive fixed point property, Amer. Math. Monthly, 76(1969), 119–132.

5. Karol Borsuk, *Theory of Retracts*, Monografie Mat., Vol. 44, Polish Scientific Publishers, Warszawa, Poland, 1967.

6. T. A. Chapman, *Lectures on Hilbert Cube Manifolds*, Conf. Board of the Math. Sci., Reg. Conf. Series in Math., No. 28, Amer. Math. Soc., Providence, R.I., 1975.

7. E. H. Connell, Properties of fixed point spaces, Proc. Amer. Math. Soc., 10(1959), 974–979.

8. C. A. Eberhart and J. B. Fugate, Approximating continua from within, Fund. Math., 72(1971), 223–231.

9. F. Burton Jones, Aposyndetic continua and certain boundary problems, Amer. J. Math., 63(1941), 545–553.

10. F. Burton Jones, Concerning aposyndetic and non-aposyndetic continua, Bull. Amer. Math. Soc., 58(1952), 137–151.

11. Ronald J. Knill, Cones, products, and fixed points, Fund. Math., 60(1967), 35–46.

12. K. Kuratowski, *Topology*, Vol. I, Academic Press, New York, N.Y., 1966.

13. K. Kuratowski, *Topology*, Vol. II, Academic Press, New York, N.Y., 1968.

14. William Lopez, An example in the fixed point theory of polyhedra, Bull. Amer. Math. Soc., 73(1967), 922–924.

15. K. D. Magill, Jr., A survey of semigroups of continuous selfmaps, Semigroup Forum, 11(1975–76), 189–282.

16. J. C. Mayer, Lex G. Oversteegen, and E. D. Tymchatyn, The Menger curve, Dissertationes Math., 252(1986).

17. Stefan, Mazurkiewicz, Sur les continus absolument indécomposables, Fund. Math., 16(1930), 151–159.

18. E. E. Moise, An indecomposable plane continuum which is homeomorphic to each of its nondegenerate subcontinua, Trans. Amer. Math. Soc., 63(1948), 581–594.

19. H. Toruńczyk, On CE-images of the Hilbert cube and characterization of Q-manifolds, Fund. Math., 106(1980), 31–40.

II

Inverse Limits of Continua

In Chapter 1, we discussed the nested intersection technique and showed how to use it to construct some interesting continua from simpler ones. In this chapter, we shall see how to use inverse limits in the same way (inverse limits are defined in 2.2). We shall gain a basic working knowledge of inverse limits and give some examples using them. As we shall soon see (2.3), inverse limits may be thought of as a special case of the nested intersection technique.

First, we briefly discuss cartesian products since inverse limits are defined as subsets of cartesian product spaces.

1. CARTESIAN PRODUCT SPACES

We shall only be concerned with finite or countable cartesian products. The *cartesian product* of the spaces X_i, $i = 1, 2, \ldots$ or $1 \leq i \leq n < \infty$, is the space of all sequences $(x_i)_{i=1}^{\infty}$ or, respectively, finite sequences $(x_i)_{i=1}^{n}$, where $x_i \in X_i$ *for each* i, with the *product topology* (i.e., the smallest topology for which the projections to the coordinate spaces X_i are all continuous [9, p. 147]), and is denoted by

$\Pi_{i=1}^{\infty} X_i$ or, respectively, by $\Pi_{i=1}^{n} X_i$ or $X_1 \times \cdots \times X_n$.

We shall always assume a cartesian product of spaces has the product topology unless otherwise stated. We also assume the Axiom of Choice [8].

We have already defined a Hilbert cube as a cartesian product of continua (1.4). We could have defined an n-cell (1.2) as a cartesian product of n arcs ($n \geq 2$). A *torus* may be defined as a cartesian product of two simple closed curves, and a *solid torus* may be defined as a cartesian product of a simple closed curve and a 2-cell. Note the following theorem.

2.1 Theorem. The finite or countable cartesian product of continua, or of nonempty compact metric spaces, is a continuum or a nonempty compact metric space, respectively.

Proof. The product space is compact by Theorem 4 of [10, p. 17], connected by Theorem 11 of [10, p. 137], and metrizable by [9, pp. 212–213]. In the finite case, it is nonempty and, in the countably finite case, it is nonempty by the Countable Axiom of Choice [8, p. 20]. This proves 2.1.

2. INVERSE LIMITS

In this section, we define inverse limits, show how to think of them as nested intersections, and obtain the important result in 2.4.

2.2 Definition of Inverse Limit. An *inverse sequence* is a ''double sequence'' $\{X_i, f_i\}_{i=1}^{\infty}$ of spaces X_i, called *coordinate spaces*, and continuous functions $f_i \colon X_{i+1} \to X_i$ called *bonding maps*. If $\{X_i, f_i\}_{i=1}^{\infty}$ is an inverse sequence, sometimes written

$$X_1 \xleftarrow{f_1} X_2 \xleftarrow{f_2} \cdots \xleftarrow{f_{i-1}} X_i \xleftarrow{f_i} X_{i+1} \xleftarrow{f_{i+1}} \cdots,$$

then the *inverse limit* of $\{X_i, f_i\}_{i=1}^{\infty}$, denoted by $\varprojlim \{X_i, f_i\}_{i=1}^{\infty}$, is the subspace of the cartesian product space $\Pi_{i=1}^{\infty} X_i$ defined by

$$\varprojlim \{X_i, f_i\}_{i=1}^{\infty} = \{(x_i)_{i=1}^{\infty} \in \prod_{i=1}^{\infty} X_i : f_i(x_{i+1}) = x_i \text{ for all } i\}.$$

The following proposition shows how we may think of inverse limits as nested intersections (conversely, see 2.15).

2.3 Proposition. Let $\{X_i, f_i\}_{i=1}^{\infty}$ be an inverse sequence. For each $n = 1, 2, \ldots$, define $Q_n(X_i, f_i)$ by

$$Q_n(X_i, f_i) = \{(x_i)_{i=1}^{\infty} \in \prod_{i=1}^{\infty} X_i : f_i(x_{i+1}) = x_i \text{ for all } i \leq n\}.$$

Then, (1)–(3) below hold:

(1) $Q_n(X_i, f_i) \supset Q_{n+1}(X_i, f_i)$ for all $n = 1, 2, \ldots$;

(2) $Q_n(X_i, f_i)$ is homeomorphic to $\prod_{i=n+1}^{\infty} X_i$ for each $n = 1, 2, \ldots$;

(3) $\varprojlim\{X_i, f_i\}_{i=1}^{\infty} = \bigcap_{n=1}^{\infty} Q_n(X_i, f_i).$

Proof. The proofs of (1) and (3) are trivial. To prove (2), fix n and define

$$h: Q_n(X_i, f_i) \rightarrow \prod_{i=n+1}^{\infty} X_i$$

by $h((x_i)_{i=1}^{\infty}) = (x_i)_{i=n+1}^{\infty}$ for each $(x_i)_{i=1}^{\infty} \in Q_n(X_i, f_i)$. It is easy to verify that h is a homeomorphism as desired for (2). This completes the proof of 2.3.

The following fundamental theorem is now an easy consequence of previous results.

2.4 Theorem. An inverse limit of continua is a continuum. Also, an inverse limit of nonempty compact metric spaces is a nonempty compact metric space.

Proof. Use 1.8 (for the second part, 1.7), 2.1, and 2.3.

We shall see how to obtain interesting continua by using inverse limits in the next section (2.8 and 2.9) and in other places in the book (e.g., 2.18, 2.27, 2.32, 10.37, and 12.22). Other examples are in, e.g., [1] and [2].

3. INDECOMPOSABLE CONTINUA FROM INVERSE LIMITS

In 1.10, we constructed an indecomposable continuum using chains. We now show how to use inverse limits to construct indecomposable continua. The idea is contained in the following definition (see [15] and [16]).

2.5 Definition of Indecomposable Inverse Sequence. An inverse
sequence $\{X_i, f_i\}_{i=1}^{\infty}$ where each X_i is a continuum is called an *indecomposable inverse sequence* provided that, for each $i = 1, 2, ...$, whenever
A_{i+1} and B_{i+1} are subcontinua of X_{i+1} such that $X_{i+1} = A_{i+1} \cup B_{i+1}$, then
$f_i(A_{i+1}) = X_i$ or $f_i(B_{i+1}) = X_i$.

It is easy to give examples of indecomposable inverse sequences (see,
e.g., 2.8 and 2.9). At this moment, however, we want to obtain 2.7. The
following lemma will be used in its proof.

2.6 Lemma. Let $\{X_i, f_i\}_{i=1}^{\infty}$ be an inverse sequence of metric spaces
with inverse limit X_{∞}. For each $i = 1, 2, ...$, let $\pi_i: X_{\infty} \to X_i$ denote the
ith projection map. Let A be a compact subset of X_{∞}. Then, $\{\pi_i(A),$
$f_i | \pi_{i+1}(A)\}_{i=1}^{\infty}$ is an inverse sequence with onto bonding maps and

(#) $\lim\limits_{\leftarrow} \{\pi_i(A), f_i | \pi_{i+1}(A)\}_{i=1}^{\infty} = A = \left[\prod\limits_{i=1}^{\infty} \pi_i(A) \right] \cap X_{\infty}.$

Proof. By 2.2, $f_i \circ \pi_{i+1} = \pi_i$ for all $i = 1, 2, ...$ It follows immediately from this that $\{\pi_i(A), f_i | \pi_{i+1}(A)\}_{i=1}^{\infty}$ is an inverse sequence with onto
bonding maps. To prove (#), first observe that

(1) $\lim\limits_{\leftarrow} \{\pi_i(A), f_i | \pi_{i+1}(A)\}_{i=1}^{\infty} = \left[\prod\limits_{i=1}^{\infty} \pi_i(A) \right] \cap X_{\infty}.$

Also, observe that

(2) $A \subset \left[\prod\limits_{i=1}^{\infty} \pi_i(A) \right] \cap X_{\infty}.$

Hence, to prove (#), it suffices by (1) and (2) to prove (3) below:

(3) $\left[\prod\limits_{i=1}^{\infty} \pi_i(A) \right] \cap X_{\infty} \subset A.$

To prove (3), let $y = (y_i)_{i=1}^{\infty} \in [\Pi_{i=1}^{\infty} \pi_i(A)] \cap X_{\infty}$. For each $j = 1, 2, ...$,
define K_j by

$K_j = A \cap \pi_j^{-1}(y_j).$

Since $y_j \in \pi_j(A)$ for each $j = 1, 2, ...$, we see that $K_j \neq \varnothing$ for each $j = 1,$
$2, ...$ By the continuity of the projections and the compactness of A,
each K_j is compact. Finally, we see that $K_j \supset K_{j+1}$ for each $j = 1, 2, ...$
since $\pi_{j+1}^{-1}(y_{j+1}) \subset \pi_j^{-1}(y_j)$, as is seen by noting that if $x \in \pi_{j+1}^{-1}(y_{j+1})$,

$\pi_{j+1}(x) = y_{j+1}$ so $f_j(\pi_{j+1}(x)) = f_j(y_{j+1})$ which, since $x \in X_\infty$ and $y \in X_\infty$, gives us that $\pi_j(x) = y_j$, i.e., $x \in \pi_j^{-1}(y_j)$. By these properties of the sets K_j, we have by the second part of 1.7 that

$$\bigcap_{j=1}^{\infty} K_j \neq \varnothing.$$

Now, let $p \in \bigcap_{j=1}^{\infty} K_j$. Clearly, then, $p \in A$ and the jth coordinate of p must be y_j for each $j = 1, 2, \ldots$. Hence, $p = y$ and, thus, $y \in A$. Therefore, we have proved (3). This completes the proof of 2.6.

2.7 Theorem. If $\{X_i, f_i\}_{i=1}^{\infty}$ is an indecomposable inverse sequence with inverse limit X_∞, then X_∞ is an indecomposable continuum.

Proof. By 2.4, X_∞ is a continuum. To show that X_∞ is indecomposable, let A and B be subcontinua of X_∞ such that $A \cup B = X_\infty$. We show that $A = X_\infty$ or $B = X_\infty$. Built into 2.5 is the fact that each bonding map f_i: $X_{i+1} \to X_i$ maps onto X_i. Hence (see 2.14), each projection π_i: $X_\infty \to X_i$ maps onto X_i. Thus, for each i, since $X_\infty = A \cup B$,

$$X_{i+1} = \pi_{i+1}(A) \cup \pi_{i+1}(B).$$

Hence (2.5), $f_i[\pi_{i+1}(A)] = X_i$ or $f_i[\pi_{i+1}(B)] = X_i$ for each i. Therefore, since $f_i \circ \pi_{i+1} = \pi_i$ for each i (2.2), we have that $\pi_i(A) = X_i$ or $\pi_i(B) = X_i$ for each i. Hence, we may assume without loss of generality that $\pi_i(A) = X_i$ for infinitely many i. Then, using the ontoness of the bonding maps and the fact that by 2.2

$$f_j \circ \cdots \circ f_k \circ \pi_{k+1} = \pi_j \qquad \text{whenever } 1 \leq j < k,$$

we see that $\pi_i(A) = X_i$ for all $i = 1, 2, \ldots$. Therefore, it now follows from (#) of 2.6 that $A = X_\infty$. This completes that proof of 2.7.

It is important to note that if each X_i in 2.7 is nondegenerate, then the inverse limit X_∞ is also nondegenerate (by 2.14).

In 2.8, 2.9, and 2.18 we show how easy it is to apply 2.7 to obtain a number of different indecomposable continua.

2.8 The p-Adic Solenoids. For each $p = 2, 3, \ldots$, let f^p: $S^1 \to S^1$ be given by $f^p(z) = z^p$ for each $z \in S^1$ (where S^1 is the unit circle in the plane, and z^p denotes the pth power of z using complex multiplication). For a given p, let

$$\sum_p = \varprojlim \{X_i, f_i\}_{i=1}^{\infty}, \text{ where each } X_i = S^1 \text{ and each } f_i = f^p.$$

Then, by 2.7, each Σ_p for $p \geq 2$ is an indecomposable continuum. It is called the *p-adic solenoid*. Some other properties of Σ_p are given in 2.16.

2.9 The Buckethandle Continuum. For each $i = 1, 2, \ldots,$ let $X_i = [0,1]$ and let $f_i : X_{i+1} \to X_i$ be defined by

$$f_i(t) = \begin{cases} 2t, & 0 \leq t \leq 1/2 \\ -2t + 2, & 1/2 \leq t \leq 1. \end{cases}$$

Then, by 2.7, $\varprojlim\{X_i, f_i\}_{i=1}^{\infty}$ is an indecomposable continuum. It is homeomorphic to the continuum pictured in Figure 4 of [10, p. 205], and is called the *Buckethandle Continuum*. See 2.18.

The examples in 2.8 and 2.9 show that we may start with very simple continua and nice maps and arrive, via inverse limits, at very complicated continua. The simplicity of the coordinate spaces and the bonding maps facilitates the study of their inverse limit (e.g., see 2.16). However, sometimes there is a drawback to the inverse limit approach in that we lose some of the geometry inherent in the type of procedure used in 1.10. Let us give one example of this. It should not be clear from the approach in 2.8 that the dyadic solenoid Σ_2 can be embedded in R^3. However, the dyadic solenoid can be thought of geometrically as the nested intersection of thinner and thinner solid tori in R^3, each one going around twice in the preceding one. It is a good exercise to carry out the details that such a nested intersection is homeomorphic to Σ_2 as defined in 2.8. On the other side of the coin, if we had defined the dyadic solenoid as such a nested intersection of solid tori, it would not be easy to see that it is a topological group (2.16). Thus: "What you gain in the peanuts, you lose in the popcorn." However, the real moral is to be familiar with both approaches so that either one may be used when appropriate.

4. ANDERSON-CHOQUET EMBEDDING THEOREM

In this section, we prove the theorem in 2.10 (which is Theorem I of [1]). It gives sufficient conditions for the embedability of an inverse limit in a space S containing the coordinate spaces of the inverse sequence. Furthermore, the embedding is obtained constructively as a specific nested intersection of subsets of S. Though the theorem has a rather lengthy and somewhat technical statement, it is not hard to prove and can be easily applied to certain constructions which yield interesting, complicated continua. Some applications are in [1], and we shall use the theorem in, e.g., 2.32 and 10.37.

2.10 Theorem. Let (S,d) be a compact metric space. Let $\{X_i, f_i\}_{i=1}^{\infty}$ be an inverse sequence where each X_i is a nonempty compact subset of S and each f_i maps onto X_i. Let

$$f_{ij} = f_i \circ \cdots \circ f_{j-1} : X_j \to X_i \qquad \text{if } j > i + 1 \text{ and } f_{ii+1} = f_i.$$

Assume (1) and (2) below:

(1) For each $\epsilon > 0$, there exists k such that for all $p \in X_k$, diameter $[\cup_{j>k} f_{kj}^{-1}(p)] < \epsilon$;

(2) For each i and each $\delta > 0$, there exists $\delta' > 0$ such that whenever $j > 1$ and $p,q \in X_j$ such that $d(f_{ij}(p), f_{ij}(q)) > \delta$, then $d(p,q) > \delta'$.

Then, $\lim\{X_i, f_i\}_{i=1}^{\infty}$ is homeomorphic to $\cap_{i=1}^{\infty} (\overline{\cup_{m \geq i} X_m})$. In particular: If $X_i \subset \overline{X_{i+1}}$ for each i, then $\lim\{X_i, f_i\}_{i=1}^{\infty}$ is homeomorphic to $\overline{\cup_{i=1}^{\infty} X_i}$.

Proof. Let $X_{\infty} = \lim\{X_i, f_i\}_{i=1}^{\infty}$. If $x = (x_i)_{i=1}^{\infty} \in X_{\infty}$, then, by (1), we see that $\{x_i\}_{i=1}^{\infty}$ is a Cauchy sequence in S and, hence, $\{x_i\}_{i=1}^{\infty}$ converges to a point $h(x)$ in S. We have thus defined a function h from X_{∞} into S. We shall verify (a)–(c) below:

(a) h is continuous;

(b) h is one-to-one;

(c) $h(X_{\infty}) = \cap_{i=1}^{\infty} (\overline{\cup_{m \geq i} X_m})$.

To prove (a), let $x = (x_i)_{i=1}^{\infty} \in X_{\infty}$ and let $\epsilon > 0$. Let $B = \{z \in S: d(z,h(x)) < \epsilon\}$. Now, choose k as guaranteed by (1). Let $\ell = k + 1$. Let

$$V = \pi_{\ell}^{-1}(B \cap X_{\ell}), \qquad \pi_{\ell} : X_{\infty} \to X_{\ell} \text{ the } \ell\text{th projection map.}$$

Since $x \in X_{\infty}$ and k satisfies (1), we see from the definition of h that $d(x_{\ell}, h(x)) < \epsilon$. Hence, $x_{\ell} \in B \cap X_{\ell}$ and, thus, $x \in V$. Next, observe that, since $B \cap X_{\ell}$ is open in X_{ℓ}, V is open in X_{∞}. Now, let $y = (y_i)_{i=1}^{\infty} \in V$. For the same reasons as for x, $d(y_{\ell}, h(y)) < \epsilon$. Also, since $y \in V$, $y_{\ell} \in B$ so $d(y_{\ell}, h(x)) < \epsilon$. Hence,

$$d(h(y), h(x)) \leq d(h(y), y_{\ell}) + d(y_{\ell}, h(x)) < 2\epsilon.$$

Therefore, it follows that we have proved (a). To prove (b), let

$$x = (x_i)_{i=1}^{\infty}, \ y = (y_i)_{i=1}^{\infty} \in X_{\infty} \qquad \text{such that } x \neq y.$$

Then, $x_n \neq y_n$ for some n. Let $\delta = d(x_n, y_n)/2$. Then, by (2), there exists δ' such that whenever $j > n$, then, since

$$d(f_{nj}(x_j), f_{nj}(y_j)) = d(x_n, y_n) = 2\delta > \delta,$$

we have $d(x_j, y_j) > \delta'$. Therefore, clearly $h(x) \neq h(y)$. This proves (b). It remains to prove (c). For convenience, let

$$Z = \bigcap_{i=1}^{\infty} \left(\overline{\bigcup_{m \ge i} X_{\infty}} \right).$$

It is clear from the definition of h that $h(X_{\infty}) \subset Z$. We prove that $Z \subset h(X_{\infty})$ by showing that $h(X_{\infty})$ is dense in Z. For this purpose, let $z \in Z$ and let $\epsilon > 0$. Let k be as guaranteed by (1). Since $z \in Z$, there exists $m \ge k$ such that $d(z,p) < \epsilon$ for some $p \in X_m$. Now, by 2.14, there exists $x = (x_i)_{i=1}^{\infty}$ in X_{∞} such that $x_m = p$. Note that, since $m \ge k$, it is clear that (1) holds with k replaced by m. Hence, it follows easily that $d(z,h(x)) < 2\epsilon$. Thus, we have proved that $h(X_{\infty})$ is dense in Z. Hence, $h(X_{\infty}) = Z$ since X_{∞} is compact (by 2.4) and h is continuous (by (a)). Therefore, we have proved (c). This completes the proof of 2.10.

Another embedding theorem for inverse limits is in 2.36 (the exercises in 2.34 and 2.35 are used in its proof).

5. $\mathcal{P}$-LIKE CONTINUA

We conclude our discussion of inverse limits by stating the theorem in 2.13 and proving the easy part. It is a compilation of several results (see [5, p. 183], [11], and [12]; also, [14, pp. 305-306]). We remark that a special case of the entire theorem is proved in 12.19, and that the ideas we use for the proof of 12.19 generalize in a straightforward manner—see the discussion following the proof of 12.19.

We need the following two definitions.

2.11 Definition of ϵ-Map. Let X and Y be metric spaces and let f: $X \to Y$. Then, f is called an ϵ-map provided that f is continuous and the diameter of $f^{-1}(f(x)) < \epsilon$ for all $x \in X$. We remark that only the metric on X is used in this definition and, in fact, there is a straightforward generalization to topological spaces by using open covers. However, we will not use this generality (for the theory of Hausdorff continua, it is necessary).

2.12 Definition of $\mathcal{P}$-Like. Let X be a compact metric space and let $\mathcal{P}$ be a given collection of compact metric spaces. Then, X is said to be $\mathcal{P}$-like provided that for each $\epsilon > 0$, there is an ϵ-map f_{ϵ} from X onto some member Y_{ϵ} of $\mathcal{P}$. See 2.23 and 2.24.

2.13 Theorem. Every continuum is an inverse limit of compact connected polyhedra with onto bonding maps. Furthermore: If X is a continuum and $\mathcal{P}$ is a collection of compact connected polyhedra, then the coordinate spaces used in the inverse limit representation of X (with onto bonding maps) may all be chosen from $\mathcal{P}$ if and only if X is $\mathcal{P}$-like.

We note that the "only if" part of the equivalence in 2.13 is easy to prove and is true even when the coordinate spaces are not polyhedra: Assume that $X = \varprojlim\{X_i, f_i\}_{i=1}^{\infty}$ where each X_i is a (nonempty) compact metric space with metric d_i and each f_i maps onto X_i. Let $\mathcal{P} = \{X_i : i = 1, 2, \ldots\}$. Let d be the metric for X defined by (comp., [9, pp. 211–213])

$$d((x_i)_{i=1}^{\infty}, (y_i)_{i=1}^{\infty}) = \sum_{i=1}^{\infty} 2^{-i} \cdot \frac{d_i(x_i, y_i)}{1 + d_i(x_i, y_i)}$$

for each $(x_i)_{i=1}^{\infty}, (y_i)_{i=1}^{\infty} \in X$. Then, the ith projection map $\pi_i : X \to X_i$ is a 2^{-i}-map for each i (and each π_i maps onto X_i by 2.14). Hence, X is $\mathcal{P}$-like.

EXERCISES

2.14 Exercise. Prove that if $\{X_i, f_i\}_{i=1}^{\infty}$ is an inverse sequence of (nonempty) spaces with onto bonding maps, then the projections π_i map the inverse limit X_{∞} onto X_i for each i; hence, if each X_i is nondegenerate, X_{∞} is nondegenerate (comp., 2.15). The first part is used in the proof of 2.7. The second part verifies the statement following the proof of 2.7.

2.15 Exercise. We have seen how inverse limits may be thought of as nested intersections (2.3). Conversely, show how every nested intersection Y,

$$Y = \bigcap_{i=1}^{\infty} Y_i, \qquad \text{where } Y_i \supset Y_{i+1} \text{ for each } i$$

can be thought of as an inverse limit of the spaces Y_i. Conclude from this that an inverse limit of nonempty spaces may be empty when the bonding maps are not onto (comp., 2.14).

2.16 Exercise. Prove that each p-adic solenoid Σ_p (2.8) is a *topological group* (meaning, an algebraic group such that the group operations of multiplication and inversion are continuous). Deduce from this that Σ_p is *homogeneous* (meaning, for any two points there is a homeomorphism of Σ_p onto Σ_p taking one of these points to the other). Prove that each nondegenerate proper subcontinuum of Σ_p is an arc. It is interesting to note that the last two properties actually characterize solenoids (Theorem 2 of [6]).

2.17 Exercise. Let $\{X_i, f_i^b\}_{i=1}^\infty$ be the inverse sequence where each $X_i = [0,1]$ and each f_i^b for $0 \le b \le 1$ is defined by

$$f_i^b(t) = \begin{cases} 2t, & 0 \le t \le 1/2 \\ (2b-2)t + 2 - b, & 1/2 \le t \le 1. \end{cases}$$

Prove that the inverse limit $X_\infty = \varprojlim\{X_i, f_i^b\}_{i=1}^\infty$ is an arc when $1/2 < b \le 1$, the $\sin(1/x)$-continuum (1.5) when $b = 1/2$, and that X_∞ contains a nondegenerate indecomposable continuum when $0 \le b < 1/2$.

2.18 Exercise. A continuous function f from X onto Y is said to be *open* (or, *interior*) provided that if U is any open subset of X, then $f(U)$ is an open subset of Y. Prove that if $\{X_i, f_i\}_{i=1}^\infty$ is an inverse sequence where each $X_i = [0,1]$ and each f_i is an open (onto) map which is not a homeomorphism, then $\varprojlim\{X_i, f_i\}_{i=1}^\infty$ is a nondegenerate indecomposable continuum. The Buckethandle Continuum (2.9) is a special case. We remark that the same result holds when each $X_i = S^1$.

2.19 Exercise. Let $X_\infty = \varprojlim\{X_i, f_i\}_{i=1}^\infty$. Let A and B be compact subsets of X_∞, and let $C = A \cap B$. For each i, let $\pi_i : X_\infty \to X_i$ denote the ith projection map and let

$$C_i = \pi_i(A) \cap \pi_i(B).$$

Prove that $C = \varprojlim\{C_i, f_i | C_{i+1}\}_{i=1}^\infty$.

This result, as well as the one in 2.6, seem to have first appeared in the literature in [3], but are due to A. D. Wallace [3, p. 233]. They have been used, for example, in the study of multicoherence [13]. The definition of multicoherence degree is in 13.32.

2.20 Exercise. Prove that an inverse limit of arcs cannot contain a simple closed curve. [Hint: Use 2.19.]

2.21 Exercise. Prove that an inverse limit of arcs, as well as an inverse limit of simple closed curves, cannot contain a simple triod (a *simple triod* is a space which is homeomorphic to the symbol T).

2.22 Exercise. Consider the situation in the diagram below where $X_\infty = \varprojlim\{X_i, f_i\}_{i=1}^\infty$, $Y_\infty = \varprojlim\{Y_i, g_i\}_{i=1}^\infty$, and each rectangle is commutative (i.e., $\varphi_i \circ f_i = g_i \circ \varphi_{i+1}$ for each i):

$$
\begin{array}{ccccccccc}
X_1 & \xleftarrow{f_1} & X_2 & \leftarrow & \cdots & \leftarrow & X_i & \xleftarrow{f_i} & X_{i+1} & \leftarrow & \cdots & X_\infty \\
\varphi_1 \downarrow & \circlearrowleft & \downarrow \varphi_2 & & & & \varphi_i \downarrow & \circlearrowleft & \downarrow \varphi_{i+1} & & & \\
Y_1 & \xleftarrow[g_1]{} & Y_2 & \leftarrow & \cdots & \leftarrow & Y_i & \xleftarrow[g_i]{} & Y_{i+1} & \leftarrow & \cdots & Y_\infty
\end{array}
$$

Define φ_∞ on X_∞ by $\varphi_\infty((x_i)_{i=1}^\infty) = (\varphi_i(x_i))_{i=1}^\infty$ for each $(x_i)_{i=1}^\infty \in X_\infty$. Prove (a)–(d) below:

(a) φ_∞ maps X_∞ into Y_∞;
(b) if each φ_i is continuous, then φ_∞ is continuous;
(c) if each φ_i is one-to-one, then φ_∞ is one-to-one;
(d) if each φ_i maps continuously onto Y_i and if each X_i is a compact metric space, then φ_∞ maps X_∞ onto Y_∞. [Hint: Fix $(y_i)_{i=1}^\infty \in Y_\infty$, let $Z_i = \varphi_i^{-1}(y_i)$ and $k_i = f_i|Z_{i+1}$ for each i, and use the second part of 2.4.]

Applications of this exercise are in 2.23, 2.25, 2.26, and, e.g., 10.37.

2.23 Exercise. Let X be the $\sin(1/x)$-continuum (1.5). Verify that X is arc-like (2.12) by showing how to define an ϵ-map from X onto $[0,1]$ for each $\epsilon > 0$. Hence, by 2.13, X can be represented as an inverse limit of arcs (with onto bonding maps). Show how to represent X in this way by considering the commutative diagram below where each $X_i = X$, each $Y_i \subset Y_{i+1}$ is a particular arc in W (1.5) beginning at $(1,\sin(1))$, each f_i is the identity map, and each g_i and φ_i is a natural horizontal projection:

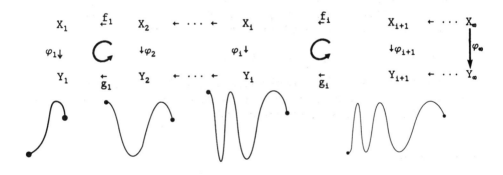

[Hint: Prove that φ_∞ (2.22) is one-to-one even though no φ_i is one-to-one.]

2.24 Exercise. Prove that the Warsaw circle (1.6) is simple closed curve-like (called *circle-like*) and not arc-like by using the definition in 2.12. Prove that the continuum constructed in 1.10 is arc-like by using 2.12.

2.25 Exercise. By using the general procedure inherent in 2.23, show that any countable cartesian product

$$\prod_{j=1}^{\infty} X_j$$

can be represented as an inverse limit of $\{Y_i, g_i\}_{i=1}^{\infty}$ with onto bonding maps g_i where

$$Y_i = \prod_{j=1}^{i} X_j, \quad \text{each } i = 1, 2, \ldots.$$

Conclude from this, e.g., that a Hilbert cube (1.4) can be represented as an inverse limit of n-cells, $n \to \infty$, with onto bonding maps.

2.26 Exercise. Let A_0 be the convex arc in the plane R^2 from $(1,0)$ to $(0,0)$ and, for each $n = 1, 2, \ldots$, let A_n be the convex arc in R^2 from $(1,0)$ to $(0,2^{-n+1})$. For each $i = 1, 2, \ldots$, let

$$X_i = A_0 \cup \left(\bigcup_{n=1}^{i+1} A_n \right), \quad Y_i = \bigcup_{n=1}^{i+1} A_n$$

and let $f_i\colon X_{i+1} \to X_i$ and $g_i\colon Y_{i+1} \to Y_i$ be the natural maps, f_i mapping A_{i+1} linearly onto A_0, g_i mapping A_{i+1} linearly onto A_i, and, in both cases, leaving all other points fixed. Prove that the two inverse limits

$$X_{\infty} = \varprojlim\{X_i, f_i\}_{i=1}^{\infty}, \quad Y_{\infty} = \varprojlim\{Y_i, g_i\}_{i=1}^{\infty}$$

are homeomorphic. [Hint: Define maps φ_i as in 2.22 which, even though they are not one-to-one, induce a one-to-one map φ_{∞}.] Let us note that the inverse limit space is (homeomorphic to)

$$\bigcup_{n=0}^{\infty} A_n.$$

With this in mind, note how, as if by magic, the arc corresponding to A_0 appears in Y_{∞}.

2.27 Exercise. A continuum X is said to be *hereditarily decomposable* provided that each nondegenerate subcontinuum of X (as well as X itself) is decomposable. All the continua we have constructed so far, except the pseudoarc (1.23), contain arcs. Obviously, hereditarily inde-

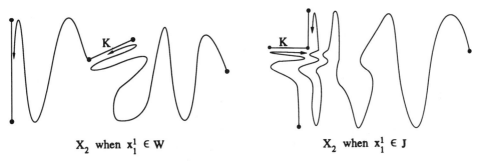

X_2 when $x_1^1 \in W$ X_2 when $x_1^1 \in J$

Figure 2.27

composable continua contain no arc. Using this fact and the existence of nondegenerate hereditarily indecomposable continua (1.23), it is easy to give an example of a decomposable continuum which contains no arc– simply take two copies of, e.g., the pseudo-arc which intersect in just one point. The purpose of this exercise is to construct an hereditarily decomposable continuum which contains no arc and to show how to do this using inverse limits–the procedure generalizes easily to produce other continua with interesting properties (see 2.32 and, e.g., [1] and [2]). As usual, the coordinate spaces X_i are constructed so that the bonding maps are natural. The reader should verify that the resulting continuum X_∞ has the desired properties.

Note the following definition: A subset D of a continuum X is said to be *continuumwise dense* in X provided that $D \cap A \neq \varnothing$ for all nonde-generate subcontinua A of X.

We start the construction by letting X_1 be the $\sin(1/x)$-continuum in 1.5. Let $p = (1/\sin(1))$, $q = (0,1)$, and $r = (0, -1)$. Let

$$D_1 = \{x_n^1 : n = 1, 2, \ldots\}$$

be a countable subset of $X_1 - \{p,q,r\}$ such that D_1 is continuumwise dense in X_1. Note that if $J = X_1 - W$ (W as in 1.5), then $\overline{D_1 \cap J} = J$. Now, let X_2 be the continuum obtained from X_1 by "replacing" the point x_1^1 with a copy Y of X_1—by this we mean that X_2 is one of the two continua drawn in Figure 2.27 below. Let K denote the arc in Y corresponding to the arc J in X_1. Let f_1 from X_2 onto X_1 be the natural map which takes K to x_1^1 and is a homeomorphism from $X_2 - K$ onto $X_1 - \{x_1^1\}$. Let s and t denote the two end points of K. Let

$$D_2 = \{x_n^2 : n = 1, 2, \ldots\}$$

be continuumwise dense in X_2 such that D_2 misses $f_1^{-1}(\{p,q,r\})$, $\{s,t\}$, and $f_1^{-1}(x_n^1)$ for all $n \geq 2$. Form X_3 from X_2 by "replacing" each of the two points x_1^2 and $f_1^{-1}(x_2^1)$ with a copy of X_1. The map f_2 from X_3 onto X_2 is defined in a manner similar to the way f_1 was defined. We obtain X_4 from X_3 by a procedure similar to the one used to obtain X_3 from X_2, this time making sure that copies of X_1 are inserted in X_3 at the first enumerated point x_1^3 of D_3 and at each of the two points $f_2^{-1}(x_2^2)$ and $(f_1 \circ f_2)^{-1}(x_3^1)$. Continuing in this fashion, we obtain the inverse sequence $\{X_i, f_i\}_{i=1}^{\infty}$. Its inverse limit X_{∞} is the desired hereditarily decomposable arcless continuum. [Hint: Roughly, X_{∞} is arcless because a nondegenerate subcontinuum A of X_{∞} must eventually project "across" an inserted copy of X_1 in some X_i, i large; thus, it can be seen by using ideas used to prove 1.12 that A is not an arc. To show that X_{∞} is hereditarily decomposable, show that each nondegenerate subcontinuum of X_{∞} has a cut point (defined in 6.1) and use 6.3.] We also note that X_{∞} is embeddable in R^2 (2.31).

2.28 Exercise. Let $X = \lim\{X_i, f_i\}_{i=1}^{\infty}$ and, for each $i = 1, 2, \ldots$, let $\pi_i \colon X \to X_i$ denote the ith projection map. Let

$$\beta = \{\pi_i^{-1}(U) \colon U \text{ is open in } X_i \text{ and } i = 1, 2, \ldots\}.$$

Prove that β, which only looks like a subbase, is actually a base for the topology on X.

2.29 Exercise. Having done the exercises in 2.23 and 2.26, can you extrapolate the technique into a general theorem? If not, it is probably because we have defined and thought of inverse limits as nested intersections. The following result gives conditions under which a compact metric space can be represented as an inverse limit of an *increasing* sequence of compact subsets.

INTERIOR APPROXIMATION THEOREM. Let Y be a compact metric space, and let $\{Y_i\}_{i=1}^{\infty}$ be a sequence of compact subsets of Y such that, for each $i = 1, 2, \ldots$, there are continuous functions g_i from Y_{i+1} onto Y_i and φ_i from Y onto Y_i such that $g_i \circ \varphi_{i+1} = \varphi_i$; if the sequence $\{\varphi_i\}_{i=1}^{\infty}$ converges uniformly to the identity map on Y, then Y is homeomorphic to $\lim\{Y_i, g_i\}_{i=1}^{\infty}$.

Prove this theorem and observe how it applies to 2.23 and 2.26 (it also applies to 2.25 when the spaces are compact metric).

The theorem appears as Lemma 4 of [4]. Simple as it is to prove, it has had profound usage in infinite-dimensional topology and continua theory (especially in the work of Curtis, Schori, and West on hyperspaces).

2.30 Exercise. In 2.23, we showed how to represent the $\sin(1/x)$-continuum as an inverse limit of arcs using 2.22. Show how to do this using 2.10 instead of 2.22.

2.31 Exercise. If the construction in 2.27 is carried out with appropriate care, then 2.10 can be applied to show that the resulting arcless continuum X_∞ is embeddable in R^2. Carry out the necessary details. We remark that since X_∞ is arc-like, its embeddability in R^2 is immediate from the general theorem in 12.20.

2.32 Exercise. Construct a nondegenerate continuum X in R^2 such that each nondegenerate subcontinuum of X separates R^2 (i.e., $R^2 - A$ is not connected for any nondegenerate subcontinuum A of X). [Hint: Use an inverse limit procedure similar to that used in 2.27, but start with Z drawn below:

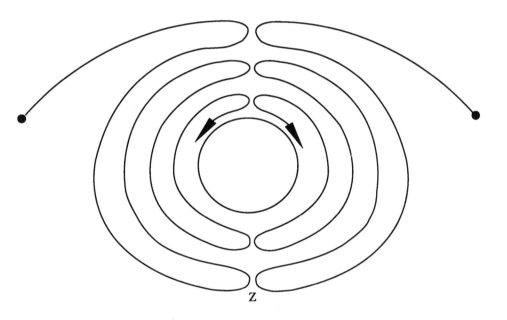

Do the "replacing" as in 2.27 but with homeomorphs of Z. Do the procedure carefully enough so that 2.10 can be applied to show that the inverse limit obtained is embeddable in R^2.]

Such an example appears in [17], but the description and verifications are much more complicated. We remark that this is another example of an hereditarily decomposable, arcless continuum (comp., 2.27).

2.33 Exercise. Let (X,d_1) and (Y,d_2) be nonempty compact metric spaces. Prove that if f is an ϵ-map from X onto Y, then there exists $\delta > 0$

such that

diameter $(f^{-1}(Z)) < \epsilon$

whenever $Z \subset Y$ and diameter $(Z) < \delta$.

2.34 Exercise. This is the first of three exercises devoted to a proof of Isbell's theorem (2.36). Prove the following result which is of a general character (comp., Lemma 2 of [7]):
Let $X = \varprojlim\{X_i, f_i\}_{i=1}^{\infty}$ where each X_i is a nonempty compact metric space with metric d_i. Let

$$f_{ij} = f_i \circ \cdots \circ f_{j-1}: X_j \to X_i \text{ if } j > i + 1 \quad \text{and} \quad f_{ii+1} = f_i.$$

Let (Y, ρ) be a complete metric space. Then, there is a sequence $\{\epsilon_i\}_{i=1}^{\infty}$, $\epsilon_i > 0$, such that if there are embeddings h_i of X_i in Y satisfying

$$\rho(h_j(x), h_i \circ f_{ij}(x)) < \delta_i/3 \quad \text{for each } x \in X_j \text{ and } j > i$$

where $\delta_i < 1/i$ and $\delta_i < \rho(h_i(y), h_i(z))$ whenever $d_i(y, z) > \epsilon_i$, then X is embedable in Y.
[Hint: Let d and π_i be as in the paragraph following 2.13 (thus, π_i is a 2^{-i}-map). Then, for each i, there exists $\eta_i > 0$ such that

$$d_i(\pi_i(x), \pi_i(y)) \geq \eta_i \quad \text{whenever } d(x, y) > 2^{-i}.$$

Let $\epsilon_i = \min\{\eta_i, 1/i\}$ for each i. Now, assume that for these numbers ϵ_i there are embeddings h_i satisfying the conditions above. Show that for each $x \in X$, $\{h_i \circ \pi_i(x)\}_{i=1}^{\infty}$ is a Cauchy sequence in Y. Then, letting

$$h(x) = \lim_{i \to \infty} h_i \circ \pi_i(x) \quad \text{for each } x \in X,$$

show that h is an embedding of X in Y.]

2.35 Exercise. This exercise does not mention inverse limits. The results are from [7, p. 78], and part (c) will be used in 2.36. Regard R^{2n} as $R_1^n \times R_2^n$ where R_1^n and R_2^n are copies of R^n. Then, a subset F of R^{2n} is said to be *flat* provided that there is a homeomorphism h of R^{2n} onto R^{2n} such that $h(F) \subset R_1^n \times \{\theta\}$ where θ denotes a point of R_2^n. Prove (a)–(c) below.

(a) If S is a closed subset of R_1^n and $f: S \to R_2^n$ is any continuous function, then the graph $G(f)$ of f is flat where

$$G(f) = \{(s, f(s)) \in R^{2n}: s \in S\}.$$

[Hint: Extend f by (i) of [9, p. 339] to a continuous function f^*: $R_1^n \to R_2^n$, and define $h: R^{2n} \to R^{2n}$ by $h(x, y) = (x, y - f^*(x))$.]

(b) If $F \subset R^{2n}$ is flat and k is any homeomorphism of R^{2n} onto R^{2n}, then $k(F)$ is flat.

(c) If S is compact and embeddable in R^n and if $f \colon S \to F$ is any continuous function from S into a flat subset F of R^{2n}, then f can be arbitrarily closely approximated by embeddings of S into flat subsets of R^{2n}. [Hint: Let d be the product metric for $R_1^n \times R_2^n$. Let $\epsilon > 0$. Let h be a homeomorphism from R^{2n} onto R^{2n} such that $h(F) \subset R_1^n \times \{\theta\}$. Since $h(f(S))$ has compact neighborhoods in R^{2n}, there exists $\delta > 0$ such that if $z \in R^{2n}$ and $d(z,h(f(s))) < \delta$, then $d(h^{-1}(z),f(s)) < \epsilon$. Let g be an embedding of S in the δ-ball about θ in R_2^n. Then, letting

$$\pi_1(x,\theta) = x \quad \text{for all } (x,\theta) \in R_1^n \times \{\theta\},$$

define $\varphi \colon S \to R^{2n}$ by

$$\varphi(s) = (\pi_1(h(f(s))),g(s)) \quad \text{for each } s \in S.$$

Show that $h^{-1} \circ \varphi$ is an embedding of S into R^{2n} within ϵ of f. Finally, use (a) and (b) to show that $h^{-1}(\varphi(S))$ is flat.]

2.36 Exercise. Prove the following embedding theorem due to J. R. Isbell (Theorem 1 of [7]):

THEOREM. If $X = \varprojlim\{X_i,f_i\}_{i=1}^\infty$ where each X_i is compact and embeddable in R^n (n fixed), then X is embeddable in R^{2n}.

[Hint: Let $\{\epsilon_i\}_{i=1}^\infty$ be as in 2.34. The proof uses (c) of 2.35 to inductively obtain embeddings h_i satisfying 2.34. Let h_1 be an embedding of X_1 in a flat subset of R^{2n}. Choose $\delta_1 < 1$ so that if $d_1(y,z) > \epsilon_1$ (d_1 the metric for X_1), then $\|h_1(y) - h_1(z)\| > \delta_1$. By (c) of 2.35, there is an embedding h_2 of X_2 in a flat subset of R^{2n} such that

$$\|h_2(x) - h_1 \circ f_1(x)\| < \delta_1/6 \quad \text{for all } x \in X_2.$$

Continue the induction so that h_j is chosen within

$$\min\{(\delta_i/3) \cdot 2^{i-j} \colon \quad i = 1, \dots, j - 1\}$$

of $h_{j-1} \circ f_{j-1}$. Then, check that this assures that h_j is within $\delta_i/3$ of $h_i \circ f_{ij}$ whenever $i < j$, as required in 2.34.]

An application of the theorem is in 12.20.

2.37 Exercise. Let $\{S_i,f_i\}_{i=1}^\infty$ be an inverse sequence where each S_i is a nonempty topological space. Assume that $i(j) < i(j + 1)$ for each $j = 1, 2, \dots$, and let

$$f_{i(j)i(j+1)}$$

$$= \begin{cases} f_{i(j)} \circ \cdots \circ f_{i(j+1)-1} \colon S_{i(j+1)} \to S_{i(j)}, & \text{if } i(j+1) > i(j)+1 \\ f_{i(j)} \colon S_{i(j+1)} \to S_{i(j)}, & \text{if } i(j+1) = i(j)+1 \end{cases}$$

Prove that $\varprojlim\{S_{i(j)}, f_{i(j)i(j+1)}\}_{j=1}^{\infty}$ and $\varprojlim\{S_i, f_i\}_{i=1}^{\infty}$ are homeomorphic. [Hint: Use 2.22.]

REFERENCES

1. R. D. Anderson and Gustave Choquet, A plane continuum no two of whose nondegenerate subcontinua are homeomorphic: an application of inverse limits, Proc. Amer. Math. Soc., 10(1959), 347–353.

2. James J. Andrews, A chainable continuum no two of whose nondegenerate subcontinua are homeomorphic, Proc. Amer. Math. Soc., 12(1961), 333–334.

3. C. E. Capel, Inverse limit spaces, Duke Math. J., 21(1954), 233–245.

4. M. K. Fort and Jack Segal, Minimal representations of the hyperspace of a continuum, Duke Math. J., 32(1965), 129–138.

5. H. Freudenthal, Entwicklungen von Räumen und ihren Gruppen, Compositio Math., 4(1937), 145–234.

6. Charles L. Hagopian, A characterization of solenoids, Pac. J. Math., 68(1977), 425–435.

7. J. R. Isbell, Embeddings of inverse limits, Ann. of Math., 70(1959), 73–84.

8. Thomas J. Jech, The Axiom of Choice, North-Holland Publ. Co., Amsterdam, Holland, 1973.

9. K. Kuratowski, Topology, Vol. I, Academic Press, New York, N.Y., 1966.

10. K. Kuratowski, Topology, Vol. II, Academic Press, New York, N.Y., 1968.

11. Ivan Lončar and Sibe Mardešić, A note on inverse sequences of ANR's, Glasnik Mat., 23(1968), 41–48.

12. Sibe Mardešić and Jack Segal, ε-mappings onto polyhedra, Trans. Amer. Math. Soc., 109(1963), 146–164.

13. Sam B. Nadler, Jr., Multicoherence techniques applied to inverse limits, Trans. Amer. Math. Soc., 157(1971), 227–234.

14. J. Nagata, *Modern General Topology*, John Wiley and Sons, Inc., New York, N.Y., 1968.

15. Jack Segal, Mapping norms and indecomposability, J. London Math. Soc., 39(1964), 598–602.

16. A. van Heemert, De R_n-adische Voortbrenging van Algemeen-Topologische Ruimten met Toepassingen op de Constructie van Niet Splitsbare Continua, Thesis, University of Groningen, 1943.

17. G. T. Whyburn, A continuum every subcontinuum of which separates the plane, Amer. J. Math., 52(1930), 319–330.

III
Decompositions of Continua

We continue the main theme of preceding chapters by examining another general method of constructing continua. We illustrate the method with specific examples and some general constructions. The method is that of upper semi-continuous decompositions. We shall first recall the definition of a decomposition space in general and see when a decomposition space of a continuum is a continuum (3.4). Then, we shall study upper semi-continuous decompositions.

1. DECOMPOSITION SPACES

3.1 Definition of Decomposition Space. Let (S,T) be a (nonempty) topological space. Let $\mathcal{D}$ be a collection of nonempty, mutually disjoint subsets of S such that $\cup \mathcal{D} = S$ (such a $\mathcal{D}$ is called a *partition* of S). Let

$$T(\mathcal{D}) = \{\mathcal{U} \subset \mathcal{D} : \cup \mathcal{U} \in T\}.$$

Note that $T(\mathcal{D})$ is a topology for $\mathcal{D}$; in fact, letting $\pi : S \to \mathcal{D}$ denote the *natural map* defined at each $x \in S$ by

$$\pi(x) = \text{the unique } D \in \mathcal{D} \text{ such that } x \in D,$$

we see that $T(\mathcal{D})$ is the largest topology for $\mathcal{D}$ such that π is continuous.

The space $(\mathcal{D}, T(\mathcal{D}))$ is called a *decomposition space* of S or, more simply, a *decomposition* of S. The topology $T(\mathcal{D})$ is called the *decomposition topology*. We emphasize that the term decomposition means a partition with the decomposition topology.

Intuitively, a decomposition is the space obtained from the original space by identifying all the points of each member of a given partition. For this reason, decompositions are frequently called *identification spaces*. They are also often called *quotient spaces* because of the intimate connection between equivalence relations and partitions in set theory.

Decompositions are an important source of examples, counterexamples, and theorems in continua theory. We will see examples of this in many places in the book. Now, however, we hasten to point out that a decomposition of a continuum X may not be a continuum even when the members of the partition are closed subsets of X (such a partition is called a *closed partition*). For example: Let $X = [-1,1]$ and let $\mathcal{D}$ be the closed partition of X given by

$$\mathcal{D} = \big\{\{x,-x\}\colon -1 < x < 1\big\} \cup \big\{\{-1\},\{1\}\big\}.$$

Then, the decomposition $(\mathcal{D}, T(\mathcal{D}))$ fails to be a continuum since it is not Hausdorff and, hence, is not metrizable. As we shall see in 3.4, not being Hausdorff is characteristic of what can go wrong.

3.2 Lemma.   If a Hausdorff space is a continuous image of a compact metric space, then it is metrizable.

Proof.   Let f be a continuous function from a compact metric space X onto a Hausdorff space Y. Then, since Y is a compact Hausdorff space, it suffices by Theorem 1 of [5, p. 241] to show that Y has a countable base. Let $\mathcal{C}$ be a countable base for X. For each finite subset $\mathcal{L}$ of $\mathcal{C}$, let

$$E(\mathcal{L}) = Y - f(X - \cup \mathcal{L}).$$

Let $\mathcal{P} = \{E(\mathcal{L})\colon \mathcal{L}$ is a finite subset of $\mathcal{C}$. Clearly, $\mathcal{P}$ is countable. Since X is compact and Y is Hausdorff, f takes closed sets to closed sets; hence, each member of $\mathcal{P}$ is an open subset of Y. Now, let U be an open subset of Y and let $q \in U$. Then, since $f^{-1}(q)$ is a compact subset of the open set $f^{-1}(U)$ and $\mathcal{C}$ is a base, there is a finite subset $\mathcal{L}$ of $\mathcal{C}$ such that

$$f^{-1}(q) \subset \cup \mathcal{L} \subset f^{-1}(U).$$

Hence, it follows easily that $q \in E(\mathcal{L}) \subset U$. Therefore, we have proved that $\mathcal{P}$ is a countable base for Y. This completes the proof of 3.2.

3.3 Theorem. A decomposition $(\mathcal{D}, T(\mathcal{D}))$ of a compact metric space X is metrizable if and only if it is Hausdorff.

Proof. Since the natural map π from X onto $\mathcal{D}$ is continuous (3.1), the "if part" of the theorem follows from 3.2. The other half of the theorem is trivial.

3.4 Theorem. A decomposition $(\mathcal{D}, T(\mathcal{D}))$ of a continuum is a continuum if and only if it is Hausdorff.

Proof. This is immediate from 3.3 and the invariance of compactness [6, p. 11] and connectedness [6, p. 128] under the continuous function π (3.1).

2. USC DECOMPOSITIONS

We shall use decompositions of compact metric spaces to construct other compact metric spaces or continua. It would be inconvenient to have to check using 3.3 each time to be sure that the decomposition is Hausdorff, hence metrizable. Therefore, we want a useful condition for this which would be easy to verify and general enough to be applicable in a lot of situations. As we shall see in 3.9, the following definition gives such a condition. Actually, the condition is both necessary and sufficient (3.26).

3.5 Definition of usc Decomposition. Let (S, T) be a topological space. A partition $\mathcal{D}$ of S is said to be *upper semi-continuous (usc)* provided that whenever $D \in \mathcal{D}$, $U \in T$, and $D \subset U$, there exists $V \in T$ with $D \subset V$ such that if $A \in \mathcal{D}$ and $A \cap V \neq \varnothing$, then $A \subset U$. Note that this condition does not take into account the decomposition topology–it is the partition that is usc. Nevertheless, when the partition is usc, we will always say the *decomposition is usc* keeping in mind that the decomposition still has the decomposition topology (3.1).

The use of the phrase "upper semi-continuous" for the condition in 3.5 is, in fact, very descriptive–see 3.25.

At this point, we head towards a proof of 3.9 and 3.10. To simplify some statements, it is convenient to have the following terminology.

3.6 Definition of $\mathcal{D}$-Saturated Set. If $\mathcal{D}$ is a decomposition of S, then any subset of S which is a union of a subcollection of $\mathcal{D}$ is said to be *$\mathcal{D}$-saturated*. Clearly, letting $\pi: S \to \mathcal{D}$ be the natural map (3.1), any $\pi^{-1}(\mathcal{C})$ for $\mathcal{C} \subset \mathcal{D}$ is $\mathcal{D}$-saturated. Also, an $A \subset S$ is $\mathcal{D}$-saturated if and only if $A = \pi^{-1}[\pi(A)]$. Hence: If V is $\mathcal{D}$-saturated and open in S, $\pi(V)$ is open in $\mathcal{D}$ (use the definitions of π and $T(\mathcal{D})$ in 3.1).

The following proposition gives two other ways to think about usc decompositions.

3.7 Proposition. Let (S,T) be a topological space, let $\mathcal{D}$ be a decomposition of S, and let $\pi: S \to \mathcal{D}$ be the natural map (3.1). Then, (1)–(3) below are equivalent.

(1) $\mathcal{D}$ is a usc decomposition;

(2) π is a closed map (i.e., π takes closed sets in S to sets which are closed with respect to $T(\mathcal{D})$);

(3) If $D \in \mathcal{D}$, $U \in T$, and $D \subset U$, then there exists $V \in T$ such that $D \subset V \subset U$ is $\mathcal{D}$-saturated.

Proof. We show that (1) implies (2), (2) implies (3), and (3) implies (1). Assume (1). Let C be a closed subset of S. We see from 3.1 that $\pi(C)$ is closed if (and only if) $\pi^{-1}[\mathcal{D} - \pi(C)]$ is open. Let

$$p \in \pi^{-1}[\mathcal{D} - \pi(C)].$$

Then, $\pi(p) \in \mathcal{D} - \pi(C)$ and, hence, $\pi(p) \subset S - C$ [since if $y \in \pi(p) \cap C$, then $\pi(y) \cap \pi(p) \neq \varnothing$ so $\pi(y) = \pi(p)$ and, thus, $\pi(p) \in \pi(C)$]. Therefore, since $S - C \in T$, there is (by 3.5) a $V \in T$ with $\pi(p) \subset V$ such that if $x \in V$, then $\pi(x) \subset S - C$. Clearly, $p \in V$. Also, $\pi(V) \subset \mathcal{D} - \pi(C)$ since if $\pi(x) \in \pi(C)$, then $\pi(x) = \pi(y)$ for some $y \in C$, so $y \in \pi(x) \cap C$, hence $\pi(x) \not\subset S - C$, therefore $x \notin V$. Thus,

$$V \subset \pi^{-1}[\mathcal{D} - \pi(C)].$$

Therefore, since $p \in V \in T$, we have proved that $\pi^{-1}[\mathcal{D} - \pi(C)]$ is open. Hence, $\pi(C)$ is closed and we have proved (2). Next, assume (2). Let $D \in \mathcal{D}$ and $U \in T$ such that $D \subset U$. Then, letting

$$V = \pi^{-1}[\mathcal{D} - \pi(S - U)]$$

it follows easily that V satisfies the conditions in (3). Clearly, (3) implies (1). This completes the proof of 3.7.

For the following simple lemma, recall that a T_1-*space* is a topological space (S,T) such that $\{x\}$ is a closed set for each $x \in S$.

3.8 Lemma. If $\mathcal{D}$ is a usc decomposition of a T_1-space (S,T), then $\mathcal{D}$ is a closed partition of S.

Proof. The easy proof is left as an exercise (3.24).

We remark that a closed partition need not be usc by the example preceding 3.2.

3.9 Theorem. Any usc decomposition of a compact metric space is metrizable.

Proof. By 3.3, it suffices to show such a decomposition is Hausdorff. Let X be a compact metric space with topology T, let $\mathcal{D}$ be a usc decomposition of X, and let $\pi: X \to \mathcal{D}$ be the natural map (3.1). To prove that $(\mathcal{D}, T(\mathcal{D}))$ is Hausdorff, let $D_1, D_2 \in \mathcal{D}$ such that $D_1 \neq D_2$. Since D_1 and D_2 are disjoint closed subsets of X (3.8) and X is normal, there exist $U_1, U_2 \in T$ such that $U_1 \cap U_2 = \varnothing$ and $D_i \subset U_i$ for each i. Since $\mathcal{D}$ is usc, 3.7 gives us $V_1, V_2 \in T$ such that $D_i \subset V_i \subset U_i$ and V_i is $\mathcal{D}$-saturated for each i. We note that $D_i \in \pi(V_i)$ for each i since $D_i \subset V_i$ and $D_i \in \mathcal{D}$. By the last comment in 3.6, $\pi(V_1)$ and $\pi(V_2)$ are open in $\mathcal{D}$. Since $U_1 \cap U_2 = \varnothing$ and $V_i \subset U_i$, $V_1 \cap V_2 = \varnothing$. Thus, since

$$\pi^{-1}[\pi(V_i)] = V_i \qquad \text{for each } i \text{ (3.6)},$$

$\pi(V_1) \cap \pi(V_2) = \varnothing$. Therefore, we have proved that $(\mathcal{D}, T(\mathcal{D}))$ is Hausdorff. This completes the proof of 3.9.

A generalization of 3.9 is in 3.36.

3.10 Theorem. Any usc decomposition of a continuum is a continuum.

Proof. Use 3.9 and the invariance of compactness [6, p. 11] and connectedness [6, p. 128] under the continuous function π (3.1).

3. EXAMPLES OF USC DECOMPOSITION

It is now appropriate to give some examples of how usc decompositions can result in interesting continua. The examples in 3.11–3.13 are very specific. The ones beginning with 3.14 are more general.

3.11 Projective n-Space. For each $n = 1, 2, \ldots$, let $\mathcal{D}$ be the partition of S^n (1.3) given by

$$\mathcal{D} = \left\{\{z, -z\}: z \in S^n\right\}.$$

It is easy to see that $\mathcal{D}$ is usc. Hence, by 3.10, the decomposition space, which is denoted by P^n, is a continuum–it is called *projective n-space*. We remark that although P^1 is a 1-sphere (3.31), the higher dimensional projective spaces have many surprising properties especially in view of how simply they are constructed. For example: P^2 contains no 2-sphere (3.31), it has the fixed point property, some simple closed curves in P^2 do not separate it, and P^2 is not embeddable in R^3 but is embeddable in R^4 (3.31). On the other hand, all the continua P^n are reasonably nice in that each P^n is an *n*-manifold.

3.12   The Moebius Strip. Let $\mathcal{D}$ be the partition of the solid square $I^2 = [0,1] \times [0,1]$ whose members are

$$\{(x,0), (1 - x,1)\} \text{ for } 0 \le x \le 1, \quad \{(x,y)\} \text{ for } 0 < y < 1.$$

It is easy to see that $\mathcal{D}$ is usc. Thus, by 3.10, the decomposition space is a continuum. It is called the *Moebius strip*, and may be thought of as an example of a one-sided surface in R^3 [2, p. 66].

3.13   The *M*-Continuum. Let X be the $\sin(1/x)$-continuum (1.5), and let $\mathcal{D}$ be the partition of X whose nondegenerate members are

$$\{(0,y), (0,1 - y)\} \quad \text{for each } y \text{ such that } 0 \le y < 1/2.$$

It follows easily that $\mathcal{D}$ is usc. Hence, by 3.10, the decomposition space is a continuum. We call it the *M-continuum* (for obvious geometrical reasons). See 3.32.

Let us note the following general, but simple, construction which will be of interest.

3.14   The Spaces *S/A*. Let (S,T) be a topological space, and let A be a nonempty closed subset of S. Let $\mathcal{D}_A$ be the partition of S given by

$$\mathcal{D}_A = \{A\} \cup \{\{z\} : z \in S - A\}.$$

It is easy to verify that $\mathcal{D}_A$ is usc. We denote the decomposition space by S/A. Intuitively, we think of S/A as being the space obtained from S by shrinking A to a point. If S is a compact metric space or a continuum, so is S/A by 3.9–3.10. Some special cases are in 3.15 and 3.16.

3.15   The Topological Cone. Let $S = X \times [0,1]$ where X is a topological space and S has the product topology. Let $A = \{(x,1) : x \in X\}$. Then, the decomposition space S/A (3.14) is called the *topological cone over X* and is denoted by $TC(X)$. The *vertex of $TC(X)$* is the point A, and the *base of $TC(X)$* is the set

$\pi(\{(x,0): x \in X\})$, π as in 3.1

although we frequently refer to X as being the base of $TC(X)$. It is easy to prove using π that $TC(X)$ is always (arcwise) connected. Hence: If X is a compact metric space, $TC(X)$ is a continuum by 3.14. In the case when X is a compact metric space, $TC(X)$ and the so-called geometrical cone are homeomorphic (3.28); hence, in this case, we will simply refer to $TC(X)$ as the *cone over X* since there will be no confusion. We remark that cones yield many examples and counterexamples in continuum theory (see, e.g., [4] and [8]). They also provide a way to think about other constructions (see, e.g., [7] and [9]–[11]). We remark that the double cone is surprisingly simple in the sense of the following result from [11]:

$TC(TC(X))$ is homeomorphic to $TC(X) \times [0,1]$.

The use of this result in [11] is particularly interesting. The result is but one of a host of "factor theorems" (it says that [0,1] can be "factored out" of the double cone). More about "factor theorems" is in 3.33.

3.16 The Topological Suspension. Starting with the topological cone $TC(X)$ with base B, the decomposition space $TC(X)/B$ (3.14) is called the *topological suspension over X* and is denoted by $TS(X)$. When X is a compact metric space, $TS(X)$ is a continuum by 3.14 and 3.15. In this case, since $TS(X)$ has a familiar geometric representation (3.29), we will simply call $TS(X)$ the *suspension over X*.

The general construction in 3.18 is one of the most important ways to obtain new spaces from old ones. First, we note the following definition for use in 3.18.

3.17 Definition of Free Union. Let (S_1,T_1) and (S_2,T_2) be topological spaces such that $S_1 \cap S_2 = \varnothing$ (without loss of generality since, for $p \neq q$, $S_1 \times \{p\}$ and $S_2 \times \{q\}$ are disjoint copies of S_1 and S_2). The *free union* of S_1 and S_2 is the topological space (S,T) where $S = S_1 \cup S_2$ and T is defined by the condition: $U \in T$ if and only if $U \cap S_i \in T_i$ for each $i = 1$, 2. The free union of S_1 and S_2 is denoted by $S_1 + S_2$.

3.18 The Attaching Spaces $S_1 \cup_f S_2$. Let (S_1,T_1) and (S_2,T_2) be topological spaces such that $S_1 \cap S_2 = \varnothing$. Let A be a nonempty closed subset of S_1, and let f be a continuous function from A into S_2. Let $\mathcal{D}$ be the partition of $S_1 + S_2$ (3.17) given by

$$\mathcal{D} = \{\{p\} \cup f^{-1}(p): p \in f(A)\} \cup \{\{x\}: x \in S_1 + S_2 - [A \cup f(A)]\}.$$

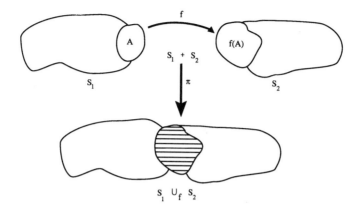

Figure 3.18

The decomposition space thus obtained is denoted by $S_1 \cup_f S_2$, as we say that S_1 *is attached by* S_2 *by* f (see Figure 3.18). Intuitively, S_1 and S_2 are sewn together along A and $f(A)$ by identifying each $a \in A$ with its image $f(a) \in S_2$. We remark that the spaces in 3.14 may easily be thought of as attaching spaces.

We leave it to the reader to verify that when S_1 and S_2 are compact metric spaces, then $\mathcal{D}$ in 3.18 is usc (3.27). This fact gives us the following result.

3.19 Theorem. If X and Y are disjoint, nonempty, compact metric spaces, then $X \cup_f Y$ is a nonempty compact metric space.

Proof. Since, clearly, $X + Y$ (3.17) is a nonempty compact metric space, the theorem follows from the statement above and 3.9.

3.20 Theorem. If X and Y are disjoint continua, then $X \cup_f Y$ is a continuum.

Proof. By 3.19, it suffices to see that $X \cup_f Y$ is connected. But this follows using the natural map (3.1)

$$\pi: X + Y \to X \cup_f Y$$

since $\pi(X)$ and $\pi(Y)$ are connected sets (by the continuity of $\pi(3.1)$)

whose union is $X \cup_f Y$ and which intersect (because for an $a \in A$, $\pi(a) = \pi(f(a))$). Therefore (Theorem 2 of [6, p. 132], $X \cup_f Y$ is connected and we have proved 3.20.

The following theorem shows how to obtain all usc decompositions of a given compact metric space. Simultaneously, it provides a useful way to obtain specific usc decompositions. Its generalization to topological spaces is in 3.23.

3.21   Theorem.   Let X be a compact metric space. If f is a continuous function from X onto a (compact) metric space Y, then, letting

$$\mathcal{D}_f = \{f^{-1}(y): y \in Y\},$$

$\mathcal{D}_f$ is a usc decomposition of X and is homeomorphic to Y. Conversely, any usc decomposition of X is a compact metric space which is a continuous image of X.

Proof.   First, suppose that $\mathcal{D}_f$ is not usc. Then, for some $f^{-1}(y_0) \in \mathcal{D}_f$, there is an open subset U of X such that $f^{-1}(y_0) \subset U$ but for which there is no V as in 3.5. It thus follows, using the compactness of $f^{-1}(y_0)$ and of $X - U$, that there are two sequences, $\{p_i\}_{i=1}^{\infty}$ and $\{q_i\}_{i=1}^{\infty}$, such that

$$p_i \to p \in f^{-1}(y_0), \qquad q_i \to q \in X - U, \qquad f(p_i) = f(q_i).$$

Hence, by the continuity of f, $f(p) = f(q)$ and, thus,

$$q \in f^{-1}(y_0) \cap (X - U)$$

which contradicts the fact that $f^{-1}(y_0) \subset U$. Therefore, $\mathcal{D}_f$ is usc. Note that $\mathcal{D}_f$ is Hausdorff by 3.9. Thus, to prove $\mathcal{D}_f$ is homeomorphic to Y, it suffices to verify the continuity of the one-to-one function h from the compact space Y onto $\mathcal{D}_f$ defined by

$$h(y) = \pi[f^{-1}(y)], \qquad \text{each } y \in Y.$$

The continuity of h follows from the "Transgression Lemma" in 3.22 (since f is a "quotient map" in the terminology of 3.22). Therefore, half of 3.21 is proved. The other half is easy since if $\mathcal{D}$ is any usc decomposition of X, $\mathcal{D}$ is a metric space (3.9) and, by using $\pi: X \to \mathcal{D}$ (3.1), we see that $\mathcal{D}$ is a continuous, hence compact, image of X. This completes the proof of 3.21.

The theorem in 3.21 may be used as follows. Let X be a compact metric space. If we want to construct, or show the non-existence of, usc decompositions of X with certain properties, we may accomplish this more easily using 3.21 and working externally by means of the continuous functions on X (or special types of them). On the other hand: If we want to show the existence, or nonexistence, of continuous images of X with certain properties, it may be best to use 3.21 and work internally by means of usc decompositions (or particular kinds of them). Let us illustrate this technique as follows. Can you imagine a usc decomposition of $I = [0,1]$ which is homeomorphic to the solid square $I \times I$? There is one by 7.9 and 3.21. In fact, given any Peano continuum Y whatsoever, there is, by 8.14 and 3.21, a usc decomposition of I which is homeomorphic to Y! We shall see other examples of this technique in action later (for example, see the discussion following the proof of 12.11).

More about the general theory of usc decompositions may be found in, e.g., [14, pp. 122–136]. We shall define and briefly discuss continuous decompositions in section 2 of Chapter XIII.

EXERCISES

3.22 Exercise. Let (S_1, T_1) and (S_2, T_2) be topological spaces. A map f from S_1 onto S_2 is called a *quotient map* (or, an *identification map*) provided that T_2 is the largest topology such that f is continuous. For example, π in 3.1 is a quotient map, as is any continuous map from a compact space onto a Hausdorff space (more generally, see (b) of 3.23). Prove the following general result, sometimes called the Transgression Lemma, which we used in the proof of 3.21 and which we shall often use to obtain maps on decomposition spaces: *Let (S_i, T_i) be topological spaces for each $i = 1, 2,$ and 3, let $f: S_1 \to S_2$ be a quotient map, and let $g: S_1 \to S_3$ be continuous; then, if g is constant on each "fiber" $f^{-1}(z)$ (i.e., $g \circ f^{-1}$ is single valued), $g \circ f^{-1}$ is continuous.*

$$
\begin{array}{ccc}
 & f & \\
S_1 & \to & S_2 \\
g \searrow & \swarrow & g \circ f^{-1} \\
 & S_3 &
\end{array}
$$

3.23 Exercise. Although this book is concerned with metric spaces, we include this exercise to give a broader perspective of 3.7 and 3.21.

Recall that a continuous function f from X onto Y is said to be a *closed map* provided that if C is any closed subset of X, then $f(C)$ is a closed subset of Y. The definition of an open map is in 2.18.

Now, assume that (S,T) and (Y,T') are nonempty topological spaces, f is a continuous function from S onto Y,

$$\mathcal{D}_f = \{f^{-1}(y): y \in Y\}$$

with the decomposition topology, $\pi: S \to \mathcal{D}_f$ is as in 3.1, and T'_f is the largest topology for Y such that f is continuous. Prove (a)–(c) below.

    (a) $T' = T'_f$ if and only if $\mathcal{D}_f$ is homeomorphic to Y by a homeomorphism $h: Y \to \mathcal{D}_f$ which can be chosen (uniquely) so that $h \circ f = \pi$. [Hint: Use 3.22.]

    (b)   If f is an open map or a closed map, then $T' = T'_f$ (thus, f is a quotient map (3.22)).

    (c)   If $T' = T'_f$, then f is a closed map if and only if $\mathcal{D}_f$ is usc. [Hint: Use 3.7.]

These results give full generality to 3.21. Also, when combined with (1) and (2) of 3.7, they show that the continuous images of a space S under closed maps and the usc decompositions of S are the same; therefore, the study of usc decompositions is really the study of continuous closed maps. Finally, the following comments about (a) are appropriate: The properties of T'_f in (a) show the real reason the term quotient map was a fitting phrase in 3.22, they show why T'_f is frequently referred to as the *quotient topology induced on Y by f*, and they show why (Y,T'_f) is called a *quotient space of S*. Indeed, we could have started in 3.1 with this seemingly more general approach. However, it seems more natural to have a decomposition space of S be defined intrinsically, i.e., in terms of S.

    3.24 Exercise. Prove 3.8. In relation to 3.8, prove the following result: If (S,T) is a topological space and $\mathcal{D}$ is a partition of S, then the decomposition space $(\mathcal{D}, T(\mathcal{D}))$ is a T_1-space if and only if $\mathcal{D}$ is a closed partition.

    3.25 Exercise. The use of the terminology "upper semi-continuous" in connection with the condition in 3.5 probably comes from the following consideration. In real analysis, a function $f: R^1 \to R^1$ is called a *usc function* provided that

$$\limsup_{t \to p} f(t) \le f(p), \qquad \text{each } p \in R^1$$

(which is the same as saying f is continuous from R^1 with the usual topology to R^1 with the topology consisting of R^1, $\varnothing$, and all sets of the form $(-\infty, x)$). Now, let $f: R^1 \to [0, \infty)$ be a function. For each $x \in R^1$, let

$$D_x = \{(x,y) \in R^2: 0 \le y \le f(x)\}.$$

Let $S = \bigcup \{D_x : x \in R^1\}$, and let $\mathcal{D} = \{D_x : x \in R^1\}$. Prove that $\mathcal{D}$ is a usc decomposition of S if and only if f is a usc function.

3.26 Exercise. Half of the following result is 3.9: A decomposition of a compact metric space is metrizable if and only if it is usc. Give the one-sentence proof (using a previous result) for the other half. Find an example of a non-usc decomposition of a metric space such that the decomposition space is an arc.

The first part of this exercise shows, e.g., that the only way a decomposition of a continuum can be a continuum is if it is usc (comp., 3.10).

3.27 Exercise. Verify the statement above 3.19 (which was used in the proof of 3.19) by a straightforward argument using sequences. Now, consider the following general situation. Let (S_1,T_1), (S_2,T_2), A, and f: $A \to S_2$ be as in 3.18. Let

$$\pi: S_1 + S_2 \to S_1 \cup_f S_2$$

be the natural map (3.1). Prove (c) below which, together with 3.9, yields 3.19 as a simple corollary. Prove (a) first, use (a) to prove (b), and then use (b) to prove (c).

(a) $\pi^{-1}[\pi(C)] = C \cup f(C \cap A) \cup f^{-1}[f(C \cap A)] \cup f^{-1}(C \cap S_2)$ for all $C \subset S_1 + S_2$.

(b) If $C \subset S_1 + S_2$ such that $C \cap S_1$ is closed in S_1, then $\pi(C)$ is closed in $S_1 \cup_f S_2$ if and only if $(C \cap S_2) \cup f(C \cap A)$ is closed in S_2.

(c) If f takes closed subsets of A onto closed subsets of S_2, then $S_1 \cup_f S_2$ is a usc decomposition of $S_1 + S_2$.

3.28 Exercise. In this exercise, we define the geometrical cone and determine its relation to the topological cone (3.15). Let $I = [0,1]$ and let

$$I^\infty = \prod_{i=1}^\infty I_i \quad \text{where each } I_i = I \,(1.4).$$

Let X be a separable metric space, and assume, without loss of generality by Theorem 1 of [5, p. 241], that $X \subset I^\infty$. Let θ denote the point in I^∞ all of whose coordinates are equal to zero. In the space $I^\infty \times I$, and for each point $(x,0)$ such that $x \in X$, form the convex segment L_x from $(\theta,1)$ to $(x,0)$,

$$L_x = \{t \cdot (\theta,1) + [1 - t] \cdot (x,0): 0 \le t \le 1\}.$$

The *geometric cone over X, $GC(X)$*, is defined by

$$GC(X) = \bigcup_{x \in X} L_x.$$

Prove that if X is compact, then $GC(X)$ and $TC(X)$ are homeomorphic. Show that if X is a countably infinite discrete space, then $GC(X)$ and $TC(X)$ are not homeomorphic.

3.29 Exercise. Formulate the appropriate definition for the *geometrical suspension*, $GS(X)$, over a separable metric space X using a construction similar to the one used in 3.28. Prove that if X is compact, then $GS(X)$ and $TS(X)$ (3.16) are homeomorphic.

3.30 Exercise. We examine how the operation of forming cones (3.15) behaves with respect to inverse limits. Let

$$X = \varprojlim \{X_i, f_i\}_{i=1}^{\infty} \qquad (2.2)$$

where each X_i is a compact metric space and each f_i maps onto X_i. Form the diagram below in which $I = [0,1]$, $j: I \to I$ is the identity map, each π_i is the natural map (3.1), and each g_i is chosen so that each rectangle is commutative (prove g_i exists and is continuous using 3.22).

Note that the inverse limit for the top row of this diagram is homeomorphic to $X \times I$. Using properties of this diagram, 2.22, and 3.22, prove that the inverse limit for the bottom row of this diagram is homeomorphic to $TC(X)$. Because of the natural way everything works, including the final homeomorphism, we summarize what has been shown by saying: *The cone over an inverse limit is the inverse limit of the respective cones*, or, more simply, *cones commute with inverse limits*. Prove an analogous assertion for suspensions.

3.31 Exercise. Recall the definition of projective n-space P^n in 3.11. Prove (a)–(e) below.
    (a) P^n is homeomorphic to the decomposition space of $R^{n+1} - \{\theta\}$,

θ = origin, obtained from

$$\mathcal{D} = \{D_v : v \in R^{n+1} - \{\theta\}\}$$

where each $D_v = \{t \cdot v : t \in R^1 - \{0\}\}$.

(b) P^n is homeomorphic to the decomposition space of B^n (1.2) obtained from

$$\mathcal{D} = \{\{x, -x\} : x \in S^{n-1}\} \cup \{\{x\} : x \in B^n - S^{n-1}\}.$$

where, for the case when $n = 1$, we assume $S^0 = \{-1, 1\}$. Hence, in particular, P^1 is a 1-sphere.

(c) P^n contains no n-sphere for $n \neq 1$.

(d) P^2 is embeddable in R^4. [Hint: Define $f : S^2 \to R^4$ by

$$f(x, y, z) = (x^2 - y^2, xy, xz, yz)$$

for each $(x, y, z) \in S^2$, and show how f induces the desired embedding.] It is more difficult to prove that P^2 is not embeddable in R^3.

(e) P^2 is homeomorphic to the attaching space $B^2 \cup_f M$ where B^2 is the 2-cell in 1.2, M is the Moebius strip in 3.12, and f is a homeomorphism from S^1 onto the manifold boundary βM,

$$\beta M = \{\pi(x, y) \in M : x = 0 \text{ or } 1\}, \qquad \pi \text{ as in 3.1.}$$

3.32 Exercise. Let X be the $\sin(1/x)$-continuum (1.5) and let Y be the M-continuum (3.13). Prove that Y is arc-like (as is X (2.23)). By the construction in 3.13, Y is a continuous image of X–prove that X is a continuous image of Y, but that X and Y are not homeomorphic. Now, let Z be the continuum in R^2 drawn in Figure 3.32 (Z is a vertical arc A together with a half-line spiraling around A and getting closer and closer to A). The continuum Z may seem to be more different from X than Y is. However, this is because of the embedding, for it is easy to show that X and Z are homeomorphic. The continua X and Y are examples of compactifications of a half-line with an arc as the remainder. Prove that all such continua are arc-like. Find examples of two such continua such that neither is a continuous image of the other. If you think we have drifted away from the main topic of this chapter, re-read the paragraph following the proof of 3.21. In the spirit of continuum theory, we are investigating usc decompositions via continuous functions.

3.33 Exercise. A space X is said to be a *factor of a space Y*, or a *Y-factor*, provided that there is a space Z such that $X \times Z$ is homeomorphic to Y. We stated a ''factor theorem'' in 3.15. Prove the following result about Q-factors where Q denotes the Hilbert cube (1.4):

If X is a Q-factor, then $X \times Q$ is homeomorphic to Q.

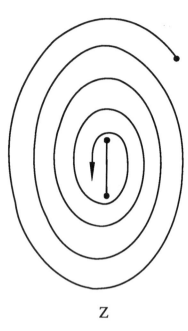

Z

Figure 3.32

[Hint: The proof is easy using the observation that a countable product of Hilbert cubes is a Hilbert cube.] We note it is known that the Q-factors are precisely the compact absolute retracts ([3], [12], [13]). This remarkable result shows, e.g., that the cartesian product of a simple triod with Q is a Hilbert cube [1].

3.34 Exercise. Prove that the usc decomposition of R^2 obtained by shrinking the y-axis to a point (3.14) is not metrizable. Compare with 3.35 and 3.36.

3.35 Exercise. Let $\mathcal{D} = \{D_t : t \in R^1\}$ be the partition of R^2 where for each $t \in R^1$, $D_t = \{(t,y) \in R^2 : y \in R^1\}$. Prove that $\mathcal{D}$ is not usc, but that the decomposition space is homeomorphic to R^1.

3.36 Exercise. Prove that if $\mathcal{D}$ is a usc decomposition of a separable metric space X such that each member of $\mathcal{D}$ is compact, then the decomposition space is a (separable) metric space. [Hint: Use (3) of 3.7 and ideas in the proof of 3.2.] Note that this result generalizes 3.9, and compare this result with 3.34.

3.37   Exercise.   Let (S,T) be a topological space, and let $(\mathcal{D}, T(\mathcal{D}))$ be a decomposition space of S. Prove that $\mathcal{D}$ is usc if and only if

$$\{D \in \mathcal{D}: D \subset U\} \in T(\mathcal{D}) \text{ for all } U \in T.$$

REFERENCES

1.   R. D. Anderson, The Hilbert cube as a product of dendrons, Notices Amer. Math. Soc., 11(1964), Abstract no. 614–649.

2.   B. H. Arnold, *Intuitive Concepts in Elementary Topology*, Prentice Hall, Inc., Englewood Cliffs, N.J., 1962.

3.   T. A. Chapman, *Lectures on Hilbert Cube Manifolds*, Conf. Board of the Math. Sci. Regional Conf. Series in Math., no. 28, Amer. Math. Soc., Providence, R.I., 1975.

4.   Ronald J. Knill, Cones, products, and fixed points, Fund. Math., 60(1967), 35–46.

5.   K. Kuratowski, *Topology*, Vol. I, Academic Press, New York, N.Y., 1966.

6.   K. Kuratowski, *Topology*, Vol. II, Academic Press, New York, N.Y., 1968.

7.   Sam B. Nadler, Jr., Continua whose cone and hyperspace are homeomorphic, Trans. Amer. Math. Soc., 230(1977), 321–345.

8.   Sam B. Nadler, Jr. and J. T. Rogers, Jr., A note a hyperspaces and the fixed point property, Colloq. Math., 25(1972), 255–257.

9.   J. T. Rogers, Jr., The cone = hyperspace property, Can. J. Math., 24(1972), 279–285.

10.   J. T. Rogers, Jr., Continua with cones homeomorphic to hyperspaces, Gen. Top. and its Appls., 3(1973), 283–289.

11.   R. M. Schori, Hyperspaces and symmetric products of topological spaces, Fund. Math., 63(1968), 77–88.

12.   H. Toruńczyk, On CE-images of the Hilbert cube and characterization of Q-manifolds, Fund. Math., 106(1980), 31–40.

13.   James E. West, Infinite products which are Hilbert cubes, Trans. Amer. Math. Soc., 150(1970), 1–25.

14.   Gordon Thomas Whyburn, *Analytic Topology*, Amer. Math. Soc. Colloq. Publ., vol. 28, Amer. Math. Soc., Providence, R.I., 1942.

IV

Limits of Sets

Many, if not all properties of continua are best examined by means of sequences of sets. We have evidence of this from preceding chapters. For example, nested sequences of sets were used to show there are indecomposable continua (1.10), to construct the Sierpiński Universal Curve (1.11), and to study inverse limits (2.3 and its consequences). From this point of view, the sequences of sets were external in the sense that they came from outside the continua to be examined. We have used internal sequences, and seen the value of doing so, but only in a very limited context (specifically, in 2.29 and the special cases referred to there). The purpose of this chapter is to lay the foundation for using internal sequences of sets to study continua. This is done by defining a notation of convergence of sets and determining its basic properties. The theory is facilitated by the use of auxiliary spaces, called hyperspaces, which we now define.

1. HYPERSPACES

4.1 Definition of Hyperspaces. For a topological space X, we let:

(1) $2^X = \{A: A$ is a nonempty closed subset of $X\}$;

(2) $C(X) = \{A \in 2^X: A$ is connected$\}$.

Now, let X be a compact metric space with metric d. For the purpose of defining an appropriate metric H_d for 2^X, we let, for each $\epsilon > 0$ and each $A \in 2^X$,

(3) $N_d(\epsilon,A) = \{x \in X: d(x,a) < \epsilon \text{ for some } a \in A\}$.

Now, for each $A, B \in 2^X$, let

(4) $H_d(A,B) = \text{glb } \{\epsilon > 0: A \subset N_d(\epsilon,B) \text{ and } B \subset N_d(\epsilon,A)\}$.

We will see in 4.2 that H_d is a metric and, in 4.6, that the topology for 2^X obtained from it is an invariant of equivalent metrics for X. The spaces 2^X and $C(X)$ with the topology obtained from H_d are called *hyperspaces* of X; H_d is called the *Hausdorff metric* induced by d. Intuitively, A and B are close together with respect to H_d provided that each point of A is close to a point of B and A and B are of approximately the same "geometric size" (definitive meaning is given to this in 4.33).

4.2 Theorem. The Hausdorff metric H_d defined in 4.1 is, in fact, a metric.

Proof. We only verify the triangle inequality, since the other properties H_d must have to be a metric are evident. First, to expedite the proof, note the following fact (which is easy to prove using the compactness of L):

(1) If $K, L \in 2^X$ and $x \in K$, then there exists $y \in L$ such that $d(x,y) \le H_d(K,L)$.

Now, let $A, B, C \in 2^X$. We show that

(*) $H_d(A,C) \le H_d(A,B) + H_d(B,C)$.

Let $\delta > 0$. Let $a \in A$. Then, by (1), there exists $b \in B$ such that $d(a,b) \le H_d(A,B)$. Now, having $b \in B$ and using (1) again, there exists $c \in C$ such that $d(b,c) \le H_d(B,C)$. Hence, using the triangle inequality for d, we have

$d(a,c) \le H_d(A,B) + H_d(B,C)$.

Thus, having started with any point $a \in A$, we have proved

$A \subset N_d(H_d(A,B) + H_d(B,C) + \delta,C)$.

A similar argument shows that

$C \subset N_d(H_d(A,B) + H_d(B,C) + \delta,A)$.

Therefore, $\delta > 0$ being arbitrary, (*) now follows. This proves 4.2.

The convergence of sets we shall use in studying continua is convergence with respect to the Hausdorff metric. We want to understand this convergence in terms of a notion of "convergence" in the original space X. We shall define convergence from this point of view in 4.9. First, we obtain an appropriate description of the Hausdorff metric topology in terms of the open sets in X. This is done in 4.5. We shall use the following notation.

4.3 Notation.   Let X be a compact metric space with topology T. For each $U \in T$, let:

(1) $\Gamma(U) = \{A \in 2^X : A \subset U\}$;

(2) $\Lambda(U) = \{A \in 2^X : A \cap U \neq \emptyset\}$.

For $U_1, \ldots, U_n \in T$, $n < \infty$, let

(3) $\langle U_1, \ldots, U_n \rangle$

$$= \left\{ A \in 2^X : A \subset \bigcup_{i=1}^{n} U_i \text{ and } A \cap U_i \neq \emptyset \text{ for each } i \right\}.$$

4.4 Lemma.   The set-theoretic relationships below in (1)–(3) hold among the sets in 4.3:

(1) $\Gamma(U) = \langle U \rangle$ and $\Lambda(U) = \langle X, U \rangle$;

(2) $\langle U_1, \ldots, U_n \rangle = \left[\Gamma \left(\bigcup_{i=1}^{n} U_i \right) \right] \cap \left(\bigcap_{i=1}^{n} \Lambda(U_i) \right)$;

and, letting $U = \bigcup_{i=1}^{n} U_i$ and $V = \bigcup_{i=1}^{m} V_i$,

(3) $\langle U_1, \ldots, U_n \rangle \cap \langle V_1, \ldots, V_m \rangle = \langle V \cap U_1, \ldots, V \cap U_n, U \cap V_1, \ldots, U \cap V_m \rangle$.

Proof.   The proof only uses the definitions in 4.3 and simple manipulations of sets. We leave the details to the reader.

4.5 Theorem.   Let X be a compact metric space with metric d and topology T. Let (using notation in 4.3):

$\mathbb{C} = \{\langle U_1, \ldots, U_n \rangle : U_i \in T \text{ for each } i = 1, \ldots, n\}$;

$\mathcal{P} = \{\Gamma(U) : U \in T\} \cup \{\Lambda(U) : U \in T\}$.

Then: $\mathbb{C}$ is a base for the topology obtained from the Hausdorff metric H_d for 2^X, and $\mathcal{P}$ is a subbase for that topology.

Proof. By (3) of 4.4 and the observation that $\langle X \rangle = 2^X$, we have that
(a) $\mathbb{C}$ is a base for some topology T_V for 2^X.
Let $[\mathcal{P}] = \{\cap \, \mathcal{L}: \mathcal{L}$ is a finite subset of $\mathcal{P}\}$. By (2) of 4.4, $\mathbb{C} \subset [\mathcal{P}]$. Since, by (3) of 4.4, $\mathbb{C}$ is closed under finite intersections, and since, by (1) of 4.4, $\mathcal{P} \subset \mathbb{C}$, clearly $[\mathcal{P}] \subset \mathbb{C}$. Hence, $[\mathcal{P}] = \mathbb{C}$. Thus, by (a), we have that
(b) $\mathcal{P}$ is a subbase for T_V.
Now, let T_H denote the topology obtained from the Hausdorff metric H_d for 2^X. In view of (a) and (b), the proof of 4.5 will be complete once we show that $T_V = T_H$. For use in the proof, let, for $A \in 2^X$ and $\epsilon > 0$,

$$B_H(A, \epsilon) = \{K \in 2^X: H_d(A, K) < \epsilon\}$$

and let, for $L, M \in 2^X$,

$$d(L, M) = \mathrm{glb}\, \{d(x, y): x \in L \text{ and } y \in M\}.$$

We first show that $T_V \subset T_H$. Let $U \in T$ such that $U \neq X$. Let $A \in \Gamma(U)$. Then, letting

$$\epsilon = d(A, X - U),$$

we see that $\epsilon > 0$ and $B_H(A, \epsilon) \subset \Gamma(U)$. Next, assume that $A \in \Lambda(U)$. Then, fixing a point $p \in A \cap U$ and this time letting

$$\epsilon = d(\{p\}, X - U),$$

we see that $\epsilon > 0$ and $B_H(A, \epsilon) \subset \Lambda(U)$. So far, we have proved that

$$\Gamma(U), \Lambda(U) \in T_H \text{ when } U \in T \text{ and } U \neq X.$$

Thus, since $\Gamma(X) = \Lambda(X) = 2^X$, we have proved that $\mathcal{P} \subset T_H$. Therefore, by (b), $T_V \subset T_H$. It remains to show that $T_H \subset T_V$. To do this, it suffices by (a) to show that for any open ball $B_H(A, \epsilon)$ in 2^X, there exist $U_1, \ldots, U_n \in T$ such that

$$A \in \langle U_1, \ldots, U_n \rangle \subset B_H(A, \epsilon).$$

Fix $A \in 2^X$ and $\epsilon > 0$. Then, since A is a nonempty compact subset of X, there is a cover of A by finitely many open subsets $U_1, \ldots, U_n$ of X such that, for each i, $U_i \cap A \neq \varnothing$ and the diameter of U_i is $< \epsilon$. Clearly, $A \in \langle U_1, \ldots, U_n \rangle$ and, using this fact, it follows easily that $\langle U_1, \ldots, U_n \rangle \subset B_H(A, \epsilon)$. Therefore, we have proved that $T_H \subset T_V$. This completes the proof of 4.5.

4.6   Corollary. Let X be a compact metric space. Then, the topology obtained from the Hausdorff metric for 2^X depends only on the topology on X (i.e.: If d and D are metrics for X each giving the topology T on X, then the topology for 2^X obtained from H_d and the one obtained from H_D are identical).

Proof. The corollary follows immediately from 4.5.

The result in 4.6 shows directly that homeomorphic compact metric spaces X and Y yield homeomorphic hyperspaces 2^X and 2^Y. Even more is true:

4.7 Theorem. Let X and Y be homeomorphic compact metric spaces, and let $h: X \to Y$ be a homeomorphism from X onto Y. Then, there is a homeomorphism $h*$ from 2^X onto 2^Y such that $h*[C(X)] = C(Y)$.

Proof. Define $h*: 2^X \to 2^Y$ by $h*(A) = h(A)$ for each $A \in 2^X$. Clearly, $h*$ maps 2^X onto 2^Y in one-to-one fashion and $h*[C(X)] = C(Y)$. To prove the continuity of $h*$ and its inverse, fix metrics d and ρ for X and Y respectively, and define D by

$$D(x_1,x_2) = \rho(h(x_1),h(x_2)), \qquad \text{each } (x_1,x_2) \in X \times X.$$

Then: D is a metric giving the topology on X, and it is easy to check that

$$H_\rho(h*(A),h*(B)) = H_D(A,B), \qquad \text{each } A, B \in 2^X,$$

provided, of course, that H_ρ and H_D are defined in terms of ρ and D (respectively) specifically as in 4.1. Hence, $h*$ is a homeomorphism. Therefore, using 4.6, we have proved 4.7.

2. LIM INF, LIM SUP, LIM

Having expressed the Hausdorff metric topology for 2^X in terms of open sets in X (4.5), we now want to give an appropriate description of convergence with respect to the Hausdorff metric by means of a notion of "convergence in X". This is done in 4.8–4.11, the accuracy of the description being verified by 4.11.

4.8 Definition of Lim Inf, Lim Sup. Let (S,T) be a topological space, and let $\{A_i\}_{i=1}^\infty$ be a sequence of subsets of S. We define lim inf A_i and lim sup A_i as follows:

lim inf $A_i = \{x \in S$: for each $U \in T$ such that $x \in U$, $U \cap A_i \neq \emptyset$ for all but finitely many $i\}$;

lim sup $A_i = \{x \in S$: for each $U \in T$ such that $x \in U$, $U \cap A_i \neq \emptyset$ for infinitely many $i\}$.

4.9 Definition of Lim. Let (S,T) and $\{A_i\}_{i=1}^\infty$ be as in 4.8, and let $A \subset S$. We write *lim* $A_i = A$ to mean

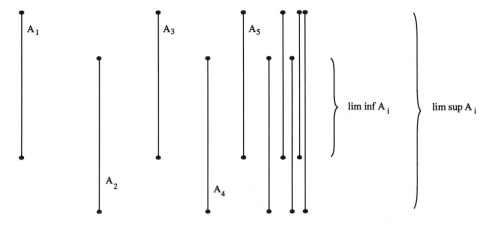

Figure 4.8

$\lim \inf A_i = A = \lim \sup A_i$.

To become more familiar with the notions introduced above, the reader may at this point want to work the exercises in 4.20 and 4.21.

We note the following simple, but useful lemma.

4.10 Lemma. Let (S,T) be a topological space, let $\{A_i\}_{i=1}^{\infty}$ be a sequence of subsets of S, and let $A \subset S$. Then, (1) and (2) below are equivalent:

(1) $\lim A_i = A$;
(2) $A \subset \lim \inf A_i$ and $\lim \sup A_i \subset A$.

Proof. The lemma is immediate from 4.8 and 4.9 since, clearly, $\lim \inf A_i \subset \lim \sup A_i$. $\blacksquare$

We are now ready to prove the following fundamental result.

4.11 Theorem. Let X be a compact metric space, and let $\{A_i\}_{i=1}^{\infty}$ be a sequence of nonempty compact subsets of X. Then, $\lim A_i = A$ if and only if $\{A_i\}_{i=1}^{\infty}$ converges to A in 2^X with respect to the Hausdorff metric H.

Proof. Assume that $\lim A_i = A$ (4.9). We first show that $A \in 2^X$. Since each $A_i \neq \varnothing$, it follows easily using the compactness of X that $\lim \sup A_i \neq \varnothing$; hence, $A \neq \varnothing$. It is easy to verify that $\lim \sup A_i$ is a closed subset of X; hence, A is closed in X. Therefore, $A \in 2^X$. We now show that $\{A_i\}_{i=1}^{\infty}$ converges to A with respect to H. We will use 4.5. Let $U_1, ..., U_n$ be open subsets of X such that $A \in \langle U_1, ..., U_n \rangle$. Since $A \cap U_j \neq \varnothing$ for each j and since $\lim \inf A_i = A$, there exists N_1 such that

(1) $A_i \cap U_j \neq \varnothing$,   all $i \geq N_1$, all $j \leq n$.

Since $A \subset \bigcup_{j=1}^n U_j$ and since $\limsup A_i = A$ and $X - \bigcup_{j=1}^n U_j$ is compact, it follows easily that there exists N_2 such that

(2) $A_i \subset \bigcup_{j=1}^n U_j$,   all $i \geq N_2$.

Now, letting $N = \max\{N_1, N_2\}$, we have, by (1) and (2), that

(3) $A_i \in \langle U_1, \ldots, U_n \rangle$,   all $i \geq N$.

Therefore, recalling 4.5, we have shown that $\{A_i\}_{i=1}^\infty$ converges to A with respect to H. To prove the converse, assume that $\{A_i\}_{i=1}^\infty$ converges to A with respect to H (note, this time, that $A \in 2^X$ automatically). We prove that $\lim A_i = A$ showing that (2) of 4.10 holds. To show that

(*) $A \subset \liminf A_i$,

let $p \in A$. Let U be an open subset of X such that $p \in U$. Since it is easy to see using the definitions in 4.1 and 4.8 that (*) holds if $A = \{p\}$, we assume for the proof of (*) that $A \neq \{p\}$. Then, $A \in \langle U, X - \{p\} \rangle$. Hence, $\langle U, X - \{p\} \rangle$ being an open subset of 2^X (4.5), there exists N such that for all $i \geq N$, $A_i \in \langle U, X - \{p\} \rangle$. Thus, $A_i \cap U \neq \varnothing$ for all $i \geq N$. Therefore, we have proved that $p \in \liminf A_i$ and, hence, we have proved (*). To show that

(#) $\limsup A_i \subset A$,

let $x \in X - A$ (if $A = X$, (#) is trivial). Let W be an open subset of X such that $A \subset W$ and $x \notin \overline{W}$. Since $A \in \langle W \rangle$ and $\langle W \rangle$ is an open subset of 2^X (4.5), there exists N such that for all $i \geq N$, $A_i \in \langle W \rangle$, i.e., $A_i \subset W$. Hence, $A_i \cap (X - \overline{W}) = \varnothing$ for all $i \geq N$. Thus, since $X - \overline{W}$ is an open subset of X and $x \in X - \overline{W}$, we see that $x \notin \limsup A_i$. Therefore, we have proved (#). This completes the proof of 4.11.

4.12   Terminology and Notation.   Let X be a compact metric space, and let $\{A_i\}_{i=1}^\infty$ be a sequence of nonempty compact subsets of X. Then, in light of 4.6 and 4.11, we say $\{A_i\}_{i=1}^\infty$ *converges* to A and write $A_i \to A$ to mean, without confusion, that $\lim A_i = A$ (4.9) or that $\lim_{i \to \infty} H(A_i, A) = 0$ where H is a Hausdorff metric for 2^X. Whether we should think of the convergence as taking place in 2^X or in X depends in general on the situation, and the context will indicate which is appropriate. We remark that our viewpoint will usually be that it takes place in X.

3. CONVERGENCE THEOREMS

As mentioned at the beginning of the chapter, we want to use sequences of subsets of a continuum X to study properties of X. For this purpose, it is important to know about the existence of convergent subsequences, and to know when the limit is connected. The results in 4.13–4.18 give basic information about this.

4.13 Theorem. If X is a compact metric space, then 2^X is compact.

Proof. Let $\mathcal{P}$ be as in 4.5. By the Alexander Subbase Lemma [6, p. 4] and by 4.5, it suffices to show that every cover of 2^X by members of $\mathcal{P}$ has a finite subcover. To this end, assume that $\mathcal{L} \subset \mathcal{P}$.

$$\mathcal{L} = \{\Gamma(U_\sigma): \sigma \in \Sigma\} \cup \{\Lambda(V_\omega): \omega \in \Omega\},$$

such that $2^X = \cup \mathcal{L}$, i.e.,

$$2^X = \left[\bigcup_{\sigma \in \Sigma} \Gamma(U_\sigma)\right] \cup \left[\bigcup_{\omega \in \Omega} \Lambda(V_\omega)\right].$$

Let $Y = X - \cup \{V_\omega: \omega \in \Omega\}$. We consider two cases.

Case 1: $Y = \varnothing$. Then, $X = \cup \{V_\omega: \omega \in \Omega\}$. Thus, since X is compact, there is a finite subset Ω_0 of Ω such that

$$X = U \{V_\omega: \omega \in \Omega_0\}.$$

Hence, it follows immediately (4.3) that

$$2^X = \bigcup_{\omega \in \Omega_0} \Lambda(V_\omega).$$

Therefore, we have a finite subcover of $\mathcal{L}$ as desired.

Case 2: $Y \neq \varnothing$. Then, from the way Y was defined, $Y \in 2^X$ and $Y \notin \Lambda(V_\omega)$ for any $\omega \in \Omega$ (4.3). Thus, since $2^X = \cup \mathcal{L}$, there exists $\alpha \in \Sigma$ such that $Y \in \Gamma(U_\alpha)$. Hence, $Y \subset U_\alpha$ (4.3). Therefore,

$$X - U_\alpha \subset X - Y = \cup \{V_\omega: \omega \in \Omega\}.$$

Thus, since $X - U_\alpha$ is compact, there is a finite subset Ω_1 of Ω such that

$$X - U_\alpha \subset \cup \{V_\omega: \omega \in \Omega_1\}.$$

Hence, it follows immediately that

$$2^X = \Gamma(U_\alpha) \cup \left[\bigcup_{\omega \in \Omega_1} \Lambda(V_\omega)\right].$$

Therefore, we have a finite subcover of $\mathscr{L}$ as desired.

From what was shown in the two cases above, we have proved 4.13.

Some pertinent information about 4.13 and its proof is in 4.24.

4.14 Corollary. Let X be a compact metric space. Then, every sequence in 2^X has a convergent subsequence. More generally: If $\{A_i\}_{i=1}^{\infty}$ is any sequence of nonempty subsets of X, then there is a subsequence $\{A_{i(j)}\}_{j=1}^{\infty}$ of $\{A_i\}_{i=1}^{\infty}$ such that $\lim A_{i(j)}$ (4.9) exists and $\lim A_{i(j)} \in 2^X$.

Proof. Since 2^X is a metric space (4.1), the first part of the corollary follows from 4.13. The second part follows by applying the first part to the sequence $\{\overline{A}_i\}_{i=1}^{\infty}$ of closures, and then using 4.11 and the easy exercise in (c) of 4.20. This completes the proof of 4.14.

Our next results concern the connectedness of $\lim A_i$. In particular, we will see in 4.17 that $C(X)$, like 2^X, is compact. This result, and its formulation in 4.18, will be used many times in subsequent chapters.

4.15 Definition of (d,ϵ)-Chain. Let (X,d) be a metric space, and let $\epsilon > 0$. A (d,ϵ)-*chain* in X is a nonempty finite subset of X together with an indexing, $\{x_1, \ldots, x_n\}$, such that

$$d(x_i, x_{i+1}) < \epsilon, \qquad \text{all } i = 1, \ldots, n - 1.$$

We say that a (d,ϵ)-chain $\{x_1, \ldots, x_n\}$, with $p = x_1$ and $q = x_n$, goes *from* p to q or (when the ordering for the indexing is unimportant) *joins* p and q. A subset Z of X is said to be (d,ϵ)-*chained* (ϵ fixed) provided that any two points of Z can be joined by a (d,ϵ)-chain in Z. A subset of X which is (d,ϵ)-chained from every $\epsilon > 0$ is said to be *d-well-chained*.

Some basic facts about the concepts introduced above are in exercises 4.22 and 4.23. We shall use (a) of 4.23 in the proof of 4.17.

4.16 Lemma. Let (X,d) be a compact metric space, and let $\{A_i\}_{i=1}^{\infty}$ be a sequence in 2^X converging to $A \in 2^X$. If each A_i is (d,ϵ_i)-chained where $\epsilon_i \to 0$ as $i \to \infty$, then $A \in C(X)$.

Proof. Suppose that A is not connected. Then, $A = K \cup L$ where K and L are nonempty, disjoint, and closed in A, hence in X. Thus, since X is normal, there are open subsets U and V of X such that $K \subset U$, $L \subset V$, and $\overline{U} \cap \overline{V} = \varnothing$ [5, p. 122]. Note from properties of U and V, and from 4.3, that $A \in \langle U,V \rangle$. Therefore, since $\langle U,V \rangle$ is open in 2^X (4.5) and $\{A_i\}_{i=1}^{\infty}$ converges to A, there exists N such that

(1) $A_i \in \langle U, V \rangle$ for all $i \geq N$.

Now, let

$$\delta = \text{glb } \{d(x,y): x \in \overline{U} \text{ and } y \in \overline{V}\}$$

and observe that $\delta > 0$ since $\overline{U}$ and $\overline{V}$ are compact and disjoint. Thus, since $\epsilon_i \to 0$ as $i \to \infty$, there exists $k \geq N$ such that $\epsilon_k < \delta$. Since, by (1), $A_k \in \langle U, V \rangle$, $A_k \cap U \neq \varnothing$ and $A_k \cap V \neq \varnothing$. But, since A_k is (d, ϵ_k)-chained and $\epsilon_k < \delta$, clearly this is impossible. Therefore, A is connected. This completes the proof of 4.16.

4.17 Theorem.   If X is a compact metric space, then $C(X)$ is compact.

Proof.   By 4.13, it suffices to show that $C(X)$ is a closed subset of 2^X. Let $A \in 2^X$ such that $A = \lim A_i$ where each $A_i \in C(X)$. Then, by (a) of 4.23, each A_i is d-well-chained. Therefore, by 4.16, $A \in C(X)$. This completes the proof of 4.17.

Since we will frequently use the following version of 4.17, we state it for convenient reference.

4.18 Corollary.   Let X be a compact metric space. Then: Every sequence of subcontinua of X has a subsequence converging to a subcontinuum of X, and, thus, every convergent sequence of subcontinua of X has a subcontinuum of X as its limit.

This completes our basic foundation for the use of sequences of sets to study continua. Applications of the ideas and results occur throughout the book.

4. GENERAL REMARKS ABOUT HYPERSPACES

In the course of the work above, we have obtained a few elementary facts about the structure of hyperspaces. Hyperspaces have been, and still are, a source of intensive investigation. An exposition of their theory may be found in [8].

In this book, the theory of hyperspaces will be examined only when it relates to or supplements other aspects of the theory of continua. At this juncture, however, a few comments about hyperspaces are worthwhile. Let X be a continuum. Then, 2^X and $C(X)$ are arcwise connected continua (5.25). We think of them as resembling cones with $F_1(X) = \{\{x\}: x \in X\}$

X

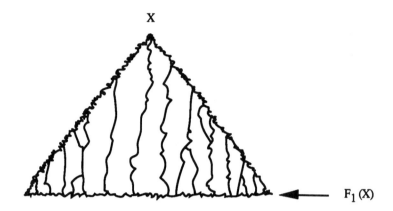

$F_1(X)$

at the bottom and the point representing X at the top (see the figure above). Clearly, $F_1(X)$ is homeomorphic to the continuum X. There is an upwards partial ordering ($\subset$), and there are horizontal levels each of which consists of all those members of the hyperspace of the same size (see 4.33).

For example, let us consider the simplest case, namely $C(I)$ where $I = [0,1]$. By defining h from $C(I)$ into R^2 by

$$h([a,b]) = (a,b) \qquad \text{for all } [a,b] \in C(I),$$

we obtain a homeomorphism of $C(I)$ onto the solid triangle T in R^2 with vertices $(0,0)$, $(0,1)$, and $(1,1)$; see the figure below. Thus, $C(I)$ is a 2-cell. In this case, a "size level" may be thought of as consisting of all the

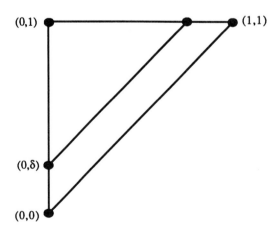

subintervals of $[0,1]$ of a fixed diameter δ (however, see (c) of 4.33). Geometrically, the "size level" for a given δ is the line segment in T of slope $= 1$ containing the point $(0,\delta)$. We note that $C(I)$ is, in fact, the cone over I, but that this is a rare occurrence (see references [7], [9], and [10] in Chapter III). You are asked to find geometric models for $C(X)$ for some other continua X in 4.29 and 4.31.

The case of 2^I is quite different. It is known, but very difficult to prove, that 2^I is a Hilbert cube [9]. Moreover, 2^X is a Hilbert cube when X is any nondegenerate Peano continuum ([1]; also proved in [13] using the characterization of the Hilbert cube we have stated in 1.4).

EXERCISES

4.19 Exercise. Prove that if (X,d) is a compact metric space and A, $B \in 2^X$, then there exist $a \in A$ and $b \in B$ such that $d(a,b) = H_d(A,B)$.

4.20 Exercise. Let (S,T) be a topological space, and let $\{E_i\}_{i=1}^\infty$ be any sequence of subsets of S. Prove (a)–(c) below concerning the concepts introduced in 4.8 and 4.9.
   (a) $\lim \inf E_i$ and $\lim \sup E_i$ are closed subsets of S.
   (b) $\lim \inf E_i = \lim \inf \overline{E}_i$ and $\lim \sup E_i = \lim \sup \overline{E}_i$.
   (c) $\lim E_i = \lim \overline{E}_i$ provided $\lim E_i$ or $\lim \overline{E}_i$ exists.

4.21 Exercise. Prove that if X is a compact metric space and if $\{E_i\}_{i=1}^\infty$ is a sequence of connected subsets of X such that $\lim \inf E_i \neq \varnothing$, then $\lim \sup E_i$ is connected. Show by an example that the conclusion may fail when X is not compact.

4.22 Exercise. Let (X,d) be a metric space. For any $Z \subset X$ and any $\epsilon > 0$, let

$$\mathbb{C}(Z,\epsilon) = \{x \in X: \text{there is a } (d,\epsilon)\text{-chain in } X \text{ from some point of } Z \text{ to } x\}.$$

Prove that $\mathbb{C}(Z,\epsilon)$ is both open and closed in X.

4.23 Exercise. The material in (a)–(d) below concerns the notions in 4.15.
   (a)   Prove that every connected metric space with metric d is d-well-chained. [Hint: Use 4.22.]
   (b)   Give an example of a complete metric space (X,d) which is d-well-chained but not connected.

(c)   Show that the property of being d-well-chained is not a topologi-
      cal invariant.

(d)   Prove that a compact metric space (X,d) is connected if and only
      if it is d-well-chained. [Note: A trivial proof is available using
      material in this chapter.]

4.24 Exercise.   The proof of 4.13 showed more than the theorem
stated. Let us examine what was shown. Let (S,T) be a topological space,
let 2^S be as in (1) of 4.1, let $\langle U_1, \ldots, U_n \rangle$ be as in (3) of 4.3, and let

$$\mathbb{C} = \{\langle U_1, \ldots, U_n \rangle : U_i \in T \text{ for each } i = 1, \ldots, n\}.$$

By (3) of 4.4 (and since $\langle S \rangle = 2^S$), $\mathbb{C}$ is a base for a topology T_V for 2^S.
The subscript V used here, as well as in the proof of 4.5, is not chosen
capriciously since, in fact, T_V is called the *Vietoris topology*. Now, from
the proof we gave for 4.13, we see that the following general result is
true: *If (S,T) is any compact topological space, then $(2^S, T_V)$ is compact*
[14]. Clearly, by 4.5, this result is a generalization of 4.13. Work (a)–(c)
below.

(a)   Prove the converse of the theorem stated above: If (S,T) is a
      topological space such that $(2^S, T_V)$ is compact, then (S,T) is com-
      pact. However, be careful–we do not assume (S,T) is a T_1-space
      (comp., Theorem 2 of [6, p. 46]), so $\{x\}$ may not be a point of 2^S
      for an $x \in S$. Also, the result is trivial when (S,T) is Hausdorff–
      why?

(b)   Prove that if (S,T) is a T_1-space such that $(2^S, T_V)$ is metrizable,
      then (S,T) is compact. [Hint: $i: S \rightarrow 2^S$, $i(x) = \{x\}$, is an embed-
      ding; so, if (S,T) is not compact, it is not countably compact.
      What next?]

(c)   Give a direct proof of 4.13 by showing how to construct a conver-
      gent subsequence of any sequence in 2^X. [Hint: Work with lim
      infs, lim sups, and a countable base for X–use 4.11.]

Facts about the Vietoris topology may be found in [5]–[7]. We mention
that an error in [7] is discussed in [12].

4.25 Exercise.   This exercise is about the relationship between the
decomposition topology (3.1) and the (relativized) Vietoris topology
(4.24). Let (S,T) be a topological space. Let $\mathcal{D}$ be a closed partition of S.
The natural set-theoretic copy of $\mathcal{D}$ in 2^S is denoted by $2^S(\mathcal{D})$, i.e.,

$$2^S(\mathcal{D}) = \{D \in 2^S : D \in \mathcal{D}\}.$$

Now, let $T(\mathcal{D})$ be the decomposition topology for $\mathcal{D}$, and let $T_V(\mathcal{D})$ denote
the subspace topology on $2^S(\mathcal{D})$ obtained from the Vietoris topology for 2^S

(which, by 4.5, is the Hausdorff metric topology when S is compact metric). Prove that

$$i: 2^S(\mathcal{D}) \rightarrow \mathcal{D}, \quad i(D) = D \text{ for each } D \in 2^S(\mathcal{D})$$

is continuous. Therefore, if we do not distinguish between $\mathcal{D}$ and $2^S(\mathcal{D})$ as sets, we see that $T(\mathcal{D}) \subset T_V(\mathcal{D})$. Most of the time, $T(\mathcal{D})$ is strictly smaller than $T_V(\mathcal{D})$–give some examples when S is compact metric and $\mathcal{D}$ is usc (3.5).

We remark that when $T(\mathcal{D}) = T_V(\mathcal{D})$, $\mathcal{D}$ is called a continuous decomposition of S. This terminology can be motivated by other considerations. This is done in Section 2 of Chapter XIII beginning with 13.6. The resulting definition in 13.9 is shown to be equivalent to $T(\mathcal{D}) = T_V(\mathcal{D})$ for compact metric spaces in 13.10.

4.26 Exercise. Let (S,T) be a T_1-space, and let S/A be as in 3.14 (with the decomposition topology). Let 2^S be as in 4.24 (with the Vietoris topology), and let Γ_A be the subspace of 2^S defined by

$$\Gamma_A = \{A \cup \{x\}: x \in S\}.$$

Prove (a) and (b) below.
- (a) The natural map $g: S \rightarrow \Gamma_A$, defined by $g(x) = A \cup \{x\}$ for each $x \in S$, is continuous.
- (b) S/A is homeomorphic to Γ_A. [Hint: Use (a) and the Transgression Lemma (3.22).]

This gives a natural way to obtain a topological copy of S/A in 2^S (comp., 4.25). We remark that if (S,T) is compact metric, we now have an explicit formula for a metric for S/A (since the Vietoris topology on Γ_A is, by 4.5, the Hausdorff metric topology). When we originally obtained the metrizability of S/A in 3.14, the proof depended, in final analysis, on 3.2 which was proved using the Urysohn metrization theorem [5, p. 241]. Finally, we remark that Γ_A plays a role in the theory of accessible points [8, pp. 395–398].

4.27 Exercise. Let X and Y be compact metric spaces, and let $f: X \rightarrow Y$ be continuous. Define $f^*: 2^X \rightarrow 2^Y$ by

$$f^*(A) = f(A), \quad \text{each } A \in 2^X.$$

Let $\hat{f} = f^*|C(X): C(X) \rightarrow C(Y)$. Prove that f^*, hence $\hat{f}$, is continuous. Clearly, if f maps X continuously onto Y, f^* maps 2^X onto 2^Y. Give an example of *continua* X and Y and a continuous function f from X onto Y such that $\hat{f}$ does not map $C(X)$ onto $C(Y)$. Is there such an example when Y is an arc?

4.28 **Exercise.** In the spirit of the comment near the end of 3.30, the result in this exercise may be summarized by saying: *Hyperspaces and inverse limits commute.* We proceed as follows. Let

$$X = \varprojlim\{X_i, f_i\}_{i=1}^{\infty} \quad (2.2)$$

where each X_i is a compact metric space, and let

$$2_{\infty}^X = \varprojlim\{2^{X_i}, f_i^*\}_{i=1}^{\infty}, \quad C_{\infty}(X) = \varprojlim\{C(X_i), \hat{f}_i\}_{i=1}^{\infty}$$

where f_i^* and $\hat{f}_i$ are as in 4.27; consider $C_{\infty}(X)$ as being contained in 2_{∞}^X by inclusion. Prove using 2.6 that there is a natural homeomorphism h from 2_{∞}^X onto 2^X such that h takes $C_{\infty}(X)$ onto $C(X)$. [Hint: To prove the continuity of h, first observe from 2.28 and 4.5 that $\mathcal{L}$, defined by

$$\mathcal{L} = \{\langle \pi_{k(1)}^{-1}(U_{k(1)}), \ldots, \pi_{k(m)}^{-1}(U_{k(m)})\rangle : U_{k(j)} \text{ is open in } X_{k(j)} \text{ and}$$
$$\pi_{k(j)} \colon X \to X_{k(j)} \text{ is the projection map for each } j = 1, \ldots, m\},$$

is a base for the Hausdorff metric topology on 2^X (X is a compact metric space by 2.4). Next, prove that $\mathcal{L}' = \mathcal{L}$ where

$$\mathcal{L}' = \{\langle \pi_i^{-1}(U_1), \ldots, \pi_i^{-1}(U_n)\rangle : U_1, \ldots, U_n \text{ are open in } X_i, \pi_i \colon X \to X_i$$
$$\text{is the projection map, and } i = 1, 2, \ldots\}.$$

Then, letting $\pi_{\infty,i} \colon 2_{\infty}^X \to 2^{X_i}$ denote the ith projection map for each i, the continuity of h follows upon verifying that

$$h^{-1}(\langle \pi_i^{-1}(U_1), \ldots, \pi_i^{-1}(U_n)\rangle) = \pi_{\infty,i}^{-1}(\langle U_1, \ldots, U_n\rangle)$$

for each $\langle \pi_i^{-1}(U_1), \ldots, \pi_i^{-1}(U_n)\rangle \in \mathcal{L}'$.]

For the case of $C(X)$, the theorem is due to Segal [10] and, for the case of 2^X, it is due to Sirota [11].

4.29 **Exercise.** In a manner somewhat similar to what we did in Section 4 when we showed $C([0,1])$ is a 2-cell, prove that $C(S^1)$ is a 2-cell.

4.30 **Exercise.** For $I = [0,1]$, prove that 2^I contains a Hilbert cube. [Hint: Make use of a collection of intervals $[a_i, b_i]$ in I where $b_{i+1} < a_i$ for each i and $a_i \to 0$ as $i \to \infty$.]

This simple fact shows that whenever X is any continuum containing an arc, 2^X contains a Hilbert cube and, hence, is infinite dimensional. Even though there are nondegenerate continua which contain no arc (1.23, 2.27, and 2.32), 2^X contains a Hilbert cube when X is *any* nondegenerate continuum (5.26).

4.31 Exercise. Let X be a simple triod (2.21). Let p be the point of order three in X. Let $C(X;p) = \{A \in C(X): p \in A\}$. Prove that $C(X;p)$ is a 3-cell, and find a homeomorph of $C(X)$ in R^3. We remark that when X is a graph, $C(X)$ has undergone detailed investigation ([2]–[4]).

4.32 Exercise. Prove that if X is a continuum, then 2^X is a continuum. [Hint: By 4.13, need only show 2^X is connected. For each $n = 1$, 2, ..., consider $X(n) = \{A \in 2^X: A \text{ has at most } n \text{ points}\}$.] It is more difficult to prove $C(X)$ is connected (5.25).

4.33 Exercise. Let X be a compact metric space. Recall from 4.1 that we used the phrase "geometric size" in our intuitive description of closeness with respect to the Hausdorff metric H_d. The purpose of this exercise is to give an appropriate, rigorous meaning to the notion of size. We do this by defining the notion of a size function. We will see that size functions always exist and that convergence with respect to H_d is as indicated in our description in 4.1. A *size function for* 2^X is a continuous function $\mu: 2^X \to R^1$ such that (i) for A, $B \in 2^X$ such that $A \subset B$ and $A \neq B$, $\mu(A) < \mu(B)$ and (ii) $\mu(\{x\}) = 0$ for all $x \in X$. A *size function for* $C(X)$ is defined similarly. Work (a)–(c) below.

(a) If (X,d) is any compact metric space, then there is a size function for 2^X. [Hint: Let $D = \{x_i: i = 1, 2, ...\}$ be a countable dense subset of X. For each i, let $f_i: X \to [0,1]$ be defined by

$$f_i(x) = 1/[1 + d(x_i,x)], \quad \text{each } x \in X$$

and let $\mu_i: 2^X \to [0,1]$ be defined by

$$\mu_i(A) = \text{diameter of } f_i(A), \quad \text{each } A \in 2^X.$$

Define $\mu: 2^X \to [0,1]$ by

$$\mu(A) = \sum_{i=1}^{\infty} 2^{-i} \cdot \mu_i(A), \quad \text{each } A \in 2^X.$$

Verify that μ is a size function.]

(b) Give definitive verification of the statement in 4.1 (referred to above) by proving the following: For any compact metric space (X,d) and any size function μ for 2^X, $A_i \to A$ in 2^X if and only if given any $\epsilon > 0$, there exists N such that for $i \geq N$

$$A_i \subset N_d(\epsilon, A)$$

and

$$|\mu(A_i) - \mu(A)| < \epsilon.$$

(c) Show that the diameter map may not be a size function for $C(X)$ even when X is an arc.

Size functions are usually called *Whitney maps*, having been discovered first by Hassler Whitney [15]. Much work has been done on them (see [8] for an accounting up until 1978). An application of size functions is in the next exercise.

4.34 Exercise. First, recall the following definitions from set theory. Let $\mathfrak{C}$ be a collection of sets. A *maximal member of* $\mathfrak{C}$ is an $F \in \mathfrak{C}$ such that no member of $\mathfrak{C}$ properly contains F. A *minimal member of* $\mathfrak{C}$ is an $E \in \mathfrak{C}$ such that no member of $\mathfrak{C}$ is properly contained in E. Prove the following result (hint: use 4.13 and (a) of 4.33):

MAXIMUM-MINIMUM THEOREM. Let X be a compact metric space. If $\mathfrak{C}$ is a nonempty closed subset of 2^X, then there is a maximal member of $\mathfrak{C}$ and a minimal member of $\mathfrak{C}$.

This result can be used in compact metric spaces in place of Zorn's Lemma or the Brouwer Reduction Theorem. Some applications of it are in 4.35, 4.36, and the proof of 8.23.

4.35 Exercise. Let Y be a continuum and let $A \subset Y$. Then, Y is said to be *irreducible about A* provided that no proper subcontinuum of Y contains A. A continuum Y is said to be *irreducible* provided that Y is irreducible about $\{p,q\}$ for some p, $q \in Y$, in which case we sometimes say Y is *irreducible between p and q*. Prove (a) and (b) below.
(a) If X is a continuum and A is a closed subset of X, then X contains a subcontinuum which is irreducible about A. [Hint: Use 4.18 in showing how to apply 4.34.]
(b) If X is a continuum and p, $q \in X$, then X contains a subcontinuum which is irreducible between p and q. Hence, every nondegenerate continuum contains a nondegenerate irreducible continuum.
Irreducible continua will be studied in some detail in Chapter XI.

4.36 Exercise. Let X and Y be compact metric spaces, and let f be a continuous function from X onto Y. Prove (a) and (b) below.
(a) There is a compact subset K of X such that $f(K) = Y$ and no closed proper subset of K maps onto Y under f. [Hint: Use 4.14 and 4.27 in showing how to apply 4.34.]
(b) If X and Y are continua, then there is a subcontinuum A of X such that $f(A) = Y$ and no proper subcontinuum of A maps onto Y under f. [Hint: Use 4.18 and 4.27 in showing how to apply 4.34.] An application is in 4.37.

4.37 Exercise. Prove that if X is a continuum, Y is a nondegenerate indecomposable continuum, and f is a continuous function from X onto Y, then X contains a nondegenerate indecomposable continuum. [Hint: Use (b) of 4.36.] For an important application of this result, see 13.57.

4.38 Exercise. This exercise is for use in the proof of 8.23. Let $I = [0,1]$. Let $\{A_i\}_{i=1}^{\infty}$ be a sequence in 2^I such that $\lim A_i = A$ where $A \neq I$ and $A_i \neq I$ for any i. Let J be a component of $I - A$ (5.1–thus, J is a maximal open or half-open interval in $I - A$). Prove there is a sequence $\{J_i\}_{i=1}^{\infty}$ of components J_i of $I - A_i$ such that $\lim \bar{J}_i = \bar{J}$. [Hint: Let $s < r < t$ where s and t are the end points of $\bar{J}$. There exists N such that $r \in I - A_i$ for all $i \geq N$. For each $i \geq N$, let J_i be the component of $I - A_i$ containing r.]

4.39 Exercise. Recall the definition of an ϵ-map (2.11) and the definitions of $\hat{f}$ and f^* (4.27). Prove that if X and Y are compact metric spaces and $f: X \to Y$ is an ϵ-map, then

$$\hat{f}: C(X) \to C(Y) \quad \text{and} \quad f^*: 2^X \to 2^Y$$

are ϵ-maps for the same ϵ as for f (provided that the Hausdorff metric H_d is as it is specifically defined in 4.1).

REFERENCES

1. D. W. Curtis and R. M. Schori, Hyperspaces of Peano continua are Hilbert cubes, Fund. Math., 101 (1978), 19–38.

2. R. Duda, On the hyperspace of subcontinua of a finite graph, I, Fund. Math., 62 (1968), 265–286.

3. R. Duda, Correction to the paper "On the hyperspace of subcontinua of a finite graph, I", Fund. Math., 69 (1970), 207–211.

4. R. Duda, On the hyperspace of subcontinua of a finite graph, II, Fund. Math., 63 (1968), 225–255.

5. K. Kuratowski, *Topology*, Vol. I, Acad. Press, New York, N.Y., 1966.

6. K. Kuratowski, *Topology*, Vol. II, Acad. Press, New York, N.Y., 1968.

7. Ernest Michael, Topologies on spaces of subsets, Trans. Amer. Math. Soc., 71 (1951), 152–182.

8. Sam B. Nadler, Jr., *Hyperspaces of Sets*, Monographs and Text-books in Pure and Applied Math., vol. 49, Marcel Dekker, Inc., New York, N.Y., 1978.

9. R. M. Schori and J. E. West, The hyperspace of the closed unit interval is a Hilbert cube, Trans. Amer. Math. Soc., 213 (1975), 217-235.

10. J. Segal, Hyperspaces of the inverse limit space, Proc. Amer. Math. Soc., 10 (1959), 706-709.

11. S. Sirota, Spectral representation of spaces of closed subsets of bicompacta, Soviet Math. Dokl., 9 (1968), 997-1000.

12. R. E. Smithson, First countable hyperspaces, Proc. Amer. Math. Soc., 56 (1976), 325-328.

13. H. Toruńczyk, On CE-images of the Hilbert cube and characterization of Q-manifolds, Fund. Math., 106 (1980), 31-40.

14. L. Vietoris, Bereiche Zweiter Ordnung, Monatshefte für Math. und Physik, 32 (1922), 258-280.

15. Hassler Whitney, Regular families of curves, Annals of Math., 34 (1933), 244-270.

V

The Boundary Bumping Theorems

Do all nondegenerate continua contain a nondegenerate proper subcontinuum? The following thoughts may be going through your mind:

Why didn't I think of this question myself? How could I have read this much and not wondered about it?

Or:

Of course they do. Let's see. If they contain an arc they do. But, what about the arcless continuum in 2.27? Oh–that continuum is decomposable, so of course it contains a nondegenerate proper subcontinuum. I see–every continuum which contains a decomposable subcontinuum contains a nondegenerate proper subcontinuum. So, what about (non-degenerate) hereditarily indecomposable continua such as, for example, the continuum in 1.23? I am not sure.

We shall show that the answer to the question in the first paragraph is "yes" in 5.5.

In Sections 1 and 2 of the chapter, we lay the foundation for a technique which has come to be known as "component juggling." In Sections 3 and 4, we obtain two important results whose proofs illustrate the technique in a straightforward way. The technique is also used in some exercises at the end of the chapter and in many proofs and exercises in subsequent chapters. Therefore, as the reader goes through the book, he will gain more understanding of and greater facility with the technique.

1. THE CUT WIRE THEOREM

Let us recall the following general and well-known definition.

5.1 Definition of Component. Let S be a space. A *component of S* is a maximal connected subset of S. If $p \in S$, then C_p, defined by

$$C_p = \cup \{A \subset S: p \in A \text{ and } A \text{ is connected}\},$$

is clearly a component of S which, since the components of S must be mutually disjoint, we call *the component of p in S* or *the component of S containing p*. More generally, if $Y \subset S$, *the component of Y in S* or *the component of S containing Y* means the maximal connected subset of S containing Y (if it exists). However, for $Y \subset S$, the phrase *component of Y* refers to a component of Y in Y where Y has the subspace topology. We remark that any component of S is a closed subset of S (since the closure of a connected set is connected), and that every connected subset of S is contained in a component of S.

Most of the work for proving the following theorem was done in 4.14–4.16 and 4.22.

5.2 Cut Wire Theorem. Let (X,d) be a compact metric space, and let A and B be closed subsets of X. If no connected subset of X intersects both A and B (equivalently, no component of X does), then $X = X_1 \cup X_2$ where X_1 and X_2 are disjoint closed subsets of X with $A \subset X_1$ and $B \subset X_2$.

Proof. Suppose that for each $i = 1, 2, \ldots$, there is a $(d, 2^{-i})$-chain K_i in X joining a point $a_i \in A$ to a point $b_i \in B$ (4.15). Then, by 4.14, there is a subsequence $\{K_{i(j)}\}_{j=1}^{\infty}$ of $\{K_i\}_{i=1}^{\infty}$ converging to a $K \in 2^X$. By 4.16, $K \in C(X)$. Also, it is easy to see that K intersects both A and B. Therefore, we have a contradiction to an assumption in our theorem. Hence, there exists $\epsilon > 0$ such that (notation as in 4.22)

$$\mathbb{C}(A,\epsilon) \cap B = \emptyset.$$

Therefore, letting $X_1 = \mathbb{C}(A,\epsilon)$ and $X_2 = X - \mathbb{C}(A,\epsilon)$, we see using the result in 4.22 that X_1 and X_2 have the desired properties. This completes the proof of 5.2.

The Cut Wire Theorem has an equivalent formulation in terms of the notion of a quasicomponent which is defined and studied briefly in 5.18. The equivalent formulation is in (h) of 5.18, and the natural analogue of 5.2 for compact Hausdorff spaces is in (i) of 5.18.

The Cut Wire Theorem is used to prove results about continua by applying it to compact subsets of continua. One such application is in the proof of 5.4.

2. BOUNDARY BUMPING THEOREMS

The results in 5.4, 5.6, and 5.7 are called Boundary Bumping Theorems because they say that, under mild conditions, a component of a set must "bump (= intersect) the boundary" of the set. The notion of (topological) boundary is defined as follows.

5.3 Definition of Boundary. Let S be a space, and let $H \subset S$. Then, the *boundary of H (in S)*, denoted by $Bd_S(H)$ or, more simply, by $Bd(H)$, is defined by

$$Bd(H) = \overline{H} \cap \overline{(S - H)}.$$

We emphasize that when we simply say *boundary of H* and write $Bd(H)$, we mean the boundary of H in the largest space under consideration.

Some elementary facts about the boundary operator are in exercise 5.17.

5.4 Boundary Bumping Theorem I. Let X be a continuum, and let U be a nonempty, proper, open subset of X. If K is a component of $\overline{U}$, then $K \cap Bd(U) \neq \emptyset$ (equivalently, since $K \subset \overline{U}$ and U is open, $K \cap (X - U) \neq \emptyset$).

Proof. Suppose that $K \cap Bd(U) = \emptyset$. Then, by 5.2, $\overline{U} = M_1 \cup M_2$ where M_1 and M_2 are disjoint closed subsets of $\overline{U}$ with $K \subset M_1$ and $Bd(U) \subset M_2$. Let

$$M_3 = M_2 \cup (X - U).$$

We show that X is not connected using M_1 and M_3. Since $\overline{U} = M_1 \cup M_2$, $X = M_1 \cup M_3$. Clearly, M_1 and M_3 are each closed in X. Since $K \neq \emptyset$ (being a component of the nonempty set $\overline{U}$) and $M_1 \supset K$, $M_1 \neq \emptyset$. Since $M_3 \supset X - U$ and $U \neq X$, $M_3 \neq \emptyset$. Now, to see that $M_1 \cap M_3 = \emptyset$, first note that since $M_1 \cap M_2 = \emptyset$,

$$M_1 \cap M_3 = M_1 \cap (X - U).$$

Thus, since $M_1 \subset \overline{U}$ and $X - U$ is closed in X,

$$M_1 \cap M_3 \subset \overline{U} \cap \overline{(X - U)} = Bd(U) \subset M_2$$

which, since $M_1 \cap M_2 = \emptyset$, shows that $M_1 \cap M_3 = \emptyset$. From the properties of M_1 and M_3 verified above, we see that X is not connected. Therefore, we have proved 5.4.

We note the following consequence of 5.4 which answers the question at the beginning of the chapter.

5.5 Corollary. Let X be a nondegenerate continuum. Then, X contains a nondegenerate proper subcontinuum. Furthermore: If A is a proper subcontinuum of X and U is an open subset of X such that $A \subset U$, then there is a subcontinuum B of X such that

$$A \subset B \neq A \quad \text{and} \quad B \subset U.$$

Proof. We prove the second part first. Let V be an open subset of X such that $A \subset V$, $\bar{V} \subset U$, and $V \neq X$. Let B be the component of $\bar{V}$ containing A (note that B exists since A is a connected subset of $\bar{V}$). Clearly, $A \subset B$ and, since $\bar{V} \subset U$, $B \subset U$. By 5.4, $B \cap (X - V) \neq \emptyset$. Thus, since $A \subset V$, $B \neq A$. This proves the second part of the corollary. The first part follows immediately from the second part by letting $p \in X$, $A = \{p\}$, and letting U be an open subset of X such that $p \in U$ and $U \neq X$.

As we will see, the following theorem is actually a consequence of the corollary above.

5.6 Boundary Bumping Theorem II. Let X be a continuum, and let E be a nonempty proper subset of X. If K is a component of E, then $\bar{K} \cap Bd(E) \neq \emptyset$ (equivalently, since $\bar{K} \subset \bar{E}$, $\bar{K} \cap \overline{(X - E)} \neq \emptyset$).

Proof. Suppose that $\bar{K} \cap \overline{(X - E)} = \emptyset$. Then (since $\bar{K} \neq \emptyset$ and $\overline{X - E} \neq \emptyset$), $\bar{K}$ is a proper subcontinuum of X. Furthermore, letting

$$U = X - \overline{(X - E)},$$

we see that $\bar{K} \subset U \subset E$ and U is open in X. Hence, by 5.5, there is a continuum B such that

$$\bar{K} \subset B \neq \bar{K} \quad \text{and} \quad B \subset U.$$

Therefore, B is a connected set properly containing K and $B \subset E$ (since $U \subset E$). This contradicts the assumption that K is a component of E. Thus, we have proved 5.6.

Our final Boundary Bumping Theorem follows immediately from the one just proved. We include it for convenient use later.

5.7 Boundary Bumping Theorem III. Let X, E, and K satisfy the hypotheses in 5.6. If E is open in X, then

$$\overline{K} \subset (X - E) \neq \varnothing, \quad \text{i.e.,} \quad \overline{K} - E \neq \varnothing.$$

If E is closed in X, then

$$K \cap (\overline{X - E}) \neq \varnothing.$$

Proof. Since, by 5.6, $\overline{K} \cap (\overline{X - E}) \neq \varnothing$, the theorem follows immediately.

We remark that a result called the Order Arc Theorem, which can be interpreted as being a continuous version of the Boundary Bumping Theorems, is in the comments below (e) of 5.25.

Many proofs in continuum theory use the following result or some closely related variation of it (such as 5.9).

5.8 Theorem. Let X be a continuum, and let $A \subset B$ be proper subcontinua of X. If K is a component of $X - B$ such that $\overline{K} - K \subset A$, then $K \cup A$ is a continuum.

Proof. By 5.7, $\overline{K} \cap B \neq \varnothing$. Thus, since $K \cap B = \varnothing$ and $\overline{K} - K \subset A$, $\overline{K} \cap A \neq \varnothing$. Therefore, $K \cup A$ is a continuum.

5.9 Corollary. Let X be a continuum, and let A be a proper subcontinuum of X. If K is a component of $X - A$, then $K \cup A$ is a continuum.

Proof. Since K is closed in $X - A$, $\overline{K} - K \subset A$. Therefore, 5.9 follows from 5.8 (letting $B = A$).

3. APPLICATION: CONVERGENCE CONTINUA

Two important notions for the study of the structure of continua are connectedness im kleinen and convergence continua. We define these notions below, and then we use boundary bumping to prove a general theorem on the existence of convergence continua (5.12).

5.10 Definition of Connected Im Kleinen. Let (S,T) be a topological space. If $p \in S$, then S is said to be *connected im kleinen at* p, written *cik at* p, provided that every neighborhood of p contains a connected neighborhood of p (neighborhood does not mean open neighborhood–G being a *neighborhood of* p means there is a $U \in T$ such that $p \in U \subset G$). It is important to note that although there is clearly no loss of generality in assuming the first neighborhood of p to be open, there *is* loss of generality in requiring the connected neighborhood of p to be open–see 5.22.

We remark that metric spaces which are cik at every point are an important class of spaces called Peano spaces. We shall devote Chapters VIII, IX, and X to the study of Peano spaces and Peano continua.

5.11 Definition of Convergence Continuum. Let S be a metric space. A nondegenerate subcontinuum A of S is called a *convergence continuum (of S)*, or a *continuum of convergence (of S)*, provided that there is a sequence $\{A_i\}_{i=1}^{\infty}$ of subcontinua A_i of S such that

$$A = \lim A_i \quad (4.9), \qquad A \cap A_i = \emptyset \text{ for each } i.$$

We remark that if S is compact, continua A_i as above may be chosen so that, in addition, they are mutually disjoint (5.23).

Some continua contain no convergence continua, e.g., an arc, a simple closed curve, a simple triod (2.21). On the other hand, any subarc of the $\sin(1/x)$-continuum X (1.5) lying on the y-axis is a convergence continuum of X. Lest you get the idea that every nondegenerate nowhere dense subcontinuum of a continuum X must be a convergence continuum of X, simply consider

$$S^1 \cup \{(x,0) \in R^2: -1 \le x \le 1\}, \qquad S^1 \text{ as in } 1.3$$

as a subcontinuum of B^2 (1.2).

We now prove the following fundamental theorem. It gives a simple sufficient condition for the existence of convergence continua as well as some useful additional information (see, e.g., 5.13).

5.12 Continuum of Convergence Theorem. Let X be a continuum, and let

$$N = \{x \in X: X \text{ is not cik at } x\}.$$

If $p \in N$, then there is a convergence continuum K of X such that $p \in K$ and $K \subset N$.

Proof. Since $p \in N$, there exists (5.10) a closed neighborhood M of p such that, letting C denote the component of p in M,

$$p \notin \text{int}(C), \qquad \text{int denoting interior in } X.$$

Hence, $p \in \overline{M - C}$. Thus, there is a sequence $\{p_i\}_{i=1}^{\infty}$ such that

(1) $p_i \to p$ as $i \to \infty$,

(2) $p_i \in M - C$ for each i.

For each $i = 1, 2, \ldots$, let C_i denote the component of p_i in M. If $C \cap C_i \neq \emptyset$ for some i, then $C \cup C_i$ would be a subcontinuum of M which, since C is a component of M, would imply $C \cup C_i \subset C$, hence $p_i \in C$ which is in contradiction to (2). Therefore:

(3) $C \cap C_i = \emptyset$ for each $i = 1, 2, \ldots$.

Now, the limit C' of a convergent subsequence of $\{C_i\}_{i=1}^{\infty}$ (4.18) would be a convergence continuum (as can be proved using 5.7, as we do below, and (3)), and $p \in C'$ (by (1)); however, C' may not be contained in N (5.24). This is remedied as follows. Let Q be a closed neighborhood of p such that

(4) $Q \subset \text{int}(M)$,   int denoting interior in X.

By (1), we assume without loss of generality that $p_i \in Q$ for each $i = 1, 2, \ldots$. Let K_i denote the component of p_i in Q for each $i = 1, 2, \ldots$. Since $Q \subset M$ (by (4)), clearly

(5) $K_i \subset C_i$ for each $i = 1, 2, \ldots$.

By 4.18, the sequence $\{K_i\}_{i=1}^{\infty}$ has a convergent subsequence $\{K_{i(j)}\}_{j=1}^{\infty}$ and its limit K is a subcontinuum of X:

(6) $K = \lim K_{i(j)}$,

(7) K is a continuum.

Since $p_{i(j)} \in K_{i(j)}$ for each j, it follows using (1) and (6) that

(8) $p \in K$.

Since each $K_{i(j)} \subset Q$ and Q is closed in X, it follows from (6) that

(9) $\overline{K} \subset Q$.

By (9) and (4), $K \subset M$. Thus, since C is the component of p in M, it follows from (7) and (8) that

(10) $K \subset C$.

By (10), (5), and (3), $K \cap K_{i(j)} = \emptyset$ for each $j = 1, 2, \ldots$. Hence, recalling (6) and (7), we will have proved K is a convergence continuum once we show K is nondegenerate. By the second part of 5.7,

$K_{i(j)} \cap (\overline{X - Q}) \neq \emptyset$ for each $j = 1, 2, \ldots$.

Hence, it follows easily using (6) that $K \cap (\overline{X - Q}) \neq \emptyset$. Thus, by (8) and the fact that Q is a neighborhood of p (so, $p \notin \overline{X - Q}$), we have that K is nondegenerate. Therefore, we have proved K is a convergence continuum. Hence, in view of (8), we will have proved 5.12 once we verify that

$K \subset N$. Suppose that there exists $x \in K - N$. Then, since $x \in K$, by (9) and (4) we have that

(11) M is a neighborhood of x

and, by (10), we have that

(12) C is the component of x in M.

By assumption $x \notin N$, i.e., X is cik at x. Hence, by (11), there is a connected neighborhood G of x such that $G \subset M$. Thus, by (12),

(13) $G \subset C$.

Since $x \in K$, by (6) and (5)

$$x \in \lim \sup K_i \subset \lim \sup C_i.$$

Thus, since G is a neighborhood of x, $G \cap C_\ell \neq \varnothing$ for some ℓ. Hence, by (13), $C \cap C_\ell \neq \varnothing$. This contradicts (3). Therefore, we have proved that $K \subset N$. This completes the proof of 5.12.

We remark that the notion in 5.11 will be used in characterizing hereditarily locally connected continua in 10.4, and that 5.12 gives half of 10.4. For now, let us note the following easy consequence of 5.12.

5.13 Corollary. A continuum X cannot fail to be cik at only one point or even at only countably many points; in fact, if X is not cik at some point, then there is a nondegenerate subcontinuum K of X such that X is not cik at any point of K.

Proof. If X is not cik at some point p, then let K be as in 5.12 (K is nondegenerate by 5.11). This proves 5.13.

4. APPLICATION: σ-CONNECTEDNESS

The following discussion leads to our next application of boundary bumping.

It is but a trivial observation that no connected space can be written as the union of more than one and only finitely many nonempty, mutually disjoint, closed subsets. Can a nondegenerate continuum be written as the union of *countably infinitely many* nonempty, mutually disjoint, closed subsets? On thinking about this question for awhile, the answer, or, at any rate, the proof for the answer may not be apparent (an example of a connected set in the plane which *can* be so written is in 5.14). It seems like the kind of question for which if the answer is "no," then a proof

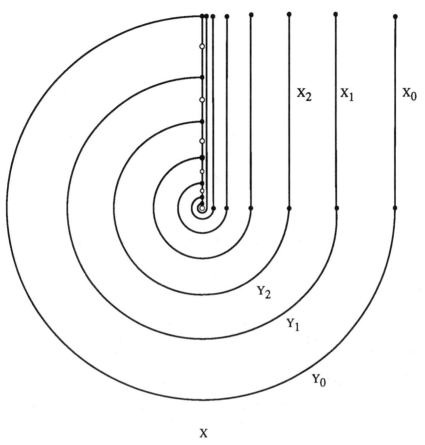

Figure 5.14 (From Ref. 4, with permission of Academic Press and Panstwowe Wydawnictwo Naukowe, Warsaw.)

may be based on the Baire category theorem [4, p. 9]. The following example suggests that such an approach may not work (although, the space is not a continuum).

5.14 Example. This is an example of a connected subset X of R^2 (see Figure 5.14) which is locally compact, hence topologically complete, and which is the union of countably infinitely many nonempty, mutually disjoint, closed, connected subsets. The space X is defined analytically as follows:

$$X = Z \cup \left[\bigcup_{n=0}^{\infty} X_n \right] \cup \left[\bigcup_{n=0}^{\infty} Y_n \right]$$

where

$$Z = \{(0,y) \in R^2 : 0 < y \leq 1 \text{ and } y \neq 3/2^n \text{ for } n = 2, 3, \ldots\}$$

and, for each $n = 0, 1, 2, \ldots,$

$$X_n = \{(2^{-n}, y) \in R^2 : 0 \leq y \leq 1\}$$
$$Y_n = \{(r,\theta) \in R^2 : r = 2^{-n} \text{ and } \pi/2 \leq \theta \leq 2\pi\} \quad \text{(polar form)}.$$

It is easy to see that X has the desired properties. The example is due to Sierpiński (see [4, p. 175] and the papers referenced there).

The example in 5.14 notwithstanding, we shall answer the question in the paragraph above 5.14 negatively. We do this is 5.16–note the following definition.

5.15 Definition of σ-Connected. A topological space (S,T) is said to be σ-*connected* provided that S is not the union of more than one and at most countably infinitely many nonempty, mutually disjoint, closed subsets.

Obviously, every σ-connected space is connected. The converse is false even for locally compact metric spaces by 5.14. However, as we now show, continua are σ-connected. The proof uses 5.5 and 1.7.

5.16 Theorem. Every continuum is σ-connected.

Proof. Suppose that a continuum X is not σ-connected. Then, since X is connected, there are countably infinitely many sets A_i satisfying

$$X = \bigcup_{i=1}^{\infty} A_i, \quad A_i \neq \varnothing, A_i \text{ closed in } X, A_i \cap A_j = \varnothing \text{ for } i \neq j.$$

Choose an open subset U of X such that $A_2 \subset U$ and $A_1 \cap U = \varnothing$; then, letting A be a component of A_2, we can apply 5.5 to obtain a subcontinuum B_1 of X such that

$$B_1 \cap A_2 \neq \varnothing, \quad B_1 \cap A_1 = \varnothing, \quad B_1 \not\subset A_2.$$

Since $B_1 \not\subset A_2$, $B_1 \cap A_j \neq \varnothing$ for some $j \neq 2$. Thus, since B_1 is connected, we see from the properties of the sets A_i that $B_1 \cap A_i \neq \varnothing$ for infinitely many i. Now, assume inductively that we have defined a subcontinuum B_n of X such that $B_n \cap A_i = \varnothing$ for all $i \leq n$ and $B_n \cap A_k \neq \varnothing$ for infinitely many k. Let

$$m = \min \{k : B_n \cap A_k \neq \varnothing\}$$

and let $\{C_i : i = 1, 2, \ldots\}$ be a one-to-one enumeration of the sets A_k which intersect B_n, with $C_1 = A_m$. Let $D_i = C_i \cap B_n$ for each $i = 1,$

2, ... Now, replacing X with B_n and A_i with D_i, we may apply 5.5 as we did above to obtain a subcontinuum B_{n+1} of B_n such that

$$B_{n+1} \cap D_2 \neq \emptyset, \quad B_{n+1} \cap D_1 = \emptyset, \quad B_{n+1} \not\subset D_2.$$

Having thus inductively defined the continua B_n for each $n = 1, 2, \ldots$, we note their pertinent properties in (1)-(3) below:

(1) B_n is nonempty and compact for each $n = 1, 2, \ldots$;
(2) $B_n \supset B_{n+1}$ for each $n = 1, 2, \ldots$;
(3) $B_n \cap A_n = \emptyset$ for each $n = 1, 2, \ldots$.

By (1), (2), and 1.7, there is a point $p \in \cap_{n=1}^{\infty} B_n$. Hence, by (3), $p \notin A_i$ for any $i = 1, 2, \ldots$. However, this is impossible since

$$X = \bigcup_{i=1}^{\infty} A_i.$$

Therefore, every continuum is σ-connected and we have proved 5.16.

For an interesting use of 5.16, see the proof of 10.4. We also note that in view of the statement of 10.4, it is initially surprising that 5.16 would play any role in its proof whatsoever.

EXERCISES

5.17 Exercise. Prove the statements in (a)-(h) below about the boundary operator Bd defined in 5.3. All sets are subsets of a given topological space S.

(a) $\overline{H} - H \subset Bd(H)$, and $Bd(H) = \overline{H} - H$ if and only if H is open in S.
(b) $Bd(H) = [H \cap \overline{(S - H)}] \cup [\overline{H} - H]$.
(c) $H \cup Bd(H) = \overline{H}$, and $\text{int}(H) \cup Bd(H) = \overline{H}$ where int denotes interior in S.
(d) $Bd(\overline{H}) \subset Bd(H)$.
(e) $Bd(H) \cup Bd(G) = Bd(H \cup G) \cup Bd(H \cap G) \cup [Bd(H) \cap Bd(G)]$.
(f) If S is connected, $H \neq \emptyset$, and $H \neq S$, then $Bd(H) \neq \emptyset$.

We note that the boundary operator is not idempotent, i.e., $Bd[Bd(H)]$ may not equal $Bd(H)$. For example, take $S = R^1$ and $H = $ rationals. However, the following formula holds:

(g) $Bd(Bd[Bd(H)]) = Bd[Bd(H)]$.

The following formula can be used to show that, for some definitions, it makes no difference whether neighborhoods or open neighborhoods are used (e.g., 10.14 and 10.57).

(h) $Bd[\text{int}(H)] \subset Bd(H)$, int $=$ interior in S.

5.18   Exercise.   In this exercise, we relate 5.2 to the notion of quasi-component (defined below after (c)). If (S,T) is a topological space, $A \subset S$, and $B \subset S$, then we say S is *connected between A and B* provided that $S \neq U \cup V$ where $A \subset U$, $B \subset V$, $U \cap V = \varnothing$, and $U, V \in T$. Obviously: If S is connected between A and B, then $A \neq \varnothing$ and $B \neq \varnothing$; also, S is connected between A and B if and only if S is connected between $\overline{A}$ and $\overline{B}$. Work (a)–(i) below.

(a)   If S is a compact metric space and S is connected between A and B, then there is a subcontinuum K of S such that $K \cap \overline{A} \neq \varnothing$ and $K \cap \overline{B} \neq \varnothing$. [Hint: Use 5.2.]

(b)   Give an example of a connected metric space S for which there are two points p, $q \in S$ such that there is no subcontinuum of S containing $\{p,q\}$. [Hint: Delete an appropriate point from the continuum in 1.5.] This shows the essential nature of the compactness assumption in (a).

(c)   For x, $y \in S$, we write $x \sim y$ to mean that S is connected between $\{x\}$ and $\{y\}$. Prove that $\sim$ is an equivalence relation.

An equivalence class with respect to the equivalence relation $\sim$ in (c) is called a *quasicomponent of S*. If Q is a quasicomponent of S and $p \in Q$, then Q is called the *quasicomponent of p (in S)* and is denoted by $[p]_\sim$.

(d) $[p]_\sim = \cap \{L: p \in L$ and L is both open and closed in $S\}$. In particular, a quasicomponent of S is closed in S.

(e)   Each component of S is contained in a quasicomponent of S.

(f)   Give an example of a metric space S such that some quasicomponent of S is not contained in any component of S.

(g)   Let S be a compact Hausdorff space, and let Q be a quasicomponent of S. If $Q \subset U$ where U is open in S, then there exists a closed and open subset L of S such that $Q \subset L \subset U$. [Hint: Show how to cover $X - U$ with finitely many closed-open sets each missing Q.]

(h)   Prove that 5.2 and the following statement are equivalent: In a compact metric space, components and quasicomponents are the same. [Hint: Assuming 5.2, use (e); the other half uses (g).] See (i).

(i)   In a compact Hausdorff space (S,T), components and quasicomponents are the same. [Hint: If not, then, by (e), some quasicomponent $Q = A \cup B$ where A and B are nonempty, disjoint, and compact (by (d)). Then, there are U, $V \in T$ with $A \subset U$, $B \subset V$, and $U \cap V = \varnothing$. Cover $S - (U \cup V)$ with $E_1, ..., E_n$ $(n < \infty)$, each $E_i \subset S - Q$, and each E_i closed and open in S. Let $E = E_1 \cup \cdots \cup E_n$, and consider $U \cup E$ and $V - E$.]

With respect to (i), some proofs use the Axiom of Choice (e.g., [2, pp. 44–47]); our hint comes from [4, pp. 168–170].

5.19 Exercise. Prove that if A is a closed proper subset of a continuum X, then the cardinality of the collection of all the components of A is less than or equal to the cardinality of the collection of all the components of $Bd(A)$. Clearly, the analogous statement with A open may be false—example?

5.20 Exercise. For any nondegenerate continuum X and any point $p \in X$, let

$$\kappa(p) = \{x \in X\text{: there is a proper subcontinuum } A \text{ of } X \text{ such that } p, x \in A\}.$$

Do (a)–(d) below.
  (a) If X is a nondegenerate continuum and $p \in X$, then $\kappa(p)$ is a dense, connected subset of X.
  (b) Let $X = [0,1]$. What is $\kappa(0)$, $\kappa(1)$, $\kappa(1/2)$? How many different sets $\kappa(p)$ are there for $p \in [0,1]$?
  (c) Let $X = S^1$ (1.3). How many different sets $\kappa(p)$ are there for $p \in S^1$?
  (d) Same as (c) except for X as in 1.5.

The sets $\kappa(p)$ defined above are called *composants* of X and, for a particular point p, $\kappa(p)$ is called the *composant of p (in X)*. Composants play a significant role in the study of indecomposable continua and, more generally, in the study of irreducible continua. We will discuss composants in much more detail in Chapter XI.

5.21 Exercise. Prove that if a compact metric space X is written as the union of countably many mutually disjoint subcontinua C_i ($i = 1, 2, \ldots$), then each C_i is a component of X.

5.22 Exercise. In 5.10, we localized the notion of connectedness by defining cik at p. Another way to localize connectedness is by means of the following condition: A topological space (S,T) is said to be *locally connected at p ($p \in S$)*, written *LC at p*, provided that each neighborhood of p contains a connected neighborhood of p which is open in S. Clearly, LC at p implies cik at p. However, the converse is false, even for continua, as can be seen on examining the continuum X in Figure 5.22. Also note the example shows that a point p at which a continuum X fails to be LC may not be a point of any convergence continuum of X (comp., 5.12). Prove (a)–(c) below.
  (a) A topological space is LC at every point if and only if each component of each open set is open.

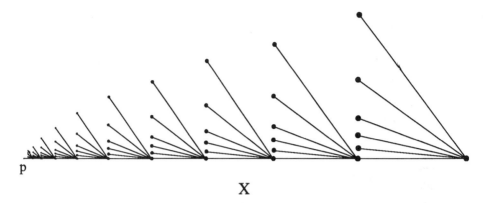

X

Figure 5.22

(b) A topological space is cik at every point if and only if it is LC at every point. [Hint: Use (a).] Hence, when assumed globally, the two notions are the same (comp., example in Figure 5.22).

(c) Let X be a continuum, and let $F = \{p \in X: X$ is not LC at $p\}$. Then, F is dense in itself (i.e., each point of F is a limit point of F) and, if $F \neq \varnothing$, F contains a nondegenerate continuum.

5.23 Exercise. Prove the statement at the end of 5.11 which asserts that if S is compact (metric), the continua A_i may be chosen so as to be mutually disjoint.

5.24 Exercise. Show by an example that the continuum C' in the proof of 5.12 may, indeed, not be contained in N.

5.25 Exercise. Let (X,d) be a continuum. We have shown that 2^X is a continuum (4.32) and that $C(X)$ is compact (4.17). We now show that $C(X)$ and 2^X are arcwise connected continua. In fact, we will show even more (see comments after (e)). A *nest* is a collection $\mathfrak{C}$ of sets such that for any C_1, $C_2 \in \mathfrak{C}$, $C_1 \subset C_2$ or $C_2 \subset C_1$. Fix $A \in C(X)$ such that $A \neq X$. Prove (a)–(e) below (H_d is specifically as defined by the formula in 4.1).

(a) If $0 < \epsilon < H_d(A,X)$, then there exists $A_\epsilon \in C(X)$ such that $A \subset A_\epsilon$ and $H_d(A,A_\epsilon) = \epsilon$. [Hint: Let $N_d(\epsilon,A)$ be as in 4.1, and use 5.4.]

(b) If $0 < \epsilon < H_d(A,X)$, then there is an (H_d,ϵ)-chain (4.15) $\mathfrak{C}_\epsilon$ in $C(X)$ from A to X such that $\mathfrak{C}_\epsilon$ is a nest. [Hint: Use (a) on A, then on A_ϵ, etc.–you get to X in only finitely many steps (why?).]

(c) There is a subcontinuum $\mathcal{L}$ of $C(X)$ such that $A, X \in \mathcal{L}$ and $\mathcal{L}$ is a nest. [Hint: Let $Z = C(X)$; Z is compact (4.17); each $\mathfrak{C}_\epsilon \in 2^Z$ for

$\mathfrak{C}_\epsilon$ as in (b); use 4.14 and 4.16 with X replaced by Z. You also must show that a limit of nests is a nest.]

(d) The continuum $\mathcal{L}$ in (c) is an arc. [Hint: Let μ be a size function (4.33). Then, $\mu|\mathcal{L}$ is a homeomorphism.] *Hence, $C(X)$ is arcwise connected.*

(e) If $B \in 2^X$ and $B \neq X$, then there is a subcontinuum $\mathcal{P}$ of 2^X such that $B, X \in \mathcal{P}$, $\mathcal{P}$ is a nest, and, hence, $\mathcal{P}$ is an arc. [Hint: Let A be a component of B, let $\mathcal{L}$ be as in (c), and consider $\mathcal{P} = \{B \cup L: L \in \mathcal{L}\}$. Prove $\mathcal{P}$ is a continuum by verifying the continuity of f: $\mathcal{L} \to \mathcal{P}$ given by $f(L) = B \cup L$ for each $L \in \mathcal{L}$. Then, use $\mu|\mathcal{P}$ as in the hint for (d).] *Hence, 2^X is arcwise connected.*

Call an arc in $C(X)$ or 2^X an *order arc* provided that it is a nest. It is easy to verify that if α is an order arc in 2^X from A to X and $A \in C(X)$, then $\alpha \subset C(X)$. From this fact and the result about $\mathcal{P}$ in (e), we have the following theorem.

ORDER ARC THEOREM. If $A \in 2^X$ and $A \neq X$, then there is an order arc α in 2^X from A to X and, if $A \in C(X)$, $\alpha \subset C(X)$.

The theorem may be viewed as a continuous version of the Boundary Bumping Theorems since it implies that we may start at any point p in a component K of $\overline{U}$ or E as in 5.4 or 5.6 (respectively), grow continuously with an order arc in $C(X)$ from $\{p\}$ to X, and, in the process, bump the boundary of U or E (respectively) while still in $\overline{K}$ [as is seen by letting α be an order arc in $C(\overline{K})$ from $\{p\}$ to $\overline{K}$, letting β be an order arc in $C(X)$ from $\overline{K}$ to X (if $\overline{K} \neq X$), and then considering $\alpha \cup \beta$]. More definitive formulations of the Order Arc Theorem are in 1.8 and 1.25 of [5].

5.26 Exercise. Prove that if X is any nondegenerate continuum, then 2^X contains a Hilbert cube and, hence, is infinite dimensional. [Hint: Let $p \in X$, and prove there is a sequence $\{A_i\}_{i=1}^\infty$ of nondegenerate mutually disjoint subcontinua in $X - \{p\}$ converging to p. By 5.24, each 2^{A_i} contains an arc. What next?]

5.27 Exercise. Let X be a continuum, let E be a nonempty proper subset of X, and let K be a component of E. Prove that K and $X - E$ are not mutually separated in X, i.e., prove that

$$\overline{K} \cap (X - E) \neq \varnothing \quad \text{or} \quad K \cap (\overline{X - E}) \neq \varnothing.$$

5.28 Exercise. Let X be a continuum, let C be a connected subset of X, and let L be a component of $X - C$. Prove that $X - L$ is connected. [Hint: Find a way to write $X - L$ so that, by using 5.27, you can see directly that $X - L$ is connected.] This result remains valid for any connected topological space X. Furthermore, the proof is straightforward using the elementary result in 6.3. Prove this general result by using 6.3.

5.29 Exercise. Prove that every nondegenerate continuum X can be written as the union of two nondegenerate, mutually disjoint, connected sets. [Hint: By 5.5, there is a nondegenerate proper subcontinuum A of X. Take two cases determined by whether or not $X - A$ is connected.] However, we remark that there are nondegenerate connected metric spaces which cannot be so written [4, p. 135]–compare with 6.20 where the sets are not required to be mutually disjoint but, instead, are required to be proper. We also note, as follows from the result in this exercise or the one in 6.20, that every nondegenerate indecomposable continuum can be written as the union of two nondegenerate, proper, connected sets.

5.30 Exercise. If X is a topological space and if A, B, $C \subset X$, then we say that C *separates A and B in X* provided that $X - C$ is not connected between A and B (5.18), i.e., $X - C = P|Q$ (6.2) with $A \subset P$ and $B \subset Q$. If any of A, B, C is known to be a one-point set $\{x\}$, we usually omit reference to it as a set–thus, e.g., we say z separates x and y in X instead of saying $\{z\}$ separates $\{x\}$ and $\{y\}$ in X.

Prove that if K is a convergence continuum in a space X and x, $y \in K$, then no subset of K can separate x and y in X. Note that it follows from this that no two points of K can be separated in X by a third point of X. Applications are in, e.g., 10.2 and 10.5.

REFERENCES

1. R. H. Bing, Concerning hereditarily indecomposable continua, Pacific J. Math., 1(1951), 43–51.

2. John G. Hocking and Gail S. Young, *Topology*, Addison-Wesley Publ. Co., Inc., Reading, Mass., 1961.

3. B. Knaster, Un continu dont tout sous-continu est indécomposable, Fund. Math., 3(1922), 247–286.

4. K. Kuratowski, *Topology*, Vol. II, Academic Press, New York, N.Y., 1968.

5. Sam B. Nadler, Jr., *Hyperspaces of Sets*, Monographs and Textbooks in Pure and Applied Math., vol. 49, Marcel Dekker, Inc., New York, N.Y., 1978.

VI

Existence of Non-Cut Points

A number of results and proofs in continuum theory depend on examining non-cut points (6.1). The existence of non-cut points, how many there may be, and where they may be situated in a continuum are important considerations. In this chapter, we prove a fundamental existence theorem in 6.6, and derive an even stronger result in 6.7. We also obtain the characterization, in terms of simple orderings, for T_1-continua with exactly two non-cut points in 6.16. Applications of these results and ideas include characterizations of special types of continua (e.g., see 6.17 and Chapters IX and X). On the other hand, using non-cut points and their location to prove results not ostensibly about non-cut points is illustrated in, for example, some of the proofs in Chapter IX (e.g., the proof of 9.19).

1. THE NON-CUT POINT EXISTENCE THEOREM

6.1 Definition of Non-Cut Point. Let (S,T) be a connected topological space, and let $p \in S$. If $S - \{p\}$ is connected, then p is called a *non-cut point of S*. If $S - \{p\}$ is not connected, then p is called a *cut point of S*.

6.2 Notation $Y = P|Q$. If Y is a topological space, we write $Y =$

$P|Q$ to mean $Y = P \cup Q$, $P \neq \emptyset$, $Q \neq \emptyset$, $P \cap Q = \emptyset$, and P and Q are both open in Y. In other words, $Y = P|Q$ means $Y = P \cup Q$ where P and Q are nonempty sets which are mutually separated in Y (recall: Two subsets of Y and *mutually separated in Y* provided that neither one contains any point of the closure of the other).

The following proposition is of general interest. It will be used in the proof of 6.6 and elsewhere.

6.3 Proposition.   Let (S,T) be a connected topological space, and let C be a connected subset of S such that

$$S - C = A|B.$$

Then: $A \cup C$ and $B \cup C$ are connected. Hence, if S and C are continua, $A \cup C$ and $B \cup C$ are continua.

Proof.   Suppose that $A \cup C = K|L$. Then, since C is connected, $C \subset K$ or $C \subset L$, say $C \subset K$. Then, $L \subset A$. Thus, $\bar{L} \cap B = \emptyset$ and $\bar{B} \cap L = \emptyset$. It now follows easily that

$$S = L|(B \cup K).$$

Therefore, we have proved the first part of 6.3. The second part is a simple consequence of the first part.

We remark that a natural generalization of 6.3 is in (c) of 6.18.
We note the following technical lemma for use in the proof of 6.6.

6.4 Lemma.   Let (S,T) be a connected topological space. Assume that $x, y \in S$ such that

$$S - \{x\} = K|L, \quad S - \{y\} = M|N.$$

If $x \in M$ and $y \in K$, then $N \cup \{y\} \subset K$.

Proof.   By 6.3, $N \cup \{y\}$ is connected. Since $x \in M$, $x \notin N \cup \{y\}$. Since $y \in K$, $[N \cup \{y\}] \cap K \neq \emptyset$. Hence, $N \cup \{y\}$ is a connected subset of $S - \{x\}$ and $N \cup \{y\}$ intersects K. Thus, since

$$S - \{x\} = K|L,$$

$N \cup \{y\}$ must be contained in K. This proves 6.4.

We shall use the Axiom of Choice, in the form of the Hausdorff Maximal Principle, in the proof of 6.6. We remark that 6.6 can be proved without the Axiom of Choice when S is metric (such a proof is sketched in 6.21). For convenience, we state the Hausdorff Maximal Principle below. A proof of its equivalence with the Axiom of Choice may be found in [1].

6.5 Hausdorff Maximal Principle (HMP). If $\mathfrak{L}$ is a collection of sets, then any nest (5.25) in $\mathfrak{L}$ is contained in a maximal nest $\mathfrak{C}$ (maximal means no nest in $\mathfrak{L}$ properly contains $\mathfrak{C}$ (4.34)).

We are now ready to prove the following important theorem. The notable metric version stating that a nondegenerate continuum must have at least two non-cut points is due to R. L. Moore (see [2, p. 177]).

6.6 Non-Cut Point Existence Theorem. Let (S,T) be a nondegenerate T_1-continuum. Assume that S has a cut point c,

$$S - \{c\} = U|V.$$

Then, there is a non-cut point of S in U and there is a non-cut point of S in V. Hence, S has at least two non-cut points.

Proof. Let N denote the set of all non-cut points of S (at this point, N may $= \varnothing$). Suppose that

(1) $N \subset V$, i.e., each $x \in U$ is a cut point of S.

By (1), and since $x \neq c$ for any $x \in U$, we may assume without loss of generality that:

(2) For each $x \in U$, $S - \{x\} = U_x|V_x$ with $c \in V_x$.

By (2) and 6.4, we have that:

(3) For each $x \in U$, $U_x \cup \{x\} \subset U$.

By (2) and 6.3, we have that:

(4) For each $x \in U$, $U_x \cup \{x\}$ and $V_x \cup \{x\}$ are connected.

Note that if $x \in U$ and $y \in U_x$, then, by (3), $y \in U$ and, hence, U_y and V_y as in (2) exist. We will prove:

(5) If $x \in U$ and $y \in U_x$, then $U_y \cup \{y\} \subset U_x$ and, thus, $x \notin U_y \cup \{y\}$.

To prove (5), let $x \in U$ and $y \in U_x$. The sets $U_y \cup \{y\}$ and $V_y \cup \{y\}$ are, by (4), connected and, since $y \neq x$ (because $y \in U_x$) and $U_y \cap V_y \neq \varnothing$, one of them is contained in $S - \{x\}$. Hence, one of (a)–(d) below must hold:

(a) $U_y \cup \{y\} \subset U_x$

(b) $V_y \cup \{y\} \subset U_x$

(c) $U_y \cup \{y\} \subset V_x$

(d) $V_y \cup \{y\} \subset V_x$

Since $y \in U_x$, $y \notin V_x$ and, hence, (c) and (d) are not possible. By (2), $c \in V_x \cap V_y$; thus, $c \notin U_x$ and $c \in V_y$. Hence, (b) is not possible. Therefore, (a) must hold. This proves (5). Next, let

$$\mathscr{L} = \{U_x \cup \{x\}: x \in U\}.$$

By HMP (6.5), there is a maximal nest $\mathfrak{C}$ in $\mathscr{L}$. Observe that each member of $\mathfrak{C}$ is closed in S (since S is a T_1-space) and that, since $\mathfrak{C}$ is a nest of nonempty sets, each nonempty finite subcollection of $\mathfrak{C}$ has nonempty intersection (since for such a $\mathscr{P} \subset \mathfrak{C}$, $\cap \mathscr{P} \in \mathscr{P}$). Thus, since S is compact, $\cap \mathfrak{C} \neq \varnothing$ (an easy exercise in complementation). Let $p \in \cap \mathfrak{C}$. By (3), $\mathfrak{C}$ is a collection of subsets of U. Hence, $p \in U$. Thus, U_p as in (2) exists. Observe that for each $U_x \cup \{x\} \in \mathfrak{C}$:

$U_p \cup \{p\} \subset U_x$ if $p \in U_x$ (by (5));

$U_p \cup \{p\} = U_x \cup \{x\}$ if $p = x$;

$p \in U_x$ or $p = x$ (since $p \in \cap \mathfrak{C}$).

Hence, we have that:

(6) $U_p \cup \{p\} \subset \cap \mathfrak{C}$.

Now, let $q \in U_p$ ($U_p \neq \varnothing$ by (2)). Then, since $p \in U$, (5) gives us that

(7) $U_q \cup \{q\}$ is a proper subset of $U_p \cup \{p\}$.

Since $q \in U$.

(8) $U_q \cup \{q\} \in \mathscr{L}$.

The facts in (6)–(8) easily yield a contradiction to the maximality of $\mathfrak{C}$. Therefore, we have proved 6.6.

A proof of 6.6 for Hausdorff continua from the point of view of partially ordered spaces is in [3].

The following corollary is significantly stronger than 6.6.

6.7 Corollary. Let (S,T) be a nondegenerate T_1-continuum. Let N denote the set of all non-cut points of S. Then, no proper connected subset of S contains N.

Proof. Suppose there is a proper connected subset Z of S such that $Z \supset N$. Let $c \in S - Z$. Then, since $c \in S - N$,

$$S - \{c\} = U|V.$$

Hence, Z being a connected subset of $S - \{c\}$, we must have the $Z \subset U$ or $Z \subset V$. Thus, since $N \subset Z$, $N \subset U$ or $N \subset V$. This contradicts 6.6. Therefore, we have proved 6.7.

In particular, 6.7 implies that every continuum is irreducible about its set of non-cut points (the definition of irreducible about a set is in 4.35).

2. LOCATION OF NON-CUT POINTS

The following result gives some information regarding the location of non-cut points.

6.8 Theorem. Let X be a continuum, let A be a proper subcontinuum of X, and let K be a component of $X - A$. Then, there is a non-cut point p of $K \cup A$ such that $p \in K$, and any such point p is also a non-cut point of X.

Proof. By 5.9, $K \cup A$ is a continuum. Hence, A being a proper subcontinuum of $K \cup A$, 6.7 guarantees that there is a non-cut point p of $K \cup A$ such that $p \in K$. Now, to see that p is a non-cut point of X, observe that

$$X - \{p\} = [(K \cup A) - \{p\}] \cup [\cup \{L \cup A: L \text{ is a component of } X - A \text{ and } L \neq K\}]$$

where $(K \cup A) - \{p\}$ is connected, each $L \cup A$ is connected (5.9), and the intersection of all these sets is nonempty (since $p \notin A$). Hence, $X - \{p\}$ is connected [2, p. 132]. This completes the proof of 6.8.

Another result which locates non-cut points is in (b) of 6.29. Some applications of these results are in Section 3 of Chapter IX and Section 1 of Chapter X.

3. SEPARATION ORDERING FOR $S(p,q)$

In view of 6.6, it is natural to inquire about the structure of those T_1-continua which have the minimum possible number of non-cut points, namely exactly two. We shall see in 6.16 that these are precisely the T_1-continua whose topology comes from a simple ordering. This result can be used to obtain an intrinsic topological characterization of the arc (6.17). We shall also obtain this characterization by a different method in Chapter IX (9.29).

The following special set $S(p,q)$ plays a significant role in our investigations. The reason will be apparent from 6.10.

6.9 Definition of $S(p,q)$. Let Z be a topological space, and let p, $q \in Z$ with $p \neq q$. Recall from 5.30 that a point $z \in Z$ is said to *separate p and q in Z* provided that

$$Z - \{z\} = A|B \qquad \text{with } p \in A \text{ and } q \in B.$$

We define $S(p,q)$ by letting

$$S(p,q) = \{z \in Z: z = p, z = q, \text{ or } z \text{ separates } p \text{ and } q \text{ in } Z\}.$$

6.10 Proposition. Let Z be a nondegenerate T_1-continuum. Then, Z has exactly two non-cut points if and only if $Z = S(p,q)$ for some p, $q \in Z$.

Proof. Assume that:

(#) Z has exactly two non-cut points p and q.

Let $c \in Z - \{p,q\}$. Then, by (#), c is a cut point of Z, i.e.,

$$Z - \{c\} = U|V \qquad (6.2).$$

Hence, by (#) and 6.6, one of p or q must be in U and the other must be in V. Thus, by 6.9, $c \in S(p,q)$. Therefore, we have proved that $Z = S(p,q)$. Conversely, assume that $Z = S(p,q)$ for some p, $q \in Z$. Then, by 6.9, the only possible non-cut points of Z are p and q. Thus, since Z has at least two non-cut points (6.6), Z has exactly two non-cut points (namely, p and q). This completes the proof of 6.10.

The following lemma and the corollary in 6.12 are technical results. Their purpose is to determine the key properties of $S(p,q)$ which will be used in the proof of 6.15.

6.11 Lemma. Let Z be a connected topological space, and let p, $q \in Z$ with $p \neq q$. Let $x, y \in S(p,q) - \{p,q\}$ with

$$Z - \{x\} = A_i|B_i, \qquad p \in A_i, q \in B_i, \qquad i = 1 \text{ and } 2$$
$$Z - \{y\} = C|D, \qquad p \in C, q \in D.$$

Then, (1) and (2) below hold:

(1) If $y \in A_i$ for some $i = 1$ or 2, then $C \cup \{y\} \subset A_1 \cap A_2$;
(2) If $y \in B_1$, then $A_1 \cup \{x\} \subset C$.

Proof. To prove (1), let $y \in A_1$. Then, $x \neq y$ so, since $C \cap D = \emptyset$, $x \notin C \cup \{y\}$ or $x \notin D \cup \{y\}$. Hence,

$$C \cup \{y\} \subset A_1 \cup B_1 \quad \text{or} \quad D \cup \{y\} \subset A_1 \cup B_1.$$

Thus, since $C \cup \{y\}$ and $D \cup \{y\}$ are connected (6.3) and $y \in A_1$,

$C \cup \{y\} \subset A_1$ or $D \cup \{y\} \subset A_1$.

Therefore, since $q \in D$ and $q \notin A_1$, we must have

(a) $C \cup \{y\} \subset A_1$.

By (a), since $x \notin A_1$, $x \notin C \cup \{y\}$. Hence, $C \cup \{y\} \subset A_2 \cup B_2$. Therefore, since $C \cup \{y\}$ is connected and $p \in C \cap A_2$, we must have

(b) $C \cup \{y\} \subset A_2$.

By (a) and (b), $C \cup \{y\} \subset A_1 \cap A_2$. Therefore, since the proof is similar starting with $y \in A_2$, we have proved (1). Now, to prove (2), let $y \in B_1$. Then, $y \notin A_1 \cup \{x\}$. Hence,

$A_1 \cup \{x\} \subset C \cup D$.

Thus, since $A_1 \cup \{x\}$ is connected (6.3),

$A_1 \cup \{x\} \subset C$ or $A_1 \cup \{x\} \subset D$.

Therefore, since $p \in A_1 \cap C$, we must have $A_1 \cup \{x\} \subset C$. This proves (2). Therefore, we have proved 6.11.

6.12 Corollary.   Let Z be a connected topological space, and let p, $q \in Z$ with $p \neq q$.

(1) For each $x \in S(p,q) - \{p,q\}$, there exist unique sets (depending only on x) P_x and Q_x such that

$$S(p,q) - \{x\} = P_x | Q_x, \, p \in P_x, \, q \in Q_x.$$

(2) If Z is a T_1-space, then P_x and Q_x in (1) are each open in the subspace topology for $S(p,q)$.

Furthermore: If $x, y \in S(p,q) - \{p,q\}$ with $x \neq y$, then

(3) $y \in P_x \cup Q_x$;
(4) if $y \in P_x$, then $P_y \cup \{y\} \subset P_x$;
(5) if $y \in Q_x$, then $P_x \cup \{x\} \subset P_y$ and, thus, $x \in P_y$.

Proof.   By (1) of 6.11, $A_1 \cap S(p,q) = A_2 \cap S(p,q)$ and, hence, we also have $B_1 \cap S(p,q) = B_2 \cap S(p,q)$. Therefore, letting

(#) $P_x = A_1 \cap S(p,q)$ and $Q_x = B_1 \cap S(p,q)$,

part (1) of 6.12 now follows. To prove part (2), simply observe from part (1) that P_x and Q_x are each open in $S(p,q) - \{x\}$ and, assuming Z is a T_1-space, $S(p,q) - \{x\}$ is open in $S(p,q)$. Part (3) is obvious. Part (4) follows immediately from (#) and (1) of 6.11 since, for C as in 6.11, $C \cap S(p,q)$ is what we now denote by P_y. Similarly, part (5) follows immediately from (#) and (2) of 6.11. This completes the proof of 6.12.

A *binary relation* for a set S is a subset of $S \times S$. We now define a binary relation $<_s$ for $S(p,q)$ called the separation ordering. The important general properties of $<_s$ are given in 6.15.

6.13   Definition of the Separation Ordering.   Let Z be a connected topological space, and let p, $q \in Z$ with $p \neq q$. The *separation ordering* $<_s$ *for* $S(p,q)$ is defined by the statements in (1)–(3) below:
   (1)   For any $z \in S(p,q) - \{p\}, p <_s z$.
   (2)   For any $z \in S(p,q) - \{q\}, z <_s q$.
   (3)   For any $x, y \in S(p,q) - \{p,q\}, y <_s x$ if and only if $y \in P_x$ [where P_x is as in (1) of 6.12].

Recall the definitions of the following general notions.

6.14   Definition of Simple Ordering and Order Topology.   A binary relation $<$ for a set Y is called a *simple ordering* provided that $<$ is (a) anti-reflexive [$y < x$ implies $y \neq x$], (b) transitive [$y < x$ and $x < t$ imply $y < t$], and (c) total [$x, y \in Y$ and $x \neq y$ imply $y < x$ or $x < y$]. Given a simple ordering $<$ on a set Y, the *order topology* for Y (induced by $<$) is the topology for Y having as a subbase all sets of the form

$$\{y \in Y: y < a\}, \quad \{y \in Y: a < y\} \quad \text{each } a \in Y.$$

In other words, all so-called open intervals (a,b),

$$(a,b) = \{y \in Y: a < y \text{ and } y < b\},$$

for a, $b \in Y$ form a base for the order topology. We note, as is easy to prove, that the order topology is always Hausdorff. This fact will be used in the proof of 6.16.

Note the following general result about $<_s$ defined in 6.13. Its proof relies on 6.12, and it will be used in the proof of 6.16.

6.15   Proposition.   The separation ordering $<_s$ for $S(p,q)$ defined in 6.13 is a simple ordering. Furthermore, if Z is a T_1-space, the separation order topology for $S(p,q)$ is contained in the subspace topology for $S(p,q)$.

   Proof.   Clearly, by 6.13, $<_s$ is a binary relation for $S(p,q)$. Now, $<_s$ is anti-reflexive since $p <_s p$ and $q <_s q$ are impossible by 6.13 and since $x \notin P_x$ by (1) of 6.12. By using (4) of 6.12, it follows easily that $<_s$ is transitive. By using (3) and (5) of 6.12, it follows easily that $<_s$ is total.

This completes the proof that $<_s$ is a simple ordering for $S(p,q)$. Now, to prove the second part of 6.15, assume Z is a T_1-space and let T_S denote the subspace topology for $S(p,q)$. Let $x \in S(p,q)$ and let

$$U = \{y \in S(p,q): y <_s x\}, \quad V = \{y \in S(p,q): x <_s y\}$$

(U and V are the subbasic sets determined by x as in 6.14). If $x = p$, then, by (1) of 6.13 and by anti-reflexiveness and transitivity of $<_s$, $U = \varnothing$ so $U \in T_S$. If $x = q$, then, by (2) of 6.13 and anti-reflexiveness of $<_s$, $U = S(p,q) - \{q\}$ so, since Z is a T_1-space, $U \in T_S$. Now, assume that $x \neq p$ and $x \neq q$. Then: By (3) of 6.13,

$$U - \{p,q\} = P_x - \{p,q\};$$

by (1) of 6.13, $p \in U$; by (1) of 6.12, $p \in P_x$ and $q \notin P_x$; and, by (2) of 6.13 and the anti-reflexiveness and transitivity of $<_s$, $q \notin U$. Hence, $U = P_x$. Thus, by (2) of 6.12, $U \in T_S$. Similar arguments show that $V \in T_S$. Therefore, by 6.14, we have proved that the separation order topology for $S(p,q)$ is contained in T_S. This completes the proof of 6.15.

4. NON-CUT POINT CHARACTERIZATION OF THE ARC

We now come to the following important characterization theorems discussed briefly at the beginning of Section 3.

6.16   Theorem.   Let (Z,T) be a nondegenerate T_1-continuum. If Z has exactly two non-cut points, then $Z = S(p,q)$ for some p, $q \in Z$ and T is equal to the separation order topology T'. Conversely: If T is the order topology obtained from some simple ordering $<$ on Z, then Z has exactly two non-cut points.

Proof.   Assume first that Z has exactly two non-cut points. Then, by 6.10, $Z = S(p,q)$ for some p, $q \in Z$. Hence, by 6.15, $T' \subset T$. Therefore, since (Z,T) is compact and (Z,T') is Hausdorff (see 6.14), $T = T'$ (since the identity map is a homeomorphism). Conversely, assume T is the order topology obtained from some simple ordering $<$ on Z. Then, since Z is compact, there are p, $q \in Z$ such that $p < x$ for all $x \in Z - \{p\}$ and $x < q$ for all $x \in Z - \{q\}$ (to obtain p, cover Z with finitely many sets of the form $\{z \in Z: a < z\}$, $a \in Z$, and let p be the minimum of the finitely many points a associated with the members of the finite cover; q is obtained similarly). Then, since T is the order topology obtained from $<$, clearly each $z \in Z - \{p,q\}$ is a cut point of Z. Hence, p and q are the only possible non-cut points of Z. Thus, by 6.6, Z has exactly two non-cut points. This completes the proof of 6.16.

Recall the definition of an arc in 1.1. This definition is not very satisfy-ing in that it is extrinsic in nature. The following theorem gives an intrin-sic characterization of the arc. A proof based on 6.16 is sketched in the exercise in 6.22. For references relating to the history of the theorem, see [2, p. 179].

6.17   Theorem.   A continuum X is an arc if and only if X has exactly two non-cut points.

Proof.   See 6.22.

In view of 6.17, it is natural to wonder about what continua have only finitely many non-cut points. This will be determined in 9.28, from which we see that such continua are extremely simple. We note that there is only one continuum with exactly three non-cut points (9.43). Let us also men-tion, as part of this theme, that continua which have only countably many non-cut points must be hlc (defined in 10.3), as follows from (b) of 6.29 and 10.4. Therefore, we see that cardinality restrictions on the set of non-cut points of continua significantly limit aspects of their structure.

In relation to the material in this chapter, especially the ideas and results beginning with 6.9, we recommend that the reader see Chapter III of [4].

EXERCISES

6.18   Exercise.   Work (a)–(c) below which concern ideas related to 6.3.
  (a)   Let (S,T) be a connected topological space. If $S = M \cup N$ where M and N are each closed, or each open, in S and $M \cap N$ is connected, then M and N are each connected.
  (b)   If a topological space Z has more than n components, $n < \infty$, then Z is the union of $n + 1$ nonempty sets $M_1, \ldots, M_{n+1}$ each two of which are mutually separated in Z. [Hint: A trivial induction.] This fact will be useful in proving (c) below.
  (c)   The following result is a generalization of 6.3. Let (S,T) be a connected topological space, and let $C_1, \ldots, C_n$ ($n < \infty$) be connected subsets of S such that

$$S - \bigcup_{i=1}^{n} C_i = A|B.$$

Then, $A \cup (\cup_{i=1}^{n} C_i)$ and $B \cup (\cup_{i=1}^{n} C_i)$ each has at most n components. [Hint: Let $Z = A \cup (\cup_{i=1}^{n} C_i)$, and suppose that Z has more than n components. Apply (b), and show that no M_i in (b) is mutually separated from B in S. What next?]

6.19 Exercise. Prove that a continuum X is indecomposable if and only if each proper subcontinuum of X is nowhere dense in X. [Hint: Use 6.3.]

6.20 Exercise. Prove that every connected topological space containing at least three points can be written as the union of two nondegenerate, proper, connected sets. [Hint: Consider whether or not there are cut points.] We note that the sets may not be able to be chosen to be mutually disjoint as they can in the case of continua (5.29)–for example, consider the one-point compactification of the rationals (a metric example is much harder [2, p. 135]).

6.21 Exercise. In this exercise, we give the main ideas for a proof of 6.6 in the case when S is metric which does not use the Axiom of Choice (i.e., does not use 6.5). You are asked to fill in the details. Assume S in 6.6 is metric. Suppose (1) in the proof of 6.6 holds. Let $D = \{p_n: n = 1, 2, \ldots\}$ be a countable dense subset of S. Let $n(1) = \min \{n: p_n \in U\}$ [$n(1)$ exists and $p_{n(1)} \in U$]. Then,
$$S - \{p_{n(1)}\} = E_1|F_1 \text{ where we may assume } c \in E_1 \text{ (}c \text{ as in 6.6)}.$$
Also, $F_1 \cup \{p_{n(1)}\} \subset U$ (6.4). Let $n(2) = \min \{n: p_n \in F_1\}$. Then,
$$S - \{p_{n(2)}\} = E_2|F_2 \text{ where we may assume } p_{n(1)} \in E_2.$$
By doing a formal induction and using 6.3 and 6.4, we obtain sets F_k and E_k such that for each $k = 1, 2, \ldots$:
  (1) $F_x \cup \{p_{n(k)}\}$ and $E_k \cup \{p_{n(k)}\}$ are continua;
  (2) $F_{k+1} \cup \{p_{n(k+1)}\} \subset F_k$ and $E_k \cup \{p_{n(k)}\} \subset E_{k+1}$;
  (3) $E_{k+1} \supset \{p_1, p_2, \ldots, p_{n(k)}\}$.
Let $F = \cup_{k=1}^{\infty} (F_k \cup \{p_{n(k)}\})$ and $E = \cup_{k=1}^{\infty} (E_k \cup \{p_{n(k)}\})$. By using (1)–(3) above and 1.7 (to see that $F \neq \varnothing$), a contradiction is obtained.

6.22 Exercise. We sketch the standard proof of 6.17 (a different proof is in 9.29). You are asked to supply the details. Assume (X, T) is a continuum with exactly two non-cut points. Then, by 6.16, $X = S(p, q)$ for some $p, q \in X$ and T is the order topology induced by the separation ordering $<_s$ (6.13–recall that $p <_s q$). Let C be a countable dense subset of $X - \{p, q\}$,

$$C = \{c_n: n = 1, 2, \ldots\} \qquad \text{where } c_i \neq c_j \text{ for } i \neq j,$$

and let D be the dyadic rationals in $(0,1)$, i.e.,

$$D = \{k/2^m: k < 2^m, k = 1, 2, ..., \text{ and } m = 1, 2, ...\}.$$

Prove (a)–(d) below.
(a) If $y <_s x$, then there exists $z \in X$ such that $y <_s z$ and $z <_s x$.
(b) There is an order isomorphism f from C onto D, i.e.: If $c_i <_s c_j$, then $f(c_i) < f(c_j)$, and $f(C) = D$. [Hint: Let $f(c_1) = 1/2$. Then, there exist first natural numbers $n(1)$ and $n(2)$ such that $c_{n(1)} <_s c_1$ and $c_1 <_s c_{n(2)}$ (why?). Let $f(c_{n(1)}) = 1/4$ and $f(c_{n(2)}) = 3/4$. Continue.]

Let $h(p) = 0$, $h(q) = 1$, and, for $x \in X - \{p,q\}$, let

$$h(x) = \text{lub}\{f(c_i): c_i \in C \text{ and } c_i <_s x\}.$$

(c) h is defined at each $x \in X$ and h is order preserving, i.e.: If $z <_s y$, then $h(z) < h(y)$.
(d) h is a homeomorphism from X onto $[0,1]$.

The other half of 6.17 is easy. We remark that there are compact connected Hausdorff spaces with exactly two non-cut points which are not metrizable and, hence, are not what we call arcs (see, e.g., J of [1, p. 164]). Appropriately, such spaces are often called *generalized arcs*.

As a by-product of what was shown above and how it was shown, prove (e) below:

(e) If J_1 and J_2 are arcs with end points p_1,q_1 and p_2,q_2, respectively, and if D_1 and D_2 are countable dense subsets of $J_1 - \{p_1,q_1\}$ and $J_2 - \{p_2,q_2\}$, respectively, then D_1 and D_2 are homeomorphic and, furthermore, there is a homeomorphism h from J_1 onto J_2 such that $h(D_1) = D_2$, $h(p_1) = p_2$, and $h(q_1) = q_2$.

6.23 Exercise. Let X be a continuum, and let A be a nonempty compact subset of X. Prove that $X - A$ is connected if and only if A is a non-cut point of X/A (3.14). A generalization is in (b) of 6.24.

6.24 Exercise. This exercise concerns non-cut points of usc decompositions (3.5). Let X be a continuum, and let $\mathcal{D}$ be a usc decomposition of X (with the decomposition topology). Work (a)–(c) below.

(a) Let $D_0 \in \mathcal{D}$. Prove that if $X - D_0$ is connected, then D_0 is a non-cut point of $\mathcal{D}$. Give an example to show that in general the converse is false (comp., (b)).
(b) Let $D_0 \in \mathcal{D}$, and assume that the members of $\mathcal{D}$ are subcontinua of X except possibly for D_0. Then, prove that D_0 is a non-cut point of $\mathcal{D}$ if and only if $X - D_0$ is connected. [Hint: Suppose $X - D_0 = C|H$ and use (3) of 3.7]. This is a generalization of 6.23.

(c) Prove that if $X = [0,1]$ and all the members of $\mathcal{D}$ are proper subcontinua of X, then $\mathcal{D}$ is an arc. [Hint: Use (b), 6.17, and, of course, 3.10]. The equivalent formulation of this result in 8.22 is used in proving 8.23.

6.25 Exercise. Let (X,T) be a continuum. A point $p \in X$ is called an *end point of* X provided that whenever $p \in U \in T$, there exists $V \in T$ such that $p \in V \subset U$ and $Bd(V)$, defined in 5.3, consists of precisely one point. This is obviously a generalization of the notion of an end point of an arc defined in 1.1. The symbol $X^{[1]}$ denotes the set of all end points of X. Work (a)–(e) below.

(a) Prove that by replacing the open sets in the definition above with neighborhoods of p, an equivalent statement is obtained, i.e., p is an end point of X if and only if p has arbitrarily small neighborhoods with one-point boundaries. [Hint: Use (f) and (h) of 5.17.]

(b) Prove that every end point of a continuum X is a non-cut point of X. We mention that the converse characterizes dendrites (10.7). Also, a better result is in 6.27.

(c) Prove that every end point of a continuum X has arbitrarily small open neighborhoods V such that $X - V$ is connected. Compare with (d).

(d) Give an example to show that if the open neighborhoods in (c) are replaced by closed neighborhoods, then (c) would be false (in spite of (a)). [Hint: Reverse the "stickers" in Figure 5.22.] Note that, in addition to its other properties, the continuum suggested by the hint is LC at the end point p–see (e).

(e) Prove that if p is an end point of a continuum X, then X is cik at p (5.10). The example in Figure 5.22 shows that X may fail to be LC at an end point p.

Some other basic facts about end points are in 6.26 and 6.27.

6.26 Exercise. Prove that if a continuum X is irreducible between p and q (4.35), then:

(a) p and q are non-cut points of X [hint: use (a) of 5.20];

(b) p and q are the only possible end points of X [hint: use (c) of 6.25].

6.27 Exercise. Prove that if X is a continuum, then, for any $H \subset X^{[1]}$ (6.25), $X - H$ is connected. [Hint: Fix $p \in X - H$, apply (b) of 4.35 to p and any $q \in X - H$, and then use (b) of 6.26.]

6.28 Exercise. Prove that if X is a continuum, then every $A \in C(X)$, and every $A \in 2^X$, is a non-cut point of $C(X)$, of 2^X respectively. [Hint: The result in the comments at the end of 5.25 will be useful.]

6.29 Exercise. If X is a connected metric space, then by a (*closed*) *cutting of* X we mean a (closed) subset C of X such that $X - C$ is not connected. Some authors define C to be a cutting of X provided that there are two points $p,q \in X - C$ such that there is no subcontinuum of $X - C$ containing $\{p,q\}$. Clearly, our notion of cutting implies this one and easy examples show that the implication is not reversible (even when X is a continuum and C is closed in X–example?). Prove (a)–(f) below assuming X is a connected, separable metric space. We remark that (b) and (d) have important applications in the proofs of some results in Chapter X (e.g., 10.7 and 10.8).

(a) If $\mathfrak{C}$ is an uncountable collection of mutually disjoint closed cuttings of X, then there exists $C \in \mathfrak{C}$ such that $X - C = U|V$ where

$$U \cap (\cup \mathfrak{C}) \neq \varnothing \neq V \cap (\cup \mathfrak{C}).$$

[Hint: If $X - C_\alpha = U_\alpha | V_\alpha$ with $U_\alpha \supset \cup \mathfrak{C} - C_\alpha$ for each $C_\alpha \in \mathfrak{C}$, then, for C_α, $C_\beta \in \mathfrak{C}$ such that $C_\alpha \neq C_\beta$, $X = (U_\alpha \cup U_\beta) \cup (V_\alpha \cap V_\beta)$; thus, $V_\alpha \cap V_\beta = \varnothing$ (why?) and, hence, X contains uncountably many mutually disjoint open sets.]

(b) If K is a convergence continuum of X (5.11), then K does not contain any uncountable collection of mutually disjoint closed cuttings of X. [Hint: Use (a).] Note that this implies that at most countably many points of K are cut points of X; thus, K contains uncountably many non-cut points of X.

(c) If Y is an uncountable collection of cut points of X, then some two points $y_1, y_2 \in Y$ are separated in X by a third point $y_3 \in Y$ (for definition, see 5.30). [Hint: Trivial by (a).]

(d) Let Y denote the set of all cut points of X. If Z is a connected subset of X, then all but at most countably many points of $Y \cap Z$ are cut points of Z.

(e) If Z is a connected subset of X, then $\bar{Z} - Z$ contains at most countably many cut points of X. [Hint: Use (d).]

(f) If X is nondegenerate, $\mathfrak{C}$ in (a) exists.

(g) Every cutting of X contains a closed cutting of X. Hence, (a) and (b) remain true without requiring the cuttings to be closed in X.

REFERENCES

1. John L. Kelley, *General Topology*, D. Van Nostrand Co., Inc., Princeton, N.J., 1955.

2. K. Kuratowski, *Topology*, Vol. II, Academic Press, New York, N.Y., 1968.

3. L. E. Ward, On the non-cut point existence theorem, Can. Math. Bull., 11(1968), 213–216.

4. Gordon Thomas Whyburn, *Analytic Topology*, Amer. Math. Soc. Colloq. Publ., vol. 28, Amer. Math. Soc., Providence, R.I., 1942.

VII
A General Mapping Theorem

The existence (or nonexistence) of continuous functions is one of the most important aspects of continuum theory. Many results are directly concerned with continuous functions or make significant use of them in their proofs. For an excellent illustration of the latter, see 13.57 and its proof.

We prove a general theorem in 7.4 which gives a unified way to obtain a number of particular mapping theorems. The theorem is easy to prove (as are the two lemmas used to prove it). Since in all the applications of 7.4 the idea of the proofs is the same, the proofs are easy to understand. Some applications are in this chapter (7.7, 7.14, 7.19, 7.20, and 7.23), and we shall use 7.4 in Chapter VIII to prove the Hahn-Mazurkiewicz Theorem (8.14).

On reading 7.4, a familiar pattern comes to mind. In Chapters I and II, we constructed continua by the nested intersection technique–now, we produce continuous functions by the same technique.

The theorem in 7.4 and the idea of using it as we do in various places is due to M. K. Fort, Jr. His only published paper about this is [3] which contains a general result implying 7.14. Except for 7.19 and 7.23, the applications of 7.4 are from the notes for Fort's topology course at the University of Georgia (1963).

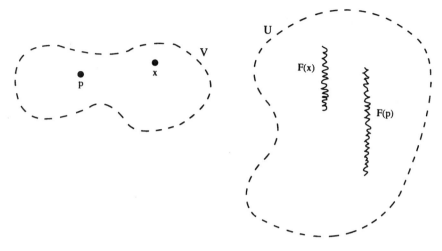

Figure 7.1

1. THE GENERAL THEOREM

The following notion will be used throughout this chapter. A dual notion will be defined in 13.6.

7.1 Definition of USC Function.   Let (X,T_1) and (Y,T_2) be topological spaces, and let (as in (1) of 4.1)

$$2^Y = \{A: A \text{ is a nonempty closed subset of } Y\}.$$

A function $F: X \to 2^Y$ is said to be *upper semi-continuous at a point $p \in X$*, written *usc at p*, provided that if $U \in T_2$ such that $F(p) \subset U$, there exists $V \in T_1$ such that $p \in V$ and $F(x) \subset U$ for all $x \in V$ (see Figure 7.1). A function $F: X \to 2^Y$ is said to be *upper semi-continuous*, written *usc*, provided that F is usc at every point of X.

The terminology in 7.1 probably came from considerations as in 3.25 or from the following easy-to-prove fact: Let $\mathcal{D}$ be a closed partition of a topological space S, and let $\pi: S \to \mathcal{D}$ be the natural map (3.1); then $\mathcal{D}$ is a usc decomposition (3.5) if and only if π, as a function from S into 2^S, is usc in the sense of 7.1.

The following two lemmas about usc functions will be used in the proof of 7.4.

7.2 Lemma.   Let (X,T_1) and (Y,T_2) be topological spaces, and let F: $X \to 2^Y$ be usc. If $F(x)$ is a singleton set $\{y_x\}$ for each $x \in X$, then the function $f: X \to Y$ defined by

$$f(x) = y_x \qquad \text{for each } x \in X$$

is continuous.

Proof. Let $p \in X$, and let $U \in T_2$ such that $f(p) \in U$. Then, since we have

$$F(p) = \{f(p)\} \subset U \text{ and } F \text{ is usc at } p,$$

by 7.1 there exists $V \in T_1$ such that $p \in V$ and $F(x) \subset U$ for all $x \in V$. Hence: $V \in T_1$, $p \in V$, and, for all $x \in V$, $\{f(x)\} \subset U$, i.e., $f(x) \in U$. Therefore, we have proved that f is continuous at p. This proves 7.2.

7.3 Lemma. Let X and Y be nonempty compact metric spaces. For each $n = 1, 2, \ldots$, let $F_n: X \to 2^Y$ be usc such that

$$F_n(x) \supset F_{n+1}(x) \qquad \text{each } x \in X, \text{ each } n = 1, 2, \ldots.$$

For each $x \in X$, let $G(x) = \bigcap_{n=1}^{\infty} F_n(x)$. Then, (1) and (2) below hold:

(1) $G: X \to 2^Y$ and G is usc;
(2) If $Y = \bigcup_{x \in X} F_n(x)$ for each $n = 1, 2, \ldots$, then $Y = \bigcup_{x \in X} G(x)$.

Proof. To prove (1), first note that, by the last part of 1.7, $G(x) \neq \varnothing$ for each $x \in X$ and, hence, $G: X \to 2^Y$. Now, to prove that G is usc, let $p \in X$ and let U be an open subset of Y such that $G(p) \subset U$. Then, by 1.7, there exists N such that $F_N(p) \subset U$. Thus, since F_N is usc at p, by 7.1 there exists an open subset V of X such that $p \in V$ and $F_N(x) \subset U$ for all $x \in V$. Therefore, since $G(x) \subset F_N(x)$ for all x, $G(x) \subset U$ for all $x \in V$. Hence, G is usc at p (7.1). Therefore, we have proved (1). To prove (2), fix $q \in Y$. Then, by the assumption in (2), for each $n = 1, 2, \ldots$ there exists $x_n \in X$ such that $q \in F_n(x_n)$. By the compactness of X, there is a subsequence $\{x_{n(i)}\}_{i=1}^{\infty}$ of $\{x_n\}_{n=1}^{\infty}$ such that $\{x_{n(i)}\}_{i=1}^{\infty}$ converges to some point $p \in X$. We now show that $q \in G(p)$. Suppose that $q \notin G(p)$. Then, letting $U = Y - \{q\}$, $G(p) \subset U$ and U is open in Y. Hence, by 1.7, there exists N such that $F_N(p) \subset U$. Thus, since F_N is usc at p, by 7.1 there exists an open subset V of X such that $p \in V$ and

(a) $F_N(x) \subset U$ for all $x \in V$.

Since $\{x_{n(i)}\}_{i=1}^{\infty}$ converges to $p \in V$, there exists $k = n(i)$ for some i such that $k \geq N$ and $x_k \in V$. Hence, by (a),

(b) $F_N(x_k) \subset U$.

Since $k \geq N$, $F_N(x_k) \supset F_k(x_k)$. Thus, since $q \in F_k(x_k)$, $q \in F_N(x_k)$. Hence, by (b), $q \in U$ which is impossible since $U = Y - \{q\}$. Therefore, we have proved that $q \in G(p)$. This completes the proof of 7.3.

Now we come to the general theorem. It says that an appropriate nested intersection of usc functions results in a continuous, onto, single-valued function between compact metric spaces.

7.4 General Mapping Theorem. Let X and Y be nonempty compact metric spaces. Assume that (1)–(4) below hold:

(1) $F_n: X \to 2^Y$ is usc for each $n = 1, 2, \ldots$;
(2) $F_n(x) \supset F_{n+1}(x)$ for each $x \in X$ and each $n = 1, 2, \ldots$;
(3) $Y = \bigcup_{x \in X} F_n(x)$ for each $n = 1, 2, \ldots$;
(4) $\lim_{n \to \infty}$ diameter $[F_n(x)] = 0$ for each $x \in X$.

Then, there is a continuous function f from X onto Y. In fact, such an f may be defined by letting

$f(x)$ = the unique point in $\bigcap_{n=1}^{\infty} F_n(x)$

for each $x \in X$.

Proof. Let $G(x) = \bigcap_{n=1}^{\infty} F_n(x)$ for each $x \in X$. Then, the formula above for f says

(#) $f(x)$ = the unique point in $G(x)$, each $x \in X$

which gives a well-defined function f by (2), (4), and the second part of 1.7. By (#), (1) of 7.3, and 7.2, $f: X \to Y$ is continuous. By (#) and (2) of 7.3, f maps X onto Y. Therefore, we have proved 7.4.

2. THE CANTOR SET

Several of the applications of 7.4 in this chapter involve the Cantor set which we now define.

7.5 Definition of Cantor Set. The *Cantor Middle-Third set* is the subspace C of $[0,1]$,

$$C = \bigcap_{i=1}^{\infty} C_i$$

where $C_1 = [0,1] - (1/3, 2/3)$ and, assuming inductively that we have defined C_i, C_{i+1} is defined by deleting from C_i the middle-third open interval of each component of C_i. A *Cantor set* is any space which is homeomorphic to C. We remark that C may be defined as the space consisting of all those numbers in $[0,1]$ which can be written in the ternary system using only the digits 0 and 2. The equivalence of these two ways of defining C follows easily since, for

$$t = \sum_{i=1}^{\infty} t_i/3^i \qquad \text{where } t_i \in \{0,1,2\} \text{ for each } i,$$

$t \in (1/3,2/3)$ if and only if t_1 must $= 1$, $t \in (1/9,2/9) \cup (7/9,8/9)$ if and only if t_2 must $= 1$ and t_1 must $= 0$ or 2, etc.

Results about Cantor sets are in the next two sections and in the exercises (7.22–7.27). The following technical lemma will be used in the proofs of 7.7 and 7.14.

7.6 Lemma. Let C be the Cantor Middle-Third set. Then:

(1) Given any integer $n \geq 1$, $C = \cup_{i=1}^{n} D_i$ where each D_i is non-empty and compact, and $D_i \cap D_j = \varnothing$ for $i \neq j$;
(2) Given any integer $m \geq 1$ and any D_i in (1), $D_i = \cup_{j=1}^{m} D_{i,j}$ where each $D_{i,j}$ is nonempty and compact, and $D_{i,j} \cap D_{i,k} = \varnothing$ for $j \neq k$.

Proof. Assume $n > 1$. From the nested intersection construction used to define C in 7.5, it is easy to find ℓ large enough so that for C_ℓ in 7.5, there are $n - 1$ points $t_1, \ldots, t_{n-1}$ in $[0,1] - C_\ell$ such that any two of these points are in different components of $X - C_\ell$ and

$$t_0 = 0 < t_1 < t_2 < \cdots < t_{n-1} < 1 = t_n.$$

Let $D_i = [t_{i-1}, t_i] \cap C$ for each $i = 1, \ldots, n$. Then, it is easy to verify that $D_1, \ldots, D_n$ satisfy (1). To prove (2), simply observe that the ideas used in proving (1) can be applied to any one of the intervals

$$[\text{glb } D_i, \text{lub } D_i], \qquad 1 \leq i \leq n$$

to obtain the desired sets $D_{i,1}, \ldots, D_{i,m}$ in (2). This completes the proof of 7.6.

3. APPLICATION: IMAGES OF CANTOR SET

For readers familiar with the standard proofs of the following theorem, you should be delighted with how easy it is to prove using 7.4. The theorem will be used to give some examples of space-filling curves (7.9).

7.7 Theorem. Every nonempty compact metric space Y is a continuous image of the Cantor Middle-Third set C.

Proof. Cover Y with finitely many nonempty compact subsets $A_1, \ldots, A_n$ each of diameter < 1 (first, by compactness, we can cover Y with finitely many nonempty open subsets $U_1, \ldots, U_n$ each of diameter < 1; then, let $A_i = \bar{U}_i$ for each $i = 1, \ldots, n$). Let $D_1, \ldots, D_n$ be as in (1) of 7.6. Define $F_1 \colon C \to 2^Y$ by letting $F_1(t) = A_i$ for each $t \in D_i$ (F_1 is well-defined on C since the sets D_i are mutually disjoint and cover C). Clearly, F_1 satisfies (1) and (3) of 7.4 [for (1), note that each D_i is open in C and F_1 is constant on each D_i–hence, F_1 is actually continuous using the Hausdorff metric on 2^Y (4.1)]. Next, for each i ($i \le i \le n$), cover A_i with finitely many nonempty compact *subsets* $A_{i,1}, \ldots, A_{i,m(i)}$ each of diameter $< 1/2$. Then, for each i ($1 \le i \le n$), let $D_{i,1}, \ldots, D_{i,m(i)}$ be as in (2) of 7.6. Define $F_2 \colon C \to 2^Y$ by letting $F_2(t) = A_{i,j}$ if $t \in D_{i,j}$. As shown for F_1, F_2 is well-defined on C and F_2 satisfies (1) and (3) of 7.4. Also, clearly, F_1 and F_2 satisfy (2) of 7.4. Continuing in this fashion (using the inductive nature of 7.6), we can define $F_n \colon C \to 2^Y$ for each $n = 1, 2, \ldots$ so that (1)–(4) of 7.4 are satisfied. Therefore, by 7.4, there is a continuous function f from C onto Y. This proves 7.7.

As a specific illustration of the technique discussed following the proof of 3.21, let us note that 7.7 determines all the usc decompositions of C:

7.8 Corollary. Every nonempty compact metric space is a usc decomposition of the Cantor Middle-Third set C (and, conversely, every usc decomposition of C is a nonempty compact metric space).

Proof. Use 7.7 and 3.21 (the converse part follows from the second part of 3.21).

Also in regards to the discussion following the proof of 3.21, let us note the following:

7.9 Examples of Space-Filling Curves. Let Y be B^n for any $n = 1$, 2, ... (1.2) or the Hilbert cube specifically defined as the countable product of unit intervals (1.4). We show how easy it is using 7.7 to obtain a continuous function from $[0,1]$ onto Y. Let C be the Cantor Middle-Third set, and let (a_i, b_i), $i = 1, 2, \ldots$, denote the components of $[0,1] - C$. Let f be a continuous function from C onto Y. We extend f to a continuous function $g \colon [0,1] \to Y$ as follows. If $t \in [0,1] - C$, then there are a unique i and a unique s such that

$$t = (1 - s) \cdot a_i + s \cdot b_i, \qquad 0 < s < 1$$

and we let

$$\varphi(t) = (1 - s) \cdot f(a_i) + s \cdot f(b_i).$$

Now, define g by

$$g(t) = \begin{cases} f(t) & \text{if } t \in C \\ \varphi(t), & \text{if } t \in [0,1] - C \end{cases}$$

By using the uniform continuity of f and the fact that $b_i - a_i \to 0$ as $i \to \infty$, it follows easily that g is continuous. Thus, since f maps C onto Y, g is a continuous function from $[0,1]$ onto Y. Such a map g is called a space-filling curve. More about space-filling curves is in 7.21. The most general result about them is proved in 8.14. We also remark that the map g defined above is analytic almost everywhere, to the consternation of some analysts when such functions g were first discovered.

4. APPLICATION: CHARACTERIZATION OF CANTOR SET

In our next application of 7.4, we shall obtain a topological characterization of Cantor sets (7.14). Note the following definition.

7.10 Definition of Totally Disconnected. A topological space S is said to be *totally disconnected* provided that $S \neq \varnothing$ and no connected subset of S contains more than one point (i.e., each component of S is a one-point set).

7.11 Lemma. Let Y be a compact totally disconnected metric space. Then, for each $\epsilon > 0$, Y can be written as the union of n [$n = n(\epsilon) < \infty$] nonempty, mutually disjoint, compact subsets $K_1, \ldots, K_n$ with diameter $(K_i) < \epsilon$ for each $i = 1, \ldots, n$.

Proof. We first prove (#) below:

(#) For each $y \in Y$, there is a closed and open subset $M(y)$ of Y such that $y \in M(y)$ and diameter $[M(y)] < \epsilon$.

To prove (#), let $y \in Y$ and let U be an open subset of Y such that $y \in U$ and diameter $(U) < \epsilon$. Let $A = \{y\}$ and let $B = Y - U$. Then, because of the assumptions about Y, we may apply 5.2-by letting $M(y)$ be the set X_1 in 5.2, we have proved (#). Now, to prove the lemma, first use the compactness of Y to cover Y with finitely many of the sets in (#), say $M(y_1), \ldots, M(y_k)$. Next, let

$$L_1 = M(y_1), \quad \text{and, for } 2 \le i \le k, L_i = M(y_i) - \bigcup_{j=1}^{i-1} M(y_j).$$

Discard all the sets L_i which are empty. Then, the remaining sets L_i are the desired sets $K_1, ..., K_n$. This completes the proof of 7.11.

The proof of 7.14 is similar to the proof of 7.7. However, to assure that the final map f is one-to-one, we need 7.11 and the following slight strengthening of part (2) of 7.6.

7.12 Lemma. Let $D_1, ..., D_n$ satisfy (1) of 7.6. Let $m \geq 2$ be a given integer. Then: For each i, $1 \leq i \leq n$, there are sets $D_{i,1} ..., D_{i,m}$ satisfying (2) of 7.6 and such that

diameter $(D_{i,j}) \leq (2/3) \cdot$ diameter (D_i),   each $j = 1, ..., m$.

Proof. Fix i, $1 \leq i \leq n$. Let $g = \text{glb } D_i$ and $\ell = \text{lub } D_i$. Since $g < \ell$ (D_i cannot be a one-point set since D_i is open in C), there exists t_1 such that

$$(2/3) \cdot g + (1/3) \cdot \ell < t_1 < (1/3) \cdot g + (2/3) \cdot \ell \quad \text{and } t_1 \notin C.$$

If $m = 2$, then $D_{i,1} = [g,t_1] \cap D_i$ and $D_{i,2} = [t_1,\ell] \cap D_i$ satisfy the desired properties. So, assume $m \geq 3$. Note that by 7.5, there are points $t < \ell$ arbitrarily close to ℓ such that $t \notin C$. Also note that since $0 < \ell \in D_i$ and D_i is open in C (7.6), there are points $s < \ell$ arbitrarily close to ℓ such that $s \in D_i$. Using these two observations, we see that there are $m - 2$ points $t_2, ..., t_{m-1}$ such that

$$t_0 = g < t_1 < t_2 < t_3 < \cdots < t_{m-1} < \ell = t_m \quad (t_1 \text{ as above})$$

$$t_j \notin C \quad \text{for any } j = 1, 2, ..., m - 1$$

$$(t_{j-1},t_j) \cap D_i \neq \varnothing \quad \text{for each } j = 2, ..., m - 1.$$

Let $D_{i,j} = [t_{j-1},t_j] \cap D_i$ for each $j = 1, ..., m$. Then, it is easy to see that $D_{i,1}, ..., D_{i,m}$ satisfy the desired properties for the lemma. This completes the proof of 7.12.

7.13 Definition of Perfect. A topological space S is said to be *perfect* provided that every point of S is a limit point of S. Some authors use the term dense in itself to describe this condition.

You should not have found the definition of a Cantor set in 7.5 any more satisfying than the definition of an arc in 1.1. We gave a topological characterization of arcs in 6.17. Now, we do the same for Cantor sets.

7.14 Theorem. A metric space Y is a Cantor set if and only if Y is compact, totally disconnected, and perfect.

Proof. Assume that *Y* is a compact, totally disconnected, perfect metric space. We construct a homeomorphism *f* from the Cantor Middle-Third set *C* onto *Y* using 7.4. For $\epsilon = 1$, let $K_1, \ldots, K_n$ be as in 7.11. Then, let $D_1, \ldots, D_n$ be as in (1) of 7.6. Define $F_1: C \rightarrow 2^Y$ by letting $F_1(t) = K_i$ for each $t \in D_i$. For the same reasons given in the proof of 7.7, F_1 is well-defined on *C* and F_1 satisfies (1) and (3) of 7.4. Fix i, $1 \leq i \leq n$. We obtain some subsets of K_i and D_i as follows. First note that since K_i is a nonempty open subset of *Y* and *Y* is perfect,

   diameter $(K_i) > 0$.

Hence, letting $\epsilon_i = (1/2) \cdot$ diameter (K_i), we have that $\epsilon_i > 0$. Thus, since K_i is compact and totally disconnected, 7.11 allows us to write K_i as the union of $m(i)$ nonempty, mutually disjoint, compact subsets $K_{i,1}, \ldots, K_{i,m(i)}$ with

   diameter $(K_{i,j}) < \epsilon_i$ for each $j = 1, \ldots, m(i)$.

Note that $m(i) \geq 2$ (since the sets $K_{i,j}$ cover K_i and are of diameter $< \epsilon_i$). Hence, there exist $D_{i,1}, \ldots, D_{i,m(i)}$ satisfying 7.12. Now, having defined

   $K_{i,1}, \ldots, K_{i,m(i)}$ and $D_{i,1}, \ldots, D_{i,m(i)}$

for each i, $1 \leq i \leq n$, define $F_2: C \rightarrow 2^Y$ by letting $F_2(t) = K_{i,j}$ for each $t \in D_{i,j}$. Then, for the same reasons as for F_1, F_2 is well-defined on *C* and F_2 satisfies (1) and (3) of 7.4. Also, since $K_{i,j} \subset K_i$ and $D_{i,j} \subset D_i$, we see that F_1 and F_2 satisfy (2) of 7.4. Continuing in this fashion (using the inductive nature of 7.12), it is easy to see how to define $F_n: C \rightarrow 2^Y$ for each $n = 1, 2, \ldots$ so that (1)–(4) of 7.4 are satisfied. Hence, by 7.4, there is a continuous function *f* from *C* onto *Y*. Furthermore, assuming *f* is the map defined by the formula in 7.4, we will show, as a consequence of the construction above, that *f* is one-to-one. To see this, let $s, t \in C$ such that $s \neq t$. Choose a positive integer *k* large enough so that

   $(2/3)^{k-1} < |s - t|$.

Let $E_1, \ldots, E_\ell$ denote the subsets of *C* used to define F_k. Since the sets $E_1, \ldots, E_\ell$ are the result of having applied 7.12 $k - 1$ times, we have that

   diameter $(E_j) < (2/3)^{k-1}$ for each $j = 1, \ldots, \ell$.

Hence, *s* and *t* cannot be in the same E_j for any $j = 1, \ldots, \ell$. Thus: Letting $K_1, \ldots, K_\ell$ denote the subsets of *Y* used to define F_k and recalling that these sets are mutually disjoint, we see from the way F_k is defined that

   $F_k(s) \cap F_k(t) = \varnothing$.

Therefore, since $f(s) \in F_k(s)$ and $f(t) \in F_k(t)$ (7.4), we have that $f(s) \neq$

$f(t)$. Hence, we have proved f is one-to-one. Thus, f is a homeomorphism from C onto Y. Therefore, Y is a Cantor set (7.5). This proves half of 7.14. The other half is trivial.

We shall see some more applications of 7.4 in the exercises below (we especially encourage the reader to do 7.20 even though the result is in 7.9). Also, as mentioned before, 7.4 will be used to prove 8.14. In the next chapter, we begin our study of locally connected spaces and lay the rest of the groundwork for the proof of 8.14.

EXERCISES

7.15 Exercise. We give some basic facts about usc functions (7.1). Prove (a)–(d) below where X and Y are topological spaces.
 (a) A function $F: X \to 2^Y$ is usc if and only if $\{x \in X: F(x) \subset U\}$ is open in X for all U open in Y or, equivalently, $\{x \in X: F(x) \cap C \neq \varnothing\}$ is closed in X for all C closed in Y.
 (b) Let f be a function from X onto Y such that $f^{-1}(y)$ is closed in X for each $y \in Y$. Define $F: Y \to 2^X$ by $F(y) = f^{-1}(y)$ for each $y \in Y$. Then, f takes closed sets in X to closed sets in Y if and only if F is usc. In particular, a continuous function f from X onto a T_1-space Y is a closed map if and only if its inverse relation $f^{-1}: Y \to 2^X$ is usc.
 (c) If $F: X \to 2^Y$ is a function such that $F(x)$ is a one-point set $\{f(x)\}$ for each $x \in X$, then F is usc if and only if $f: X \to Y$ is continuous (comp., 7.2). Thus, (1)–(4) of 7.4 are *equivalent* to the existence of a continuous function from one nonempty compact metric space X onto another such space Y! The value of 7.4 is, of course, its explicit method for constructing such maps.
 (d) If X is compact, $F: X \to 2^Y$ is usc, and $F(x)$ is compact for each $x \in X$, then $\bigcup_{x \in X} F(x)$ is compact.
Some other general facts about usc functions are in 7.16, 7.17, [4, pp. 173–182], and [5, pp. 57–64].

7.16 Exercise. Let X and Y be compact metric spaces, let $F: X \to 2^Y$ be a function, and let $p \in X$. Prove that F is usc at p if and only if whenever $x_i \to p$ as $i \to \infty$, lim sup $F(x_i) \subset F(p)$. Comp., 3.25.

7.17 Exercise. Let X and Y be compact metric spaces, let $F: X \to 2^Y$ be a function, and let

$$\mathfrak{L} = \{(x,y) \in X \times Y: y \in F(x)\}.$$

Prove that F is usc if and only if $\mathcal{L}$ is closed in $X \times Y$.

7.18 Exercise. Set-valued functions which are usc but not (necessarily) continuous arise quite naturally. For example, in light of the comment following 7.1, most of Chapter III is devoted to studying such a function. Another example is in (b) of 7.15. In this exercise we give a different type of example. Let (X,d) be a compact metric space with the fixed point property (1.19). Let $\mathfrak{C}$ denote the space of all continuous functions from X into X with the uniform metric ρ,

$$\rho(f,g) = \text{lub } \{d(f(x),g(x)): x \in X\}, \quad f, g \in \mathfrak{C}.$$

For each $f \in \mathfrak{C}$, let $F(f) = \{x \in X: f(x) = x\}$. Prove that $F: \mathfrak{C} \to 2^X$ is usc and, by taking $X = [0,1]$, give an example to show that F is not continuous (where 2^X has the Hausdorff metric).

See [2] and [7]. We remark that the upper semi-continuity of F and various properties of the functions f at which F is continuous have applications in areas such as stability of solutions of differential equations and continuity of roots of analytic maps.

7.19 Exercise. A *compactification* of a space S is a compact space Z containing a topological copy S' of S as a dense subset; $Z - S'$ is called the *remainder* of the compactification.

Prove that if Z_1 and Z_2 are any two metric compactifications of a countable discrete space having a Cantor set as the remainder, then Z_1 is homeomorphic to Z_2. [Hint: Use 7.4.] An application is in 7.26.

The result remains valid on replacing the Cantor set with any compact metric space [8, p. 85].

7.20 Exercise. In 7.9, we obtain some space-filling curves using 7.7. It is particularly instructive to construct a space-filling curve for $I \times I$, $I = [0,1]$, by making direct use of 7.4. This will illustrate very well many of the main ideas used in proving 8.14. We indicate the procedure, and ask the reader to carry it out. Divide $I \times I$ into four congruent closed squares $A_1, \ldots, A_4$ where the indexing assures that $A_i \cap A_{i+1}$ is a common edge of A_i and A_{i+1} for $i = 1, 2, 3$. Define $F_1: I \to 2^{I \times I}$ as follows:

$$F_1(0) = A_1, \quad F_1(1) = A_4,$$
$$F_1(t) = A_i, \quad (i - 1)/4 < t < i/4 \text{ and } i = 1, \ldots, 4,$$
$$F_1(i/4) = A_i \cup A_{i+1}, \quad i = 1, 2, 3.$$

Next, divide each A_i into four congruent closed squares $A_{i,1}, \ldots, A_{i,4}$ where the indexing assures that the intersections

$A_{i,j} \cap A_{i,j+1}$, $1 \le i \le 4$ and $i \le j \le 3$

$A_{i,4} \cap A_{i+1,1}$, $1 \le i \le 3$

are, in each case, a common edge of the two squares being intersected. Define F_2 on $[0,1/4]$ as follows:

$F_2(0) = A_{1,1}$, $F_2(1/4) = A_{1,4} \cup A_{2,1}$,

$F_2(t) = A_{1,j}$, $(j - 1)/16 < t < j/16$ and $j = 1, ..., 4$,

$F_2(j/16) = A_{1,j} \cup A_{1,j+1}$, $j = 1, 2, 3$.

Define F_2 on $(1/4,2/4]$ analogously, etc. Finish defining F_2, and see why F_1 and F_2 satisfy (1)-(3) of 7.4 (note that this time, in contrast to the proofs of 7.7 and 7.14, F_1 and F_2 are not continuous). Then, define F_3 and check its properties carefully to be sure you understand why and how the process continues to obtain F_n for each $n = 1, 2, ...$ satisfying (1)-(4) of 7.4. Then, by 7.4, there is a continuous function f from I onto $I \times I$.

7.21 Exercise. We enlarge our examples of space-filling curves (7.9) as follows. A metric space (Y,d) is said to be *uniformly locally arcwise connected*, written *ULAC*, provided that for each $\epsilon > 0$ there exists $\delta > 0$ such that if $d(x,y) < \delta$ and $x \ne y$, then there is an arc $A \subset Y$ such that A has end points x and y and diameter $(A) < \epsilon$. Prove that if Y is any ULAC continuum, then there is a continuous function from $[0,1]$ onto Y. [Hint: Modify the proof of 7.9 using the uniform continuity of f: $C \to Y$.]

It is easy to see that any compact connected polyhedron is ULAC, so, by the result in this exercise, we have a wider range of space-filling curves than in 7.9. In fact, any Peano continuum is ULAC (8.30). If we proved this and then applied the result in the present exercise, we would have a proof of 8.14. This is the way 8.14 is often proved, more specifically: 8.23 is proved first, then 8.30 is proved using a consequence of 8.23, and then 8.14 is proved. However, sometimes there are errors in the proof of 8.23-see [1]. Our proof of 8.14 will be different, and we shall use 8.14 to prove 8.23.

7.22 Exercise. A topological space S is said to be *homogeneous* provided that for any two points $p, q \in S$, there is a homeomorphism h from S onto S such that $h(p) = q$. Prove that a Cantor set is homogeneous. [Hint: The proof is a consequence of a simple consideration when indexing in the proof of 7.14.] Thus: Even though some points of the Cantor Middle-Third set C may seem to be ''different'' than other points (e.g., 1/3 and 1/4), this is due to the embedding of C in $[0,1]$ and not to any property of C itself.

7.23   Exercise.   Prove that every compact, totally disconnected metric space can be topologically embedded in the Cantor Middle-Third set. [Hint: Use 7.4.] We remark that there are uncountably many mutually nonhomeomorphic compact, totally disconnected metric spaces [6].

7.24   Exercise.   Prove that every nonempty open subset U of a nondegenerate continuum X contains a Cantor set. [Hint: The construction in 7.5 can be mimicked in U by thinking of C_i in 7.5 as being a union rather than a complement–then, apply 7.14 to the intersection.]

7.25   Exercise.   This exercise concerns the recognition of Cantor sets using 7.14. Prove (a)–(d) below.
   (a)   The finite or countable cartesian product of Cantor sets is a Cantor set.
   (b)   The countably infinite cartesian product of the two-point discrete space $\{0,1\}$ with itself is a Cantor set.
   (c)   If X is a compact metric space, then X is a Cantor set if and only if 2^X is a Cantor set.
   (d)   Let $X = \{0\} \cup \{1/n: n = 1, 2, ...\}$, usual topology from R^1, and let $\mathbb{C} = \{A \in 2^X: 0 \in A\}$. Then, $\mathbb{C}$ is a Cantor set. An application is in 7.26.

7.26   Exercise.   Let $X = \{0\} \cup \{1/n: n = 1, 2, ...\}$ with the usual topology from R^1. Prove that 2^X is homeomorphic to $C \cup M$ where C is the Cantor Middle-Third set and M is the set of all the midpoints of the components of $[0,1] - C$. [Hint: Use 7.19 and (d) of 7.25.]
   The result is a special case of the Theorem in [8, p. 88].

7.27   Exercise.   Let C be the Cantor Middle-Third set, and let (a_i,b_i), $i = 1, 2, ...$, denote the components of $[0,1] - C$. Let

$$\mathcal{D} = \left\{ \{a_i,b_i\}: i = 1, 2, ... \right\} \cup \left\{ \{t\}: t \in C - \bigcup_{i=1}^{\infty} \{a_i,b_i\} \right\}$$

Prove that $\mathcal{D}$ is a usc decomposition of C. What familiar space is the decomposition space homeomorphic to?

REFERENCES

1. B. J. Ball, Arcwise connectedness and the persistence of errors, Amer. Math. Monthly, 91(1984), 431–433.

2. M. K. Fort, Jr., Essential and nonessential fixed points, Amer. J. Math., 72(1950), 315–322.

3. M. K. Fort, Jr., One-to-one mappings onto the Cantor set, J. of the Indian Math. Soc., 25(1961), 103–107.

4. K. Kuratowski, *Topology*, Vol. I, Academic Press, New York, N.Y., 1966.

5. K. Kuratowski, *Topology*, Vol. II, Academic Press, New York, N.Y., 1968.

6. S. Mazurkiewicz and W. Sierpiński, Contribution á la topologie des ensembles dénombrables, Fund. Math., 1(1920), 17–27.

7. Barrett O'Neill, Essential sets and fixed points, Amer. J. Math., 75(1953), 497–509.

8. A. Pelczyński, A remark on spaces 2^X for zero-dimensional X, Bull. Pol. Acad. Sci., 13(1965), 85–89.

Part Two
SPECIAL CONTINUA AND MAPS

VIII

General Theory of Peano Continua

The proper background has now been given to study special classes of continua. This is the first of five chapters devoted to such a study.

In this chapter, we develop the basic structure of Peano spaces and Peano continua. In the context of continuum theory, Peano continua are, in a sense, the most fundamental special class of continua to study in that they are defined merely by making connectedness a local property (8.1). Our results center on the existence of space-filling curves and arcs. In particular, we shall see that Peano continua are well behaved in that they can be written as the union of finitely many arbitrarily small Peano sub-continua, they are arcwise and locally arcwise connected, and they are characterized by the property of being the continuous metric images of the closed interval [0,1]. We shall obtain a number of consequences of these results in the main body of the chapter and in the exercises at the end. We remark that some special types of Peano continua will be examined in the next two chapters.

1. PEANO SPACES AND PROPERTY S

We begin with the following definition.

8.1 Definition of Peano Space and Peano Continuum. A metric space X is called a *Peano space* provided that for each $p \in X$ and each neighborhood N of p, there is a connected open subset U of X such that $p \in U \subset N$. Thus, by 5.22, a metric space X is a Peano space if and only if (i) X is LC at every point, (ii) each component of each open subset of X is open in X [(a) of 5.22], or (iii) X is cik at every point [(b) is 5.22]. A *Peano continuum* is a Peano space which is a continuum.

Different authors attach different meanings to the term Peano space. For some authors, Peano space means Peano continuum; for others, a Peano space is defined to be a Hausdorff continuous image of [0,1]. The definitions in 8.1 are the most convenient for our purposes.

The following notion is due to Sierpiński [14]. It will play a crucial role in determining the structure of Peano spaces and Peano continua.

8.2   Definition of Property S.   Let (X,d) be a metric space. A non-empty subset Y of X is said to have *property S* provided that for each $\epsilon > 0$, there are finitely many connected subsets $A_1, \ldots, A_n$ of Y such that

$$Y = \bigcup_{i=1}^{n} A_i, \qquad \text{diameter } (A_i) < \epsilon \text{ for each } i = 1, \ldots, n.$$

We shall see in 8.4 that, for compact metric spaces, having property S is equivalent to being a Peano space. We remark that in general, without the presence of compactness, property S is not a topological invariant [e.g., consider $(0,1)$ and R^1 with their usual metrics].

8.3   Theorem.   If a metric space (X,d) has property S, then (X,d) is a Peano space.

Proof.   By (iii) of 8.1, it suffices to show that X is cik at every point. Let $p \in X$ and let $\epsilon > 0$. Then, since X has property S, there are finitely many connected sets $A_1, \ldots, A_n$ such that

$$X = \bigcup_{i=1}^{n} A_i, \qquad \text{diameter } (A_i) < \epsilon/2 \text{ for each } i = 1, \ldots, n.$$

Let $G = \cup \{A_i : p \in \overline{A}_i\}$. Clearly,

G is connected, diameter $(G) < \epsilon$

and, since $p \notin \overline{X - G}$, G is a neighborhood of p. Therefore, we have proved that X is cik at p (5.10). This completes the proof of 8.3.

8.4   Theorem.   A nonempty compact metric space (X,d) is a Peano space if and only if (X,d) has property S. Thus, in particular, a continuum

(X,d) is a Peano continuum if and only if for each $\epsilon > 0$, X is the union of finitely many subcontinua each having diameter $< \epsilon$.

Proof. With regard to the first part, half is 8.3 and the other half is simply a matter of using 8.1 and compactness to cover X with finitely many connected open subsets of diameter $< \epsilon$ for any given $\epsilon > 0$. The second part is immediate from the first part using that the closure of a connected set is connected. This completes the proof of 8.4.

It is important to notice that the subcontinua guaranteed by the second part of 8.4 may not be Peano continua. Whereas it is true that they may be so chosen, this fact is by no means trivial to prove. Obtaining this result is our next goal (8.10). It is one of the most important results about the structure of Peano continua. For example, it is a crucial ingredient in the proof of the Hahn–Mazurkiewicz Theorem (8.14).

8.5 Proposition. Let (X,d) be a metric space, and let $Y \subset X$ such that Y has property S. Then, for any Z such that

$$Y \subset Z \subset \bar{Y},$$

Z has property S and, hence, Z is a Peano space.

Proof. Let $\epsilon > 0$. Then, letting $A_1, \ldots, A_n$ be as in 8.2 and letting B_i be the closure of A_i in Z for each i, the first part of the theorem follows easily (each B_i is connected by Corollary 3 (ii) of [8, p. 132]). The second part of the theorem is immediate from the first part and 8.3.

The following technical definition contains the key idea for proving 8.10.

8.6 Definition of $S(\epsilon)$-Chain. Let (X,d) be a metric space and let $\epsilon > 0$. An $S(\epsilon)$-*chain* is a nonempty, finite, indexed collection $\mathcal{L} = \{L_1, \ldots, L_n\}$ of subsets of X satisfying (1)–(3) below:

(1) $L_i \cap L_{i+1} \neq \varnothing$ for each $i = 1, \ldots, n-1$ (weak chain);
(2) L_i is connected for each $i = 1, \ldots, n$;
(3) diameter $(L_i) < \epsilon \cdot 2^{-i}$ for each $i = 1, \ldots, n$.

If $\mathcal{L} = \{L_1, \ldots, L_n\}$ is an $S(\epsilon)$-chain, then each $L_i \in \mathcal{L}$ is called a *link* of $\mathcal{L}$; if $x \in L_1$ and $y \in L_n$, then we say that $\mathcal{L}$ is an $S(\epsilon)$-*chain from x to y*. If $A \subset X$, then we define $S(A,\epsilon)$ as follows:

$$S(A,\epsilon) = \{y \in X: \text{there is an } S(\epsilon)\text{-chain from some point of } A \text{ to } y\}.$$

The important properties of the sets $S(A,\epsilon)$ are given in 8.7 and 8.8. Briefly, these sets provide a way to obtain small, connected, open subsets having property S in any metric space with property S.

8.7 Proposition. If a metric space (X,d) has property S, then, for any nonempty subset A of X and any $\epsilon > 0$, $S(A,\epsilon)$ has property S.

Proof. Fix $\delta > 0$. We show that $S(A,\epsilon) = \bigcup_{i=1}^{n} B_i$, some $n < \infty$, where each B_i is connected and has diameter $< \delta$ (8.2). To do this, first choose a positive integer k such that

(1) $\displaystyle\sum_{i=k}^{\infty} \epsilon \cdot 2^{-1} < \delta/4.$

Next, let

$\quad K = \{y \in S(A,\epsilon):$ there is an $S(\epsilon)$-chain with at most k links from some point of A to $y\}.$

Since (X,d) has property S, there is a finite cover of X by connected sets each having diameter $< \epsilon \cdot 2^{-k-1}$ (8.2). Let $E_1, \ldots, E_n$ denote the members of this cover which intersect K (if none of them intersects K, then $K = \varnothing$ and, thus, since $A \subset K$, $A = \varnothing$, a contradiction). Note that (2)–(5) hold:

(2) $K \subset \bigcup_{i=1}^{n} E_i$;
(3) $E_i \cap K \neq \varnothing$ for each i;
(4) E_i is connected for each i;
(5) diameter $(E_i) < \epsilon \cdot 2^{-k-1}$ for each i.

To prove (6) below, fix some $i = 1, \ldots, n$. By (3), there is an $S(\epsilon)$-chain $\{L_1, \ldots, L_t\}$ with $t \le k$ from a point of A to a point of $E_i \cap K$. Then, we see from (4), (5), and 8.6 that

$\quad \{L_1, \ldots, L_t, L_{t+1} = E_i\}$

is an $S(\epsilon)$-chain from a point of A to any point of E_i. Hence, we have proved that

(6) $E_i \subset S(A,\epsilon)$ for each $i = 1, \ldots, n$.

For each i, $1 \le i \le n$, let $\mathcal{P}_i$ denote the collection of sets M satisfying (7)–(10) below:

(7) $M \subset S(A,\epsilon)$;
(8) $M \cap E_i \neq \varnothing$;
(9) M is connected;
(10) diameter $(M) < \delta/4.$

Now, let

$\quad B_i = \bigcup \mathcal{P}_i$ for each $i = 1, \ldots, n$.

Observe that any given E_i satisfies (7)–(10) [(7) by (6), (8) since $E_i \neq \varnothing$, (9) by (4), and (10) by (1) and (5)]. Hence,

(11) $E_i \subset B_i$ for each $i = 1, \ldots, n$.

Now, we show that $B_1, \ldots, B_n$ have the desired properties stated at the beginning of the proof. By (4), (8), and (9), each B_i is connected. It follows easily from (1), (5), (8), and (10) that each B_i has diameter $< \delta$. By (7), $B_i \subset S(A, \epsilon)$ for each $i = 1, \ldots, n$. Hence, it remains to prove that

(#) $S(A, \epsilon) \subset \bigcup_{i=1}^{n} B_i$.

To prove (#), let $y \in S(A, \epsilon)$. Note that by (2) and (11), $K \subset \bigcup_{i=1}^{n} B_i$. Hence, to prove (#), assume $y \notin K$. Since $y \in S(A, \epsilon)$, there is an $S(\epsilon)$-chain $\mathcal{L} = \{L_1, \ldots, L_m\}$ from a point of A to y. Since $y \notin K$, $m > k$. Let

$$H = \bigcup_{i=k}^{m} L_i.$$

Clearly, by the definition of K, $L_k \subset K$. Hence, by (2), $L_k \cap E_i \neq \varnothing$ for some i. We show that $H \subset B_i$ by showing that H satisfies (7)–(10). By the definition of $S(A, \epsilon)$ in 8.6, clearly

$\bigcup \mathcal{L} \subset S(A, \epsilon)$.

Hence, clearly H satisfies (7). Since $L_k \cap E_i \neq \varnothing$, H satisfies (8). By (1) and (2) of 8.6, H satisfies (9). By (1) of 8.6,

$$\text{diameter}\,(H) \leq \sum_{i=k}^{m} \text{diameter}\,(L_i)$$

and, hence, by (3) of 8.6,

$$\text{diameter}\,(H) \leq \sum_{i=k}^{m} \epsilon \cdot 2^{-i};$$

thus, by (1), H satisfies (10). Now, having proved that H satisfies (7)–(10), we have that $H \subset B_i$. Hence, recalling that $y \in L_m$, we have that $y \in B_i$. Therefore, we have proved (#). This completes the proof of 8.7.

8.8 Lemma. Let (X, d) be a metric space, let A be a nonempty subset of X, and let $\epsilon > 0$. Then, (1)–(3) below hold:

(1) diameter $[S(A, \epsilon)] \leq$ diameter $(A) + 2\epsilon$;
(2) if A is connected, then $S(A, \epsilon)$ is connected;
(3) if (X, d) has property S, then $S(A, \epsilon)$ is an open subset of X.

Proof. Part (1) follows easily using the fact that, by (1) and (3) of 8.6, every $S(\epsilon)$-chain has diameter $< \epsilon$. Part (2) follows easily using the

fact that, by (1) and (2) of 8.6, every $S(\epsilon)$-chain is connected. To prove
part (3), let $y \in S(A,\epsilon)$. Then (8.6), there is an $S(\epsilon)$-chain $\{L_1, \ldots, L_n\}$
from some point of A to y. By 8.3, there is a connected open subset U of X
such that $y \in U$ and

diameter $(U) < \epsilon \cdot 2^{-n-1}$.

Clearly, then,

$\{L_1, \ldots, L_n, L_{n+1} = U\}$

is an $S(\epsilon)$-chain from a point of A to any point of U. Hence, $U \subset S(A,\epsilon)$.
Therefore, we have proved part (3).

The following theorem shows that for any metric space (X,d) having
property S, there is a sequence of "finite subdivisions of smaller and
smaller mesh," each "subdivision" consisting of finitely many connected sets and being a refinement of the preceding ones. The analogy
with triangulations of polyhedra is obvious and inescapable even though
(for various choices) two of our connected sets may not intersect nicely.

8.9 Theorem. If a metric space (X,d) has property S, then, for any
$\epsilon > 0$, X is the union of finitely many connected sets each of which has
property S and is of diameter $< \epsilon$; furthermore, these sets may be chosen
to be open in X or closed in X.

Proof. By 8.2, $X = \bigcup_{i=1}^{n} A_i$, some $n < \infty$, where each A_i is connected and of diameter $< \epsilon/3$. By 8.7 and 8.8, the sets

$S(A_i,\epsilon/3), \quad i = 1, \ldots, n$

are open in X [(3) of 8.8] and satisfy the desired properties. Hence, by
8.5, the closed sets

$\overline{S(A_i,\epsilon/3)}, \quad i = 1, \ldots, n$

also satisfy the desired properties. This completes the proof of 8.9.

Now we come to the result we have wanted to obtain since proving 8.4.
As you can see from the proof, all the work has been done.

8.10 Theorem. If (X,d) is a Peano continuum, then, for any $\epsilon > 0$,
X is the union of finitely many Peano continua each of which is of diameter $< \epsilon$.

Proof. By 8.4, (X,d) has property S. Hence, by 8.9, X is the union of
finitely many connected sets $A_1, \ldots, A_n$ each of which has property S, is of

diameter $< \epsilon$, and is closed in X. Therefore, each A_i being a continuum, the result follows on applying 8.3 to each A_i. This proves 8.10.

2. THE HAHN-MAZURKIEWICZ THEOREM

In this section, we use 8.10 and 7.4 to show that every Peano continuum is a continuous image of [0,1] (8.14). We then derive some simple consequences of this theorem. In the next section, we use the theorem to prove the Arcwise Connectedness Theorem (8.23).

First, we note the following concept which was actually introduced as part of 8.6.

8.11 Definition of Weak Chain. A *weak chain* is a nonempty, finite, indexed collection $\mathcal{L} = \{L_1, \ldots, L_n\}$ of sets such that

$$L_i \cap L_{i+1} \neq \varnothing \qquad \text{for each } i = 1, \ldots, n - 1.$$

Let $\mathcal{L} = \{L_1, \ldots, L_n\}$ be a weak chain. Then, we say $\mathcal{L}$ is a *weak chain from L_1 to L_n* and, if $x \in L_1$ and $y \in L_n$, we also say $\mathcal{L}$ is a *weak chain from x to y*. Each $L_i \in \mathcal{L}$ is called a *link* of $\mathcal{L}$.

The following easy-to-prove lemma is used in the proof of 8.13.

8.12 Lemma. If $\mathbb{C} = \{C_1, \ldots, C_m\}$ and $\mathcal{L} = \{L_1, \ldots, L_n\}$ are weak chains with $C_1 = L_1$, then there is a weak chain $\mathcal{P}$ from C_1 to L_n such that $\mathcal{P} = \mathbb{C} \cup \mathcal{L}$.

Proof. Simply let $\mathcal{P} = \{P_1, \ldots, P_{2m+n}\}$ where $P_i = C_i$ for $1 \leq i \leq m$, $P_{m+i} = C_{m-i+1}$ for $1 \leq i \leq m$, and $P_{2m+i} = L_i$ for $1 \leq i \leq n$. Then, it is easy to see that $\mathcal{P}$ has the desired properties.

8.13 Lemma. Let S be a connected topological space, and let $p, q \in S$. If $\mathbb{C}$ is a finite collection of nonempty closed subsets of S covering S, then the entire collection $\mathbb{C}$ can be indexed so as to form a weak chain from p to q.

Proof. Since $\cup \mathbb{C} = S$, there exists $C_1 \in \mathbb{C}$ such that $p \in C_1$. Let $\mathbb{C}_0 = \{C \in \mathbb{C}: \text{there is a weak chain from } C_1 \text{ to } C \text{ whose links are members of } \mathbb{C}\}$. We first show that

(*) $\mathbb{C}_0 = \mathbb{C}$.

To prove (*), we consider the sets $A = \cup \mathbb{C}_0$ and $B = \cup (\mathbb{C} - \mathbb{C}_0)$. It is easy to see that any member of $\mathbb{C}$ which intersects a member of $\mathbb{C}_0$ is itself a member of $\mathfrak{C}_0$; hence, $A \cap B = \varnothing$. Since A and B are each a finite union of closed subsets of S, A and B are each closed in S. Since $\cup \, \mathfrak{C} = S$, $A \cup B = S$. Since $C_1 \in \mathfrak{C}_0$ and $C_1 \neq \varnothing$, $A \neq \varnothing$. Thus, from the established properties of A and B and by the connectedness of S, it must be that $B = \varnothing$. Therefore, since $\varnothing \notin \mathfrak{C}$, $\mathfrak{C} \subset \mathfrak{C}_0$. Hence, we have proved (*). Now, it is easy to see how 8.13 follows from (*) by repeated applications of 8.12. This completes the proof of 8.13.

We have seen examples of space-filling curves in 7.9, 7.20, and 7.21. We now show that *every* Peano continuum is a continuous image of [0,1]. We encourage the reader to peruse 7.20 before proceeding to the proof of 8.14–it will then be clear that 8.10 and 8.13 are the natural ingredients necessary for generalizing the process in 7.20.

8.14 Hahn-Mazurkiewicz Theorem.   Every Peano continuum is a continuous image of the closed interval [0,1].

Proof.   The proof uses 8.10, 8.13, and 7.4. Let (X,d) be a Peano continuum. By 8.10 and 8.13, there is a weak chain $\{A_1, \ldots, A_n\}$ of Peano subcontinua of X covering X such that each A_i is of diameter < 1. Write [0,1] as the union of n nondegenerate closed subintervals $I_1, \ldots, I_n$ of the form

$$I_i = [t_{i-1}, t_i], \quad 1 \leq i \leq n, \text{ where } t_0 = 0 < t_1 < t_2 < \cdots < t_n = 1.$$

Define $F_1 \colon [0,1] \to C(X)$ [(2) of 4.1] by

$$F_1(t) = \begin{cases} A_i, & \text{if } t \in I_i - \{t_1, \ldots, t_{n-1}\} \\ A_i \cup A_{i+1}, & \text{if } t = t_i \text{ and } 1 \leq i \leq n-1 \\ A_1, & \text{if } t = 0 \text{ (only needed if } n = 1) \\ A_n, & \text{if } t = 1 \text{ (only needed if } n = 1) \end{cases}$$

It is easy to verify that F_1 satisfies (1) and (3) of 7.4. Next, let $p_1 \in A_1$, $p_i \in A_i \cap A_{i-1}$ for $2 \leq i \leq n$, and $p_{n+1} \in A_n$. By 8.10 and 8.13 (applied to each A_i), there is for each i, $1 \leq i \leq n$, a weak chain $\{A_1^i, \ldots, A_{m(i)}^i\}$ of Peano subcontinua of A_i from p_i to p_{i+1} covering A_i such that each A_j^i is of diameter $< 1/2$. For each i, $1 \leq i \leq n$, write I_i as the union of $m(i)$ nondegenerate closed subintervals $I_1^i, \ldots, I_{m(i)}^i$ of the form

$$I_j^i = [t_{j-1}^i, t_j^i], \quad 1 \leq j \leq m(i),$$
$$\text{where } t_0^i = t_{i-1} < t_1^i < t_2^i < \cdots < t_{m(i)}^i = t_i.$$

Define $F_2 \colon [0,1] \to C(X)$ by

$$F_2(t) = \begin{cases} A_j^i, & \text{if } t \in I_j^i - \{t_0^i, \ldots, t_{m(i)}^i\} \\ A_j^i \cup A_{j+1}^i, & \text{if } t = t_j^i \text{ and } 0 < j < m(i) \\ A_{m(i-1)}^{i-1} \cup A_1^i, & \text{if } t = t_0^i \text{ and } 2 \leq i \leq n \\ A_1^1, & \text{if } t = 0 \\ A_{m(n)}^n, & \text{if } t = 1 \end{cases}$$

It is easy to verify that F_2 satisfies (1) and (3) of 7.4 and that F_1 and F_2 satisfy (2) of 7.4. Continuing in this fashion, it is easy to see how to define F_n for each $n = 1, 2, \ldots$ so that (1)–(4) of 7.4 are satisfied. Therefore, by 7.4, there is a continuous function from $[0,1]$ onto X. This completes the proof of 8.14.

The property of being a continuous image of $[0,1]$ actually characterizes Peano continua, as we shall see in 8.18.

8.15 Lemma. Let S_1 and S_2 be topological spaces, and let f be a continuous function from S_1 onto S_2. If C is a component of S_2, then $f^{-1}(C)$ is a union of some components of S_1.

Proof. It is easy to verify that $f^{-1}(C) = \cup \{K: K$ is a component of S_1 and $K \cap f^{-1}(C) \neq \varnothing\}$. This proves 8.15.

A continuous image of a Peano space may not be a Peano space (even if the range space is metric). In fact, it is easy to find a one-to-one continuous function from $[0,1)$ onto the Warsaw circle (1.6). However, note the following result.

8.16 Proposition. If f is a (continuous) closed map of a Peano space X onto a metric space Y, then Y is a Peano space.

Proof. We show that Y satisfies (ii) of 8.1. Let C be a component of an open subset U of Y. Then, by 8.15, $f^{-1}(C)$ is a union of some components of $f^{-1}(U)$. Thus, since $f^{-1}(U)$ is open in X and X satisfies (ii) of 8.1, $f^{-1}(C)$ is open in X. Hence, $X - f^{-1}(C)$ is closed in X. Thus, since f is a closed map and

$$f[X - f^{-1}(C)] = Y - C,$$

$Y - C$ is closed in Y. Hence, C is open in Y. Therefore, we have proved that Y satisfies (ii) of 8.1. This proves 8.16.

We remark that 8.16 shows that a metrizable usc decomposition of a Peano space is a Peano space (use 3.7).

8.17 Corollary. Any continuous Hausdorff image of a Peano continuum is a Peano continuum.

Proof. Use 3.2 and 8.16.

Now, we have the following characterization theorem.

8.18 Theorem. A Hausdorff continuum X is a Peano continuum if and only if X is a continuous image of $[0,1]$.

Proof. Use 8.14 and 8.17.

There are fairly simple continua such that no one of them is a continuous image of any other (you were asked to find two such continua in 3.32). However, this cannot happen with Peano continua:

8.19 Theorem. Any two nondegenerate Peano continua are continuous images of one another. In fact: If X and Y are nondegenerate Peano continua, then, for any n points ($n < \infty$)

$$x_1, \ldots, x_n \in X \quad \text{and} \quad y_1, \ldots, y_n \in Y,$$

there is a continuous function f from X onto Y such that $f(x_i) = y_i$ for each $i = 1, \ldots, n$.

Proof. Clearly, it suffices to prove the second part of the theorem. Start with a continuous function α from $[0,1]$ onto Y (8.14). For each i, $1 \le i \le n$, let $t_i \in [0,1]$ such that $\alpha(t_i) = y_i$. Next, let $p, q \in X - \{x_1, \ldots, x_n\}$ with $p \ne q$. Define

$$\beta: \{p,q,x_1, \ldots, x_n\} \to [0,1]$$

by letting $\beta(p) = 0$, $\beta(q) = 1$, and $\beta(x_i) = t_i$ for each $i = 1, \ldots, n$. By the Tietze Extension Theorem [7, p. 127], extend β to a continuous function $\gamma: X \to [0,1]$, and note that γ maps X onto $[0,1]$ since X is connected and $0,1 \in \gamma(X)$. Now, letting $f = \alpha \circ \gamma$, we see that f has the desired properties. Therefore, we have proved 8.19.

Let us note that we have determined all the usc decompositions of any given Peano continuum.

8.20 Corollary. Let X be a given nondegenerate Peano continuum. Then, a topological space Y is a usc decomposition of X if and only if Y is a Peano continuum.

Proof. The "if part" follows by using the first part of 8.19 and then applying the first part of 3.21. The "only if part" follows by using the second part of 3.21 and then applying 8.17.

3. ARCS IN PEANO CONTINUA

Our next main result is that Peano continua are arcwise connected (8.23). The proof is done by using 8.14 (specifically, 8.19) and by using an auxiliary result about monotone maps in 8.22. The following brief discussion is included to motivate the terminology in 8.21.

Recall that a function f from R^1 to R^1 is called monotone provided that $f(s) \leq f(t)$ for all $s \leq t$ or $f(s) \geq f(t)$ for all $s \leq t$. It is easy to see that a continuous function $f\colon R^1 \to R^1$ is monotone if and only if $f^{-1}(t)$ is connected for all $t \in R^1$. Thus, the following definition is a natural topological generalization of the familiar notion of monotone real-valued functions of a real variable.

8.21 Definition of Monotone Map. Let S_1 and S_2 be topological spaces. A function $f\colon S_1 \to S_2$ is said to be *monotone* provided that f is continuous and $f^{-1}(y)$ is connected for each $y \in S_2$.

8.22 Proposition. If Y is a nondegenerate Hausdorff space and $f\colon [0,1] \to Y$ is a monotone map onto Y, then Y is an arc.

Proof. Clearly, Y is compact and, by 3.2, Y is a metric space. Hence, $\mathcal{D}_f$ has the properties stated in 3.21. Since $\mathcal{D}_f$ is usc and since f is monotone and Y is nondegenerate, we see that (c) of 6.24 applies to show that $\mathcal{D}_f$ is an arc. Therefore, Y is an arc (3.21). This proves 8.22.

We have seen many examples of continua which are not arcwise connected. In 2.27, we even saw an example of an hereditarily decomposable continuum which contains no arc whatsoever. We now prove the fundamentally important fact that all Peano continua are arcwise connected. We shall then show that Peano continua are also locally arcwise connected (8.25). We remark that in many books the arcwise and local arcwise connectedness of Peano continua is proved before proving 8.14, and then 8.14 is proved using these results (in the manner indicated in 7.21). However, in some of these books there is a basic flaw in the proofs of the arcwise connectedness theorem. We refer the reader to [1] for a discussion of the error and how to correct it. The idea of using 8.14 and 8.22 to prove the following result is due to J. L. Kelley.

8.23 Arcwise Connectedness Theorem. Every (nondegenerate) Peano continuum is arcwise connected.

Proof. Let X be a nondegenerate Peano continuum, and let p, $q \in X$ with $p \neq q$. By the second part of 8.19, there is a continuous function f from $I = [0,1]$ onto X such that $f(0) = p$ and $f(1) = q$. Let

$$\mathbb{C} = \{A \in 2^I : p, q \in f(A) \text{ and if } s \text{ and } t \text{ are the end points of the closure of a component } J \text{ of } I - A, \text{ then } f(s) = f(t)\}.$$

We shall apply the Maximum-Minimum Theorem (4.34) to $\mathbb{C}$. To be able to do this, we must show (*) and (**) below hold:

   (*) $\mathbb{C} \neq \varnothing$
   (**) $\mathbb{C}$ is closed in 2^I.

Since $I \in \mathbb{C}$, (*) holds. To prove (**), let $A_i \in \mathbb{C}$ for each $i = 1, 2, \ldots$ such that $A_i \to A$ for some $A \in 2^I$ (4.12). Since $A_i \to A$, $f(A_i) \to f(A)$ by 4.27. Thus, since p, $q \in f(A_i)$ for each i, we see easily that p, $q \in f(A)$. Since $I \in \mathbb{C}$, we may assume for the rest of the proof of (**) that $A \neq I$ and, hence, we may also assume that $A_i \neq I$ for any i. Now, let J be a component of $I - A$ and let $s < t$ be the end points of $\bar{J}$. By 4.38, there is a sequence $\{J_i\}_{i=1}^{\infty}$ of components J_i of $I - A_i$ such that $\bar{J}_i \to \bar{J}$. Each $\bar{J}_i = [s_i, t_i]$ for some s_i, $t_i \in I$. Since $\bar{J}_i \to \bar{J}$, we see easily that $s_i \to s$ and $t_i \to t$. Thus, by the continuity of f,

$$f(s_i) \to f(s) \quad \text{and} \quad f(t_i) \to f(t).$$

Since $A_i \in \mathbb{C}$ for each i, $f(s_i) = f(t_i)$ for each i. Hence, $f(s) = f(t)$. Now, having verified all the necessary properties, we have proved that $A \in \mathbb{C}$. Therefore, we have proved (**). By (*) and (**), we may apply 4.34 thus obtaining a minimal member M of $\mathbb{C}$. Note that p, $q \in f(M)$ since $M \in \mathbb{C}$. We will show that $f(M)$ is an arc by representing $f(M)$ as a monotone image of I and then applying 8.22. First, we verify that M has the following property:

   (#) If s, $t \in M$, $s < t$, and $f(s) = f(t)$, then $M \cap [s,t] = \{s,t\}$.

To prove (#), first let $u = \text{glb}(M)$ and $v = \text{lub}(M)$. Since $M \in \mathbb{C}$ and since $f(0) = p$ and $f(1) = q$, we see that $f(u) = p$ and $f(v) = q$ [since if $u > 0$, $[0,u)$ is a component of $I - M$, so $f(0) = f(u)$; similarly, $q = f(v)$]. Now, let s and t satisfy the hypothesis of (#). Then, using that u, $v \in M - (s,t)$ to know that

$$p, q \in f[M - (s,t)],$$

it is clear that $M - (s,t) \in \mathbb{C}$. Hence, by the minimality of M, we see that

$M \cap (s,t) = \emptyset$. Thus, $M \cap [s,t] = \{s,t\}$ and, therefore, we have proved (#). Now, we define a monotone map g from I onto $f(M)$ as follows. If $r \in M$, let $g(r) = f(r)$. If $r \in I - M$, $r \in J$ where J is a component of $I - M$; then, letting s be an end point of $\bar{J}$, let $g(r) = f(s)$. Since $M \in \mathfrak{C}$, g is well-defined. Since $g|M = f|M$ and since every value of g is in $f(M)$, $g(I) = f(M)$. An easy sequence argument using the continuity of f shows that g is continuous. To show g is monotone, let $z \in g(M)$. By (#) and since $g|M = f|M$, there are at most two points of M in $g^{-1}(z)$. If

$$M \cap g^{-1}(z) = \{s\},$$

then $g^{-1}(z) = \{s\}$, $[0,s]$, or $[s,1]$. If

$$M \cap g^{-1}(z) = \{s,t\} \qquad \text{with, say, } s < t,$$

then, noting in this case that $s = u > 0$ and $t = v < 1$ are *each* impossible (by the minimality of M), we see that $g^{-1}(z) = [s,t]$. This completes the proof that g is monotone. Now, since $p,q \in f(M) = g(M) \subset g(I)$ and, by 8.22, $g(I)$ is an arc, it follows that we have proved 8.23.

We shall now show how to use 8.23 and some other results to prove that Peano continua are locally arcwise connected. We note the following definition.

8.24 Definition of Locally Arcwise Connected. Let (S,T) be a topological space. If $p \in S$, then S is said to be *locally arcwise connected at p*, written *LAC at p*, provided that every neighborhood of p contains an arcwise connected neighborhood of p. The space S is said to be *locally arcwise connected*, written *LAC*, provided that S is LAC at every point. See 8.43.

8.25 Theorem. Every open subset of a Peano continuum is LAC. In particular, every Peano continuum is LAC.

Proof. Let U be an open subset of a Peano continuum X, and assume for the proof that $U \neq \emptyset$. Let $p \in U$ and let $\epsilon > 0$. Then, by 8.9, there is a connected open subset V of X such that $p \in V$, $\bar{V} \subset U$, V has property S, and diameter $(V) < \epsilon$. By 8.5, $\bar{V}$ has property S. Hence, by 8.3 and since $\bar{V}$ is a continuum, $\bar{V}$ is a Peano continuum. Thus, by 8.23, $\bar{V}$ is arcwise connected. Hence, $\bar{V}$ being a neighborhood of p contained in U and of diameter $< \epsilon$, we have proved that U is LAC at p (8.24). Therefore, we have proved 8.25.

With respect to 8.25 and 8.23, we remark that a continuum can be arcwise connected without being LAC (e.g., 1.6). However, as the proof of the following result shows, every LAC continuum is arcwise connected.

8.26 Theorem. Any connected open subset of a Peano continuum is arcwise connected.

Proof. Let U be a connected open subset of a Peano continuum X, and assume for the proof that $U \neq \varnothing$. Fix $p \in U$, and define E as follows:

$$E = \{p\} \cup \{x \in U: \text{there is an arc in } U \text{ from } p \text{ to } x\}.$$

Since $p \in E$, $E \neq \varnothing$. Note that since U is open in X, U is LAC by 8.25. Hence, a simple argument shows that E and $U - E$ are each open in U. Thus, since U is connected and $E \neq \varnothing$, we must have that $U = E$. Therefore, 8.26 follows easily.

8.27 Corollary. Any separable, locally compact, connected Peano space Z is arcwise connected.

Proof. Since 8.23 applies if Z is compact, we assume for the proof that Z is not compact. Then, the one-point compactification Z^* of Z, $Z^* = Z \cup \{\infty\}$ where $\infty \notin Z$, is a continuum by Theorem 5 of [8, p. 43]. Since Z is open in Z^*, it is easy to see that Z^* is cik at each point of Z. Hence, by 5.13, Z^* is cik at every point. Thus, by (iii) of 8.1, Z^* is a Peano continuum. Therefore, since Z is a connected open subset of Z^*, Z is arcwise connected by 8.26. This proves 8.27.

We remark that the following more general result than 8.27 can be proved: Every connected, topologically complete Peano space is arcwise connected. We will not prove this result (a correct proof can be found in some of the books referenced in [1]). We mention that without the completeness assumption the result would be false (see 8.29).

We also call the reader's attention to [16], which contains a natural generalization of 8.23.

Finally, we note that much work has been done in relation to 8.14 and 8.23 in the setting of locally connected Hausdorff continua using generalized arcs (defined in 6.22). Since we shall not cover this topic in this book, we refer the reader to [9], [3], [13], and the many references contained therein.

EXERCISES

8.28 Exercise. Prove that arcwise connectedness is a continuous invariant for Hausdorff spaces; more specifically, prove that if Y is a Hausdorff space, X is arcwise connected, and f is a continuous function from X onto Y, then Y is arcwise connected. [Hint: All the work has been done–simply quote the two relevant results from the present chapter.]

8.29 Exercise. We give an example of a nondegenerate connected Peano space X in R^2 such that X contains no arc (see the comment following the proof of 8.27). Let Y denote a nondegenerate arcless continuum in R^2 (e.g., take Y to be the pseudo-arc (1.23) or the continuum X_∞ in 2.27 (see 2.31)). Let $I^2 = [0,1] \times [0,1]$. For each $n = 1, 2, \ldots$, divide I^2 into n^2 congruent solid squares and let $\mathbb{C}_n$ denote the set of these squares. For each $A \in \mathbb{C}_n$, there is a topological copy Y_A of Y such that $Y_A \subset A$ and Y_A intersects each of the four sides of A in exactly one point, that point being the midpoint of the given side (show why this can be done). Let

$$X = \cup \; \{Y_A \colon A \in \mathbb{C}_n \text{ for some } n\}.$$

Prove that X has the desired properties. Moreover, show that X is ULC (defined in 8.42). The example is due to R. L. Moore [11].

8.30 Exercise. Prove that every Peano continuum is ULAC (7.21).

8.31 Exercise. We have seen in 8.26 that any connected open subset of a Peano continuum is arcwise connected. Give an example of a connected open subset U of a Peano continuum X such that $\overline{U}$ is not arcwise connected. [Hint: First, let M be the result of fattening up W in 1.5, being sure to narrow the width of the fattening as you tend towards the y-axis. Then, to get X, add countably many appropriate arcs to $\overline{M}$ using R^3.]

8.32 Exercise. The following definition and results are motivated in part by 8.31. Let X be a topological space, let $Z \subset X$, and let $p \in Z$. We say that p is *arcwise accessible from $X - Z$* provided that there is an arc in $(X - Z) \cup \{p\}$ having p as an end point. Do (a)-(c) below.

(a) Prove that if X is a Peano continuum and U is an open subset of X such that U has property S, then every point of $Bd(U)$ is arcwise accessible from U. [Hint: Let $p \in Bd(U)$. Use 8.9 to show there are connected open subsets V_i of U, $i = 1, 2, \ldots$, such that V_i has property S, $V_i \supset V_{i+1}$, diameter $(V_i) < 1/i$, and $p \in Bd(V_i)$. Then, use 8.26 to construct a Peano continuum.]

(b)   Prove that if X is a LAC topological space and U is an open subset of X, then the set of all points of $Bd(U)$ which are arcwise accessible from U is dense in $Bd(U)$.

(c)   Give an example of a dense, connected, open subset U of B^2 (1.2) for which there is a point $p \in Bd(U)$ such that p is not arcwise accessible from U. Comp., (a).

8.33 Exercise.   Prove that if X is a connected Peano space, E is a nonempty proper subset of X, and C is a component of E, then $\overline{C} \cap Bd(E) \neq \varnothing$. Comp., 5.6.

8.34 Exercise.   Prove that a locally compact, connected metric space X is a Peano space if and only if whenever K is a compact subset of X and U is an open set in X such that $K \subset U$, then all but finitely many components of $X - K$ lie in U. [Hint: For the "if" part, it suffices to show (ii) of 8.1 holds for any open set V such that $\overline{V}$ is compact. Suppose a component Q of such a V is not open. Then, there is $q \in Q \cap \overline{(V - Q)}$. Let $K = Bd(V)$. There is U open in X such that $K \subset U$ and $q \notin \overline{U}$. Observe that each component of V is a component of $X - K$. For the "only if" part, assume the hypotheses about X and let K and U be as in the condition, $K \neq \varnothing$ and $K \neq X$. There is W open such that $K \subset W \subset U$ and $\overline{W}$ is compact. Let $\mathscr{L} = \{L: L \text{ is a component of } X - K \text{ and } L \not\subset U\}$. Use 8.33 to see that $L \cap Bd(W) \neq \varnothing$ for each $L \in \mathscr{L}$. Then, using (ii) of 8.1, conclude that $\mathscr{L}$ is finite.]

8.35 Exercise.   Prove that if X is a Peano space, then components and quasicomponents are the same. Comp., (i) of 5.18.

8.36 Exercise.   Prove that if X is a Peano space, $Y \subset X$, and Z is a connected subset of Y such that Z is open relative to Y, then there is a connected open subset V of X such that $Z = Y \cap V$. [Hint: A one-line proof does it.]

8.37 Exercise.   Let X be a Peano space, and let A and B be closed subsets of X such that $X = A \cup B$. Prove that if $A \cap B$ is a Peano space, then A and B are Peano spaces. [Hint: First, it suffices to show A is LC at each $p \in A \cap B$ (why?). To do this, let U be open in X with $p \in U$. Then, the component C of p in $A \cap B \cap U$ is open in $A \cap B \cap U$. Use 8.36 to find a connected open subset V of U such that $C = A \cap B \cap V$. Finally, show that $V \cap A$ is connected by letting $S = V, M = V \cap A, N = V \cap B$ and showing that (a) of 6.18 can be applied.]

8.38 Exercise. Prove that any continuous function f from any closed subset A of $[0,1]$ into a Peano continuum X can be extended to a continuous function $g\colon [0,1] \to X$. [Hint: Use 8.30.]

8.39 Exercise. Prove (a) and (b) below.
(a) If A is any compact subset of a Peano continuum Y, then there is a Peano subcontinuum X of Y such that $A \subset X$ and $X - A$ is the union of at most countably many arcs. [Hint: Use 7.7, 8.38, and, eventually, 8.17.]
(b) Any continuum is a nested intersection of a sequence of Peano continua, i.e., given any continuum X, there are Peano continua X_i, $i = 1, 2, \ldots$, such that $X_i \supset X_{i+1}$ for each i and $\cap_{i=1}^{\infty} X_i$ is homeomorphic to X. [Hint: Consider X to be in the Hilbert cube I^∞ [6, p. 241]; then, in $I^\infty \times I^\infty$, find countably many arcs A_i such that each A_i intersects X precisely in its two end points, any two arcs A_i and A_j are disjoint outside of X, and, for each n, $X \cup (\cup_{i=n}^{\infty} A_i)$ is a Peano continuum.]

One consequence of (a) is that any compact metric space A of dimension ≥ 1 can be embedded in a Peano continuum of the same dimension (first, embed A in the Hilbert cube Y [7, p. 241] and apply (a), and then use the Sum Theorem [4, p. 30]). An application of (b) is in 12.42.

8.40 Exercise. Let X be a nondegenerate Peano continuum. Prove (a) and (b) below.
(a) If X is not an arc, then X contains a simple closed curve or a simple triod. [Hint: For $p \neq q$, non-cut points of X (6.6), and A an arc with end points p and q (8.23), find a way to use 8.26.]
(b) If X contains no simple triod, then X is an arc or a simple closed curve. [Hint: Use (a) and 8.23.]

Applications are in, e.g., 8.41 and 9.5.

8.41 Exercise. Using 2.12 and (a) of 8.40, prove that a Peano continuum is arc-like if and only if it is an arc. We note that this result is immediate from 2.13 (which we did not prove), 2.20, 2.21, and 8.40. We also note that a better result is in 12.6.

8.42 Exercise. A metric space (X,d) is said to be *uniformly locally connected*, written *ULC*, provided that for each $\epsilon > 0$, there is a $\delta = \delta(\epsilon) > 0$ such that if $x, y \in X$ and $d(x,y) < \delta$, then there is a connected subset C of X such that $x, y \in C$ and diameter $(C) < \epsilon$ (comp., 7.21). Work (a)–(e) below.
(a) If (X,d) is ULC, then (X,d) is a Peano space.

(b)   A compact metric space (X,d) is a Peano space if and only if (X,d) is ULC.

(c)   Give an example of a metric space having property S (hence, a Peano space (8.3)) which is not ULC.

(d)   Give an example of a metric space which is ULC and which does not have property S. Comp., (e).

(e)   If (X,d) is a compact metric space, then, using the subspace metric, any ULC subspace Z of X has property S. [Hint: Let $\epsilon >$ 0. Let $\delta = \delta(\epsilon/3)$ be as in the definition of ULC for Z. Let $\{z_i: i = 1, 2, \ldots\}$ be a countable dense subset of Z. For each i, let $A_i = \cup \{C: C \subset Z, z_i \in C, C$ is connected, and diameter $(C) < \epsilon/3\}$. Show that $Z = \cup_{i=1}^{n} A_i$ for some $n < \infty$.]

8.43   Exercise.   In 8.24, we localized the notion of arcwise connectedness. Another way to do this is to say that a topological space S is *strongly locally arcwise connected at p* $(p \in S)$, written SLAC at p, provided that each neighborhood of p contains an arcwise connected neighborhood of p which is open in S. Thus, LAC at p is to SLAC at p as cik at p is to LC at p. As a consequence of 8.26, Peano continua are SLAC at every point. The example in Figure 5.22 shows that LAC at p does not imply SLAC at p. Prove that a topological space is LAC at every point if and only if it is SLAC at every point. [Hint: First prove an analogue of (a) of 5.22 using *arc components* (meaning, maximal arcwise connected sets).]

8.44   Exercise.   A space X is said to be *semi-locally-connected at p* $(p \in X)$, written SLC at p, provided that each neighborhood of p contains a neighborhood V of p (see (a)) such that $X - V$ has only finitely many components. A space X is said to be *semi-locally-connected*, written SLC, provided that X is SLC at every point. Observe that the continuum in Figure 5.22 is SLC at the point p but not LC at p. Also observe that, by (c) of 6.25, any continuum is SLC at any end point. Work (a)–(e) below.

(a)   Prove that requiring the neighborhood V in the definition above to be open is no restriction.

(b)   Give an example of a continuum which is LC at a point p but not SLC at p. [Hint: Add an appropriate arc to 1.5.] Comp., (d).

(c)   Give an example of a continuum which is SLC at a point p but not cik at p. [Hint: Add two appropriate arcs to 1.5.]

(d)   Prove that, in spite of (b), every Peano continuum is SLC. [Hint: Follows easily using only 8.1.] An application is in 8.45.

(e)   Prove that a continuum is SLC if and only if it is aposyndetic (defined in 1.22).

We remark that the class of SLC continua is an important extension of the class of Peano continua partly because a lot of cyclic element theory can be done in the setting of SLC continua (see [15]). The result in (e) is due to Jones [5].

8.45 Exercise. Let X be a Peano continuum, and let p be a non-cut point of X. Prove that for each $\epsilon > 0$, there is a connected open subset U of X such that $p \in U$, diameter $(U) < \epsilon$, and $X - U$ is connected. [Hint: First use (d) of 8.44, then 8.26, and finally 5.28.]

8.46 Exercise. Prove that if f is a continuous function from a compact metric space X onto a metric space Y, then f is monotone (8.21) if and only if $f^{-1}(E)$ is connected for all connected subsets E of Y. The example following the proof of 8.15 illustrates the necessity of requiring X to be compact. The result will be used often (e.g., in 8.47, (a) of 9.45, and 10.52).

8.47 Exercise. Prove that if $X = \varprojlim\{X_i, f_i\}_{i=1}^{\infty}$ where each X_i is a Peano continuum (or is SLC (8.44)) and each f_i is a monotone map onto X_i, then X is a Peano continuum (or is SLC). [Hint: Show first that if $Z = \varprojlim\{Z_i, g_i\}_{i=1}^{\infty}$ where each Z_i is a continuum and each g_i is a monotone map onto Z_i, then each projection $\pi_i : Z \to Z_i$ is a monotone map onto Z_i (use 2.4, 2.6, 2.14, and 8.46). Then, use 2.28 and 8.46.]

8.48 Exercise. Prove that X is a Peano continuum if and only if 2^X is a Peano continuum. Prove the same result with 2^X replaced by $C(X)$.

8.49 Exercise. A topological space S is said to be *contractible* provided there is a continuous function $h : S \times [0,1] \to S$ such that for each $x \in S$, $h(x,0) = x$ and $h(x,1) = p$ for some given $p \in S$. Such a map h is called a *contraction*. Intuitively, S being contractible means S can be continuously deformed to a point over a finite time interval (regarding this intuitive statement, see 12.62). Prove the following result:

THEOREM. If X is a Peano continuum, then $C(X)$ and 2^X are contractible.

[Hint: Let μ be a size function for $C(X)$ (4.33). Define k on $X \times [0,\mu(X)]$ by

$$k(x,t) = \cup \{B \in C(X) : x \in B \text{ and } \mu(B) = t\}.$$

Using the Order Arc Theorem near the end of 5.25, and using 8.30 and (b) of 4.33, it follows that k maps into $C(X)$ and does so continuously. Now, define h on $2^X \times [0,\mu(X)]$ by

$$h(A,t) = \cup \{k(a,t): a \in A\}$$

and check that h gives a contraction of 2^X and that the restriction of h to $C(X) \times [0, \mu(X)]$ gives a contraction of $C(X)$.]

The theorem is due to Wojdyslawski [17], but the hint is from [6]. Much more is now known–see [12].

8.50 Exercise. Let us say that a metric space (X,d) is *boundedly connected* provided that for any x, $y \in X$, we can find a bounded connected subset Z of X such that x, $y \in Z$. If (X,d) is boundedly connected, then, for any x, $y \in X$, let

$$\rho_r(x,y) = \text{glb} \{\text{diameter } (Z): Z \text{ is a bounded connected subset of } X \text{ and } x, y \in Z\}.$$

Work (a)–(c) below.
  (a) Prove that if (X,d) is boundedly connected, then ρ_r is a metric and the d-topology is contained in the ρ_r-topology.
  (b) Prove that a metric space (X,d) is a connected Peano space if and only if (X,d) is boundedly connected and the d-topology is equal to the ρ_r-topology.
  (c) Prove that a connected metric space is a Peano space if and only if it is homeomorphic to a space in which every open ball is connected.
For continua, a much stronger result than (b) or (c) is true ([2] or [10]).

8.51 Exercise. Prove that a continuum which is the union of two Peano continua must be a Peano continuum. Compare, 10.38.

REFERENCES

1. B. J. Ball, Arcwise connectedness and the persistence of errors, Amer. Math. Monthly, 91(1984), 431–433.

2. R. H. Bing, Partitioning a set, Bull. Amer. Math. Soc., 55(1949), 1101–1110.

3. J. L. Cornette and B. Lehman, Another locally connected Hausdorff continuum not connected by ordered continua, Proc. Amer. Math. Soc., 35(1972), 281–284.

4. Witold Hurewicz and Henry Wallman, *Dimension Theory*, Princeton Univ. Press, Princeton, N.J., 1948.

5. F. Burton Jones, Aposyndetic continua and certain boundary problems, Amer. J. Math., 63(1941), 545–553.

6. J. L. Kelley, Hyperspaces of a continuum, Trans. Amer. Math. Soc., 52(1942), 22–36.

7. K. Kuratowski, *Topology*, Vol. I, Academic Press, New York, N.Y., 1966.

8. K. Kuratowski, *Topology*, Vol. II, Academic Press, New York, N.Y., 1968.

9. Sibe Mardešić, On the Hahn-Mazurkiewicz theorem in nonmetric spaces, Proc. Amer. Math. Soc., 11(1960), 929–937.

10. E. E. Moise, Grille decomposition and convexification theorems for compact locally connected continua, Bull. Amer. Math. Soc., 55(1949), 1111–1121.

11. R. L. Moore, A connected and regular point set which contains no arc, Bull. Amer. Math. Soc., 32(1926), 331–332.

12. Sam B. Nadler, Jr., *Hyperspaces of Sets*, Monographs and Textbooks in Pure and Applied Math., vol. 49, Marcel Dekker, Inc., New York, N.Y., 1978.

13. J. Nikiel, Images of arcs—a nonseparable version of the Hahn–Mazurkiewicz theorem, Fund. Math., 129(1988), 91–120.

14. W. Sierpiński, Sur une condition pour qu'un continu soit une courbe jordanienne, Fund. Math., 1(1920), 44–60.

15. Gordon Thomas Whyburn, *Analytic Topology*, Amer. Math. Soc. Colloq. Publ., vol. 28, Amer. Math. Soc., Providence, R.I., 1942.

16. G. T. Whyburn, On n-arc connectedness, Trans. Amer. Math. Soc., 63(1948), 452–456.

17. M. Wojdyslawski, Sur la contractilité des hyperespaces des continus localement connexes, Fund. Math., 30(1938), 247–252.

IX
Graphs

We began our study of special classes of continua in Chapter VIII by examining Peano continua in general. Recall that we chose to begin with Peano continua because of the fundamental way in which they are defined (i.e., by merely localizing connectedness). Later, we saw that one of their most important and pervasive properties is their arcwise connectivity. In view of this, it can be said that a basic class of Peano continua consists of those which are simply a finite union of arcs. In this chapter, we study a subclass of these continua called graphs (9.1). We obtain many characterizations of graphs, and give a number of applications. In our applications, we characterize particular continua. For example: We give four characterizations of simple closed curves [9.6, 9.31, and (a) and (c) of 9.44], two characterizations of trees (9.28 and 9.42), and we show that there are exactly five continua which become disconnected upon the removal of any three of their points (9.33).

1. THE NOTIONS OF GRAPH AND ORDER OF A IN X

9.1 Definition of Graph. A *graph* is a continuum which can be written as the union of finitely many arcs any two of which are either disjoint or intersect only in one or both of their end points.

We remark that some authors use the term linear graph, finite graph, or one-dimensional compact connected polyhedron to describe what we have called a graph in 9.1. Other authors use the term graph to mean a more abstract object (for an introductory treatment of abstract graphs, see, e.g., Chapters 10 and 11 of [2]).

It is easy to verify that a graph is a Peano continuum–moreover, every subcontinuum of a graph is a Peano continuum (9.4) and, in fact, is a graph (9.10.1).

We note that the union of two intersecting graphs, even arcs, may not be a graph, as the continuum drawn below shows.

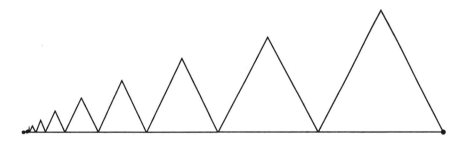

However, let us note the following fundamentally important fact.

9.2 Proposition. If X and Y are graphs such that $X \cap Y$ is nonempty and finite, then $X \cup Y$ is a graph.

Proof. The easy proof is left as an exercise (9.37).

For any set A, we let $|A|$ denote the cardinality of A. The following notion is not only important in the theory of graphs, but also, for example, in the classification of curves.

9.3 Definition of Order of A in X. Let (X,T) be a topological space, and let $A \subset X$. Let β be a cardinal number. We say that A *is of order less than or equal to β in X*, written

$$\text{ord}(A,X) \leq \beta,$$

provided that for each $U \in T$ such that $A \subset U$, there exists $V \in T$ such that

$$A \subset V \subset U \quad \text{and} \quad |Bd(V)| \leq \beta.$$

We say that A *is of order β in X*, written

$$\text{ord}(A,X) = \beta,$$

provided that ord$(A,X) \leq \beta$ and ord$(A,X) \nleq \alpha$ for any cardinal number $\alpha < \beta$. If $A = \{p\}$, then we frequently write ord(p,X) instead of ord$(\{p\},X)$ and say p is of order ... instead of saying $\{p\}$ is of order .... Clearly, ord$(p,X) = 1$ if and only if p is an end point of X (6.25).

2. CHARACTERIZATIONS OF GRAPHS IN TERMS OF ORDER

We give three characterizations of graphs in terms of the notion of order (9.10, 9.12, and 9.13). In the process, we obtain other results, e.g., a characterization of simple closed curves in terms of order (9.6).

9.4 Proposition. If X is a continuum such that ord$(x,X) < \aleph_0$ for all $x \in X$, then every subcontinuum of X is a Peano continuum. Hence, every subcontinuum of a graph is a Peano continuum (comp., 9.10.1).

Proof. Assume X is a continuum such that ord$(x,X) < \aleph_0$ for all $x \in X$. Then, it is immediate from 5.11 that X contains no convergence continuum. Hence, clearly, no subcontinuum of X contains a convergence continuum. Thus, by 5.12, every subcontinuum of X is cik at every point. Therefore, by 8.1, every subcontinuum of X is a Peano continuum. This proves the first part of 9.4. The second part follows from the first part since it is easy to see using 9.1 that if X is a graph, ord$(x,X) < \aleph_0$ for all $x \in X$. This completes the proof of 9.4.

The following proposition determines those nondegenerate continua each point of which is of order ≤ 2.

9.5 Proposition. If X is a nondegenerate continuum, then ord$(x,X) \leq 2$ for all $x \in X$ if and only if X is an arc or a simple closed curve.

Proof. Assume that ord$(x,X) \leq 2$ for all $x \in X$. Then, by 9.4, X is a Peano continuum and, clearly, X cannot contain a simple triod. Therefore, by (b) of 8.40, X is an arc or a simple closed curve. This proves half of 9.5. The other half is trivial.

Note the following characterization of simple closed curves. We shall give some others in 9.31 and 9.44.

9.6 Corollary. A continuum X is a simple closed curve if and only if each point of X is of order 2 in X.

Proof. Use 9.5.

Similarly, a continuum X is an arc if and only if there exist p, $q \in X$ with $p \neq q$ such that $\text{ord}(p,X) = 1 = \text{ord}(q,X)$ and $\text{ord}(x,X) = 2$ when $x \in X - \{p,q\}$. This result can be proved without assuming X is compact (see [7, p. 293]).

9.7 Lemma. Let X be a Peano continuum, and let $p \in X$ such that $\text{ord}(p,X) = n < \aleph_0$. Then, there is a countable local base $\{B_i: i = 1, 2, ...\}$ at p such that, for each $i = 1, 2, ..., B_i$ is a connected open subset of X and $|Bd(B_i)| = n$.

Proof. By 9.3, there is a countable local base $\{U_i: i = 1, 2, ...\}$ at p such that, for each $i = 1, 2, ...,$
   (1) U_i is open in X;
   (2)   diameter $(U_i) < 1/i$;
   (3) $|Bd(U_i)| = n$.
For each $i = 1, 2, ...,$ let V_i denote the component of U_i containing p. By (1) and by (ii) of 8.1, we see that:
   (4) $Bd(V_i) \subset Bd(U_i)$ for each $i = 1, 2, ...;$
   (5) $\{V_i: i = 1, 2, ...\}$ is a local base at p.
By (3) and (4), $|Bd(V_i)| \leq n$ for each $i = 1, 2, ...$ Also, since $\text{ord}(p,X) = n$, we see from (2) and (5) that $|Bd(V_i)| < n$ for only finitely many i. Hence, there exists N such that

$$|Bd(V_i)| = n \qquad \text{for all } i \geq N.$$

Thus, letting $B_i = V_{N+i}$ for each $i = 1, 2, ...,$ it follows that the sets B_i have the desired properties. This proves 9.7.

The following definition generalizes the notion of a simple triod (defined in 2.21).

9.8 Definition of Simple n-Od. For a given integer $n \geq 3$, a *simple n-od* is a space which is homeomorphic to the cone over an n-point discrete space (3.15; also, see 3.28). If Z is a simple n-od, then the unique point of Z which is of order ≥ 3 in Z is called the *vertex* of Z. Clearly, a simple n-od is a graph and a simple 3-od is what in 2.21 we called a simple triod.

The following lemma is a very special case of Menger's n-Beinsatz Theorem [7, p. 277].

9.9 Lemma. Let X be a continuum with exactly one point p of order ≥ 3 in X. If $\mathrm{ord}(p, X) = n < \aleph_0$, then p is the vertex of a simple n-od which is a neighborhood of p in X.

Proof. By 9.4, X is a Peano continuum. Hence, by 9.7, there is a countable local base $\{B_i : i = 1, 2, \ldots\}$ at p such that, for each $i = 1$, $2, \ldots$, B_i is a connected open subset of X and $|Bd(B_i)| = n$. If, for each i, there were a point $x_i \in Bd(B_i)$ such that x_i is not a limit point of $X - B_i$, then $\{B_i \cup \{x_i\} : i = 1, 2, \ldots\}$ would be a collection of open subsets of X forming a local base at p such that

$$|Bd(B_i \cup \{x_i\})| = n - 1, \quad \text{each } i = 1, 2, \ldots$$

which would contradict that $\mathrm{ord}(p, X) = n$. Hence, there exists k such that

(1)   Each $q_j \in Bd(B_k)$, $1 \leq j \leq n$, is a limit point of $X - B_k$.

Note (2)–(4) below:

(2) $\overline{B}_k$ is arcwise connected (9.4 and 8.23);
(3) X is LAC at each q_j (9.4 and 8.25);
(4) $\mathrm{ord}(x, X) \leq 2$ for all $x \neq p$.

By (1), (3), and (4), we have:

(5)   Each q_j must be an end point of any arc in $\overline{B}_k$ to which q_j belongs.

By (2), there is an arc A_j in $\overline{B}_k$ with end points p and q_j for each $j = 1, \ldots,$ n. It follows easily from (4) and (5) that

$$A_\ell \cap A_m = \{p\} \quad \text{when } \ell \neq m.$$

Hence, $\bigcup_{j=1}^{n} A_j$ is a simple n-od with vertex p. Thus, since $\mathrm{ord}(p, X) = n$, it follows easily using (2), (4), and (5) that $\overline{B}_k = \bigcup_{j=1}^{n} A_j$. This completes the proof of 9.9.

We are now ready to prove our first characterization of graphs.

9.10 Theorem. A continuum X is a graph if and only if (1) and (2) below both hold:

(1) $\mathrm{ord}(x, X) < \aleph_0$ for all $x \in X$;
(2) $\mathrm{ord}(x, X) \leq 2$ for all but finitely many $x \in X$.

Proof. Clearly (using 9.1), any graph X satisfies (1) and (2). To prove the converse, first note that if Y is a continuum with no points of order ≥ 3 in Y, then, by 9.5, Y is a graph. Assume inductively that if Y is any continuum satisfying (1) and (2) with at most k points of order ≥ 3 in Y, then Y is a graph. Now, let Z be a continuum satisfying (1) and (2) with exactly $k + 1$ points p_i, $1 \le i \le k + 1$, of order ≥ 3 in Z. Then, since Z satisfies (1), we may apply 9.4 to see that Z is a Peano continuum. Hence, there is a connected open subset U of Z such that $p_i \in U$ and $p_i \notin \overline{U}$ for any $i \ne 1$. Note that $\overline{U}$ is a continuum and p_1 is the only point of $\overline{U}$ of order ≥ 3 in $\overline{U}$. Thus, letting

$$n = \text{ord}(p_1, \overline{U})$$

(note: $n < \aleph_0$ since Z satisfies (1)), we have by 9.9 that p_1 is the vertex of a simple n-od which is a neighborhood of p_1 in $\overline{U}$ and, hence, in Z. Thus, clearly, there is a connected open subset V of Z such that $p_1 \in V$, $\overline{V}$ is an n-od, and $|Bd(V)| = n$. Since $Bd(Z - V) = Bd(V)$, we have by 5.19 that $Z - V$ has at most n components $K_1, ..., K_j$. Since $p_1 \notin K_i$ for any i, we have by the inductive assumption that each K_i is a graph. Also, by 5.6 (with $X = Z$ and $E = Z - V$), each $K_i \cap \overline{V} \ne \varnothing$ and, since $K_i \cap \overline{V} \subset Bd(V)$, each $K_i \cap \overline{V}$ is finite. Thus (since $\overline{V}$ is a graph), we may apply 9.2 to see that $\overline{V} \cup K_1$ is a graph. Hence, since

$$(\overline{V} \cup K_1) \cap K_2 = \overline{V} \cap K_2,$$

we may apply 9.2 to see that $(\overline{V} \cup K_1) \cup K_2$ is a graph. Continuing in this fashion, j applications of 9.2 give us that Z is a graph. Therefore, by induction, we have proved that if a continuum satisfies (1) and (2), then it is a graph. This completes the proof of 9.10.

9.10.1 Corollary. Every subcontinuum of a graph is a graph.

Proof. Use 9.10.

The following lemma will be used in the proofs of our next two characterizations (9.12 and 9.13). We remark that the assumption that X be a Peano continuum in the second part of the lemma is necessary (9.39).

9.11 Lemma. Let X be a continuum such that there exist points $x_i \in X$, $i = 1, 2, ...$, satisfying

$$\text{ord}(x_i, X) \ge 3 \quad \text{for each } i \text{ and } x_i \ne x_j \text{ when } i \ne j.$$

Then: There is a subcontinuum K of X such that

$$\text{ord}(K, X) \ge \aleph_0$$

and, if X is a Peano continuum, there is a subcontinuum L of X such that

$$|L^{[1]}| \ge \aleph_0 \quad (L^{[1]} = \text{the set of end points of } L \text{ (6.25)}).$$

Proof. If X is not Peanian, then, by 5.12, there exists $p \in X$ such that $\text{ord}(p, X) \ge \aleph_0$ and the first part of the lemma follows taking $K = \{p\}$. Hence, for the purpose of proving the lemma, we may assume that X is a

Peano continuum. We also assume without loss of generality that $\{x_i\}_{i=1}^{\infty}$ converges to a point $x \in X$ and that $x_i \neq x$ for all i. Then (8.1), there are connected open subsets U_i of X such that

$$x_i \in U_i, \text{ diameter } (U_i) < 1/i, \quad U_i \cap U_j = \varnothing \text{ for } i \neq j.$$

One of the situations in (a) or (b) below holds:

    (a)   There is an arc A in X such that $x_i \in A$ for infinitely many i.

    (b)   No arc in X contains x_i for infinitely many i.

Assume first that (a) holds, without loss of generality, for all i. Since $\text{ord}(x_i, X) \geq 3$ and $\text{ord}(x_i, A) \leq 2$ for each i, there exists $p_i \in U_i - A$ for each i. By 8.26, there is, for each i, an arc A_i in U_i such that A_i has end points x_i and p_i. Let

$$K = A \quad \text{and} \quad L = A \cup \left(\bigcup_{i=1}^{\infty} A_i \right).$$

It is easy to see that K and L have the desired properties. Next, assume that (b) holds. Then, since X is LAC at x (8.25) and $\{x_i\}_{i=1}^{\infty}$ converges to x, it follows easily that there are arcs B_n, $n = 1, 2, \ldots$, with end points x and $x_{i(n)}$ such that

$$\text{diameter } (B_n) < 1/n \quad \text{and} \quad x_{i(n)} \notin B_j \text{ for any } j \neq n.$$

For each n, let C_n be a proper subarc of B_n such that x is one of the end points of C_n and $C_n \cap U_{i(n)} \neq \varnothing$. Let

$$K = \bigcup_{n=1}^{\infty} C_n = L.$$

Again, it is easy to see that K and L have the desired properties. This completes the proof of 9.11.

The following theorem is due to J. L. Kelley, though it was actually stated and proved in 5.2 of [5] for the case when X is a Peano continuum. We note that Kelley used the theorem to prove that a Peano continuum X has a finite-dimensional hyperspace $C(X)$ if and only if X is a graph (5.4 of [5]).

9.12 Theorem. A continuum X is a graph if and only if $\text{ord}(A,X) < \aleph_0$ for all subcontinua A of X.

Proof. The "only if" part is easy using 9.1. To prove the "if" part, assume that X is a continuum such that $\text{ord}(A,X) < \aleph_0$ for all subcontinua A of X. Then, clearly, X satisfies (1) of 9.10 and, by the first part of 9.11,

X satisfies (2) of 9.10. Therefore, by 9.10, X is a graph. This completes the proof of 9.12.

We have characterized graphs in terms of the order of points (9.10) and the order of subcontinua (9.12). Though the following characterization involves the order of points, it is of a slightly different nature and will be used to obtain a completely different type of characterization (9.24). We also remark that the condition that X be a Peano continuum is necessary (see 9.39).

9.13   Theorem.   A Peano continuum X is a graph if and only if each subcontinuum of X has only finitely many end points.

Proof.   By using 9.10.1, the "only if" part follows immediately from 9.1. To prove the "if" part, assume that X is a Peano continuum such that each subcontinuum of X has only finitely many end points. Then, by the second part of 9.11, X satisfies (2) of 9.10. Suppose that X does not satisfy (1) of 9.10, i.e., $\text{ord}(p,X) \geq \aleph_0$ for some $p \in X$. By (2) of 9.10, there is a connected open subset U of X such that $p \in U$, $\text{ord}(x,X) \leq 2$ for all $x \in U - \{p\}$, and diameter $(U) < 1$. Let $p_1 \in U - \{p\}$. By 8.26, there is an arc A_1 in U with end points p and p_1. Assume inductively that we have defined n arcs A_i, $1 \leq i \leq n$, with end points p and p_i such that for each i

   diameter $(A_i) < 1/i$, $A_i \cap A_j = \{p\}$ whenever $i \neq j$.

Then, let V be a connected open subset of U such that $p \in V$, $p_i \notin V$ for any i, and diameter $(V) < 1/(n + 1)$. Note that since $V \subset U$, $\text{ord}(x,X) \leq 2$ for all $x \in V - \{p\}$. Thus, by using the properties in the inductive assumption and noting that $\text{ord}(p,X) > n$, we see that there exists $p_{n+1} \in V - \bigcup_{i=1}^{n} A_i$. By 8.26, there is an arc A_{n+1} in V with end points p and p_{n+1}. Since $\text{ord}(x,X) \leq 2$ for all $x \in V - \{p\}$ and since $p_i \notin A_{n+1}$ for any i, it follows easily that $A_{n+1} \cap A_i = \{p\}$ for each $i = 1, ..., n$. Now, using the properties of the inductively defined arcs A_i, $i = 1, 2, ...$, it is easy to see that $\bigcup_{i=1}^{\infty} A_i$ is a subcontinuum of X with infinitely many end points. This is a contradiction. Hence, X satisfies (1) of 9.10. Therefore, since we have already shown X satisfies (2) of 9.10, X is a graph. This completes the proof of 9.13.

3.   DISCONNECTION NUMBER

The classical topological characterization of a simple closed curve is that it is the only continuum which becomes disconnected upon the removal of any two points. We shall prove this result later (9.31). For now, we note that the *statement* of the result leads naturally to the following question: What are the continua which become disconnected upon the removal of any n points where $n \leq \aleph_0$ is fixed? We devote most of the rest of this chapter to answering this question and giving some applications. In particular, we shall show there is such an n for a continuum X if and only if X is a graph (9.24), and we shall derive a formula for computing the smallest such n for any given graph (9.34). Our applications are concerned with characterizations of special continua.

We formalize the idea discussed above as follows:

9.14 Definition of Disconnection Number. Let X be a connected space. A cardinal number $n \leq \aleph_0$ is called a *disconnection number for X* provided that whenever $A \subset X$ such that $|A| = n$, then $X - A$ is not connected. We write $D(X) \leq \aleph_0$ to mean there is a disconnection number for X. When $D(X) \leq \aleph_0$, we let $D^s(X)$ denote the smallest disconnection number for X.

We note the following three simple propositions about disconnection numbers for continua.

9.15 Proposition. If X is a continuum such that $D(X) \leq \aleph_0$, then $D^s(X) \geq 2$.

Proof. The proposition follows immediately from the fact that X has a non-cut point (6.6).

9.16 Proposition. If X is a continuum, then $D(X) \leq \aleph_0$ if and only if $\aleph_0$ itself is a disconnection number for X.

Proof. Assume that $D(X) \leq \aleph_0$. Let $n = D^s(X)$, and assume for the purpose of proof that $n < \aleph_0$. Let $A \subset X$ such that $|A| = \aleph_0$. Let $B \subset A$ such that $|B| = n$. Then, $X - B$ is not connected. Let K_1 and K_2 be two components of $X - B$. Then, by the first part of 5.7, $\overline{K_i} \cap B \neq \varnothing$ for each $i = 1$ and 2. Hence, K_1 and K_2 must be nondegenerate. Thus, since K_1 and K_2 are connected, K_1 and K_2 are uncountable. Hence,

$$K_1 - A \neq \varnothing \quad \text{and} \quad K_2 - A \neq \varnothing$$

and, therefore, it follows that $X - A$ is not connected. This proves half of 9.16. The other half is trivial.

9.17 Proposition. Let X be a continuum. If an integer n is a disconnection number for X, then so is any integer $j \geq n$.

Proof. The proof is similar to the proof of 9.16 and, thus, is omitted.

Now, let us note the following simple result about graphs.

9.18 Proposition. If X is a nondegenerate graph, then $D(X) \leq \aleph_0$ and, in fact, $D^s(X) < \aleph_0$.

Proof. By 9.1, $X = \cup_{i=1}^k A_i$ $(1 \leq k < \infty)$ where each A_i is an arc and any two of these arcs are either disjoint or intersect only in one or both of their end points. Let

$$n = \begin{cases} k + 1, & \text{if } X \text{ is not an arc} \\ 3, & \text{if } X \text{ is an arc.} \end{cases}$$

Then, it is easy to verify that X becomes disconnected upon the removal of any set of n distinct points. This proves 9.18.

Our next goal is to prove the converse of 9.18. This is obtained in 9.24. The proof is based on the following five results.

9.19 Lemma. Let X be a continuum such that $D(X) \leq \aleph_0$. Let Z be a proper subcontinuum of X, and let m denote the cardinality of the set of components of $X - Z$. Then, $m < D^s(X)$ and, thus, $m < \aleph_0$.

Proof. If $m \geq \aleph_0$, let $\mathbb{C}$ be a collection of $\aleph_0$ of the components of $X - Z$; if $m < \aleph_0$, let $\mathbb{C}$ be the collection of all the components of $X - Z$. Whichever is the case, let $g = |\mathbb{C}|$ and let C_i, $1 \leq i \leq g$, be a one-to-one enumeration of the members C_i of $\mathbb{C}$. For each $C_i \in \mathbb{C}$, let $K_i = C_i \cup Z$. By 5.9, each K_i is a continuum. Hence, Z being a proper subcontinuum of each K_i, 6.8 shows that we may choose a non-cut point q_i of K_i such that $q_i \in C_i$ for each i. Let $Q = \{q_i : 1 \leq i \leq g\}$, and note for use later that $|Q| = g$ (since, for $i \neq j$, q_i and q_j were chosen in different components of $X - Z$). We define $\mathcal{L}$ and $\mathcal{P}$ as follows (the set $\mathcal{P}$ may be empty):

$$\mathcal{L} = \{K_i - \{q_i\} : 1 \leq i \leq g\}$$
$$\mathcal{P} = \{C \cup Z : C \text{ is a component of } X - Z \text{ and } C \notin \mathbb{C}\}.$$

Each member of $\mathcal{L}$ is connected and contains Z, and, if $\mathcal{P} \neq \varnothing$, the same is true (by 5.9) of each member of $\mathcal{P}$. Hence, by Corollary 3(i) of [7, p. 132],

$$(\cup \mathcal{L}) \cup (\cup \mathcal{P}) \text{ is connected.}$$

Now, observe that

$$(\cup \mathcal{L}) \cup (\cup \mathcal{P}) = X - Q.$$

Hence, $X - Q$ is connected. Therefore, since $|Q| = g$, g is not a disconnection number for X. Thus, since $g \leq \aleph_0$, we have by 9.16 and 9.17 that $g < D^s(X)$. Therefore, since $g < \aleph_0$ implies $g = m$, we have that $m < D^s(X)$. This completes the proof of 9.19.

9.20 Lemma.   Let X be a continuum such that $D(X) \leq \aleph_0$. Then, X does not contain a nowhere dense nondegenerate subcontinuum.

Proof.   Suppose that X contains a nowhere dense nondegenerate subcontinuum Z. By 9.19, $X - Z$ has only finitely many components $C_1, \ldots, C_m$. Since Z is nowhere dense in X,

$$(1) \quad Z = \bigcup_{i=1}^{m} (\overline{C}_i \cap Z).$$

Therefore, by the Baire theorem (Theorem 6 of [7, p. 9]), there exists k such that $\overline{C}_k \cap Z$ contains a nonempty open subset W of Z. Let V be a nonempty proper open subset of Z such that $\overline{V} \subset W$. Let

$$K = \cup \{\overline{C}_i \cap V : |\overline{C}_i \cap V| \leq \aleph_0\}$$

(note: K may be empty). Since $|K| \leq \aleph_0$, there exists $A \subset V - K$ such that $|A| = \aleph_0$ (Theorem 6 of [7, p. 9]). Let

$$L = (\overline{C}_k - A) \cup (Z - V).$$

We show that L is connected. Since $A \subset Z$, $\overline{C}_k - A$ is connected by Corollary 3(ii) of [7, p. 132]. Let Q be a component of $Z - V$. Since Q is closed in Z, we see by 5.6 that $Q \cap \overline{V} \neq \varnothing$. Thus, since $\overline{V} \subset \overline{C}_k$, $Q \cap \overline{C}_k \neq \varnothing$. Since $A \subset V$, $Q \cap A = \varnothing$. It now follows that since $\overline{C}_k - A$ is connected, $(\overline{C}_k - A) \cup Q$ is a connected subset of L. Therefore, since Q was any component of $Z - V$, we see using Corollary 3(i) of [7, p. 132] that L is connected. Next, we prove (2) below:

$$(2) \quad L \cup \left[\bigcup_{i=1}^{m} (\overline{C}_i - A)\right] \text{ is connected.}$$

We prove (2) by showing that $L \cup (\overline{C}_i - A)$ is connected for each $i = 1, \ldots, m$. Fix i. By the first part of 5.7, $\overline{C}_i \cap Z \neq \varnothing$. Suppose that $\overline{C}_i \cap Z \subset A$. Then, since $|A| = \aleph_0$ and $V \subset Z$, $|\overline{C}_i \cap V| \leq \aleph_0$. Hence $\overline{C}_i \cap V \subset K$. Since $\overline{C}_i \cap Z \neq \varnothing$, there is a point $p \in \overline{C}_i \cap Z$. Thus, since we have

$$\overline{C}_i \cap Z \subset A, \quad A \subset V, \quad \overline{C}_i \cap V \subset K$$

we see that $p \in A \cap K$, which is a contradiction since $A \cap K = \varnothing$.

Therefore, $\overline{C}_i \cap Z \not\subset A$. Hence, there exists $q \in \overline{C}_i \cap Z$ such that $q \notin A$. Since $\overline{C}_k \supset V$, we see that $q \in L$ (since if $q \in V$, then $q \in \overline{C}_k - A$). Thus, $L \cap (\overline{C}_i - A) \neq \emptyset$. Therefore, since L and $\overline{C}_i - A$ are connected (the latter because $A \subset Z$–use Corollary 3(ii) of [7, p. 132]), we have that $L \cup (\overline{C}_i - A)$ is connected. From this fact for each $i \leq m$ and by Corollary 3(i) of [7, p. 132], (2) now follows. Observe that by using (1) and the fact that $A \subset V \subset Z$, it follows easily that

$$(3) \quad L \cup \left[\bigcup_{i=1}^{m} (\overline{C}_i - A) \right] = X - A.$$

Thus, since $|A| = \aleph_0$, we have by (2) and (3) that $\aleph_0$ is not a disconnection number for X. Therefore, since $D(X) \leq \aleph_0$, we have a contradiction to 9.16. This completes the proof of 9.20.

9.21 Corollary. If X is a continuum such that $D(X) \leq \aleph_0$, then X is a Peano continuum.

Proof. By 9.20 and the definition of 5.11, X does not contain a convergence continuum. Therefore, by 5.12 and (iii) of 8.1, X is a Peano continuum. This proves 9.21.

We remark that 9.21 can be proved without 9.20 by using 5.12 and (b) of 6.29 to show that if X is not a Peano continuum, then $D(X) \not\leq \aleph_0$. Such a proof is in 4.1 of [10].

9.22 Lemma. Let X be a continuum such that $D(X) \leq \aleph_0$. If Z is a nondegenerate subcontinuum of X, then $D(Z) \leq \aleph_0$.

Proof. Assume for the purpose of proof that $Z \neq X$. Let $A \subset Z$ such that $|A| = \aleph_0$. Suppose that $Z - A$ is connected. By 9.19, $X - Z$ has only a finite number m of components $C_1, \ldots, C_m$. By the first part of 5.7, $\overline{C}_i \cap Z \neq \emptyset$ for each i. Hence, there exists $p_i \in \overline{C}_i \cap Z$ for each i. Let:

$H_i = C_i \cup \{p_i\}$ for each i;

$M = (Z - A) \cup \{p_1, \ldots, p_m\}$;

$B = A - \{p_1, \ldots, p_m\}$ (note: B may equal A).

By using Corollary 3(ii) of [7, p. 132], we see that each H_i is connected and that M is connected (for the case of M, note that M lies between $Z - A$ and $\overline{Z - A}$ since $|A| = \aleph_0$ implies, by Theorem 6 of [7, p. 9], that $\overline{Z - A} = Z$). Thus, since $H_i \cap M \neq \emptyset$ for each i, we have by Corollary 3(i) of [7, p. 132] that

(1) $\left(\bigcup_{i=1}^{m} H_i\right) \cup M$ is connected.

Next, observe that

(2) $\left(\bigcup_{i=1}^{m} H_i\right) \cup M = X - B$

By (1), (2), and the fact that $|B| = \aleph_0$ (since $|A| = \aleph_0$ and $m < \aleph_0$), we have that $\aleph_0$ is not a disconnection number for X. However, since $D(X) \leq \aleph_0$, this is a contradiction to 9.16. Hence, $Z - A$ is not connected. Therefore, we have proved that $D(Z) \leq \aleph_0$. This proves 9.22.

The following corollary, together with 9.21, will allow us to apply 9.13 to obtain our main result (9.24).

9.23 Corollary. If X is a continuum such that $D(X) \leq \aleph_0$, then each subcontinuum of X has only finitely many end points.

Proof. Let Z be a subcontinuum of X and let J denote the set of end points of Z. For the purpose of proof, assume that Z is nondegenerate (since if $|Z| = 1$, then $|J| = 0$). Then, by 9.22, $D(Z) \leq \aleph_0$. Suppose that $|J| \geq \aleph_0$. Then, there exists $H \subset J$ such that $|H| = \aleph_0$. By 6.27, $Z - H$ is connected. Thus, since $D(Z) \leq \aleph_0$ and $|H| = \aleph_0$, we have a contradiction to 9.16. Hence, $|J| < \aleph_0$ and, therefore, we have proved 9.23.

Now, we come to our characterization of graphs in terms of disconnection numbers. In connection with this result, the reader should see 9.34, which gives a formula for $D^s(X)$, and the comments following it.

9.24 Theorem. Let X be a nondegenerate continuum. Then, (1)–(3) below are equivalent:

(1) X is a graph.
(2) $D(X) \leq \aleph_0$.
(3) Some integer n is a disconnection number for X, i.e., $D^s(X) < \aleph_0$.

Proof. By 9.18, (1) implies (2) and (3). Assuming (2), 9.21 and 9.23 show that we may apply 9.13 to obtain (1). Clearly, (3) implies (2). This completes the proof of 9.24.

Let us note that we have proved in 9.24 that if a continuum X becomes disconnected upon the removal of every countably infinite subset, then there is a fixed integer n such that X becomes disconnected upon the removal of every subset of cardinality n. This was proved by reducing the implication to graphs [i.e., showing (2) implies (1) in 9.24]–a direct proof seems elusive.

4. APPLICATIONS OF THE CHARACTERIZATION BY DISCONNECTION NUMBER

We now give some applications of 9.24. Note the following definition.

9.25 Definition of Tree. A *tree*, or *acyclic graph*, is a graph which contains no simple closed curve.

The following two propositions will be used in the proofs of several of our applications. Recall the definition of end point in 6.25.

9.26 Proposition. Let X be a nondegenerate graph, and let $p \in X$ such that p does not belong to any simple closed curve in X. Then, p is a non-cut point of X if and only if p is an end point of X.

Proof. Assume that p is a non-cut point of X. Let $\epsilon > 0$. By 9.40 (and the hypothesis about p in 9.26), there is an open subset W of X such that $p \in W$ and W does not contain any point of any simple closed curve in X. By the second part of 9.4, X is a Peano continuum. Hence, by 8.45, there is a connected open subset U of X such that
$p \in U$, $U \subset W$, $\overline{U} \neq X$, diameter $(U) < \epsilon$, and $X - U$ is connected. Suppose that $|Bd(U)| \geq 2$. Then, by (b) of 8.32, there exist $x, y \in Bd(U)$ with $x \neq y$ such that x and y are arcwise accessible from U. Thus, since U is arcwise connected (8.26), it follows that there is an arc A in $U \cup \{x,y\}$ such that A has end points x and y. By the second part of 9.4, $X - U$ is a Peano continuum. Hence, by 8.23, there is an arc B in $X - U$ such that B has end points x and y. Clearly, $A \cup B$ is a simple closed curve and $(A \cup B) \cap W \neq \emptyset$, a contradiction to our choice of W. Thus, $|Bd(U)| = 1$. Therefore, it follows that we have proved p is an end point of X. This proves half of 9.26. The other half is in (b) of 6.25. Therefore, we have proved 9.26.

9.27 Proposition. Let X be a nondegenerate tree. Then, a point $p \in X$ is a non-cut point of X if and only if p is an end point of X.

Proof. The result is an immediate consequence of 9.26.

Our first application of 9.24 is the following characterization of trees. It shows that those continua with only finitely many non-cut points have a particularly simple structure (comp., comments in the paragraph following 6.17).

9.28 Theorem. A continuum X is a tree if and only if X has only finitely many non-cut points.

Proof. Assume that X is a nondegenerate continuum whose set F of non-cut points is finite. We first show that $|F| + 1$ is a disconnection number for X. Let $A \subset X$ such that $|A| = |F| + 1$. Then, there exists $p \in A$ such that p is a cut point of X. Hence, there are at least two components of $X - \{p\}$ and both of them are uncountable (since, by the first part of 5.7, both of them are nondegenerate). Thus, since $|A| < \aleph_0$ and $p \in A$, we see that $X - A$ is not connected. Therefore, we have proved that $|F| + 1$ is a disconnection number for X. Thus, by 9.24, X is a graph. Suppose that X contains a simple closed curve S. Let

$$B = \{z \in S: \text{ord}(z,X) > 2\}.$$

We see that any point $p \in S - B$ is a non-cut point of X since $S - \{p\}$ is connected and since, for any $x \in X - S$, there is an arc A in X from x to a point $b \in B$ such that $A \cap S = \{b\}$. In other words, $S - B \subset F$. However, since $|B| < \aleph_0$ by (2) of 9.10, this contradicts our assumption that $|F| < \aleph_0$. Therefore, X is a tree by 9.25. Conversely, assume that X is a nondegenerate tree. Then, by 9.13, X has only finitely many end points. Thus, by 9.27, X has only finitely many non-cut points. This completes the proof of 9.28.

We remark that the "if" part of 9.28 can also be proved using (b) and (d) of 6.29, 5.12, 6.7, and 8.23.

The following result was first stated in 6.17, and the usual proof was sketched in 6.22. The proof we now give is especially simple because 9.28 reduces the result to one about trees.

9.29 Corollary. A continuum X is an arc if and only if X has exactly two non-cut points.

Proof. Assume that X is a continuum with exactly two non-cut points p and q. By 9.28, X is a tree. Hence, X is a graph. Since graphs are arcwise connected (9.1), there is an arc A in X from p to q. Thus, by 6.7, $A = X$ and, therefore, X is an arc. This proves half of 9.29. The other half is trivial using the definition of an arc (1.1). Therefore, we have proved 9.29.

For another application of 9.28 similar to the one just done, see 9.43.
Our next applications of 9.24 are in 9.31 and 9.33. To facilitate their proofs, we give the following general and useful proposition.

9.30 Proposition. Let X be a nondegenerate graph and let $D^s(X) = n$ (by 9.15 and 9.18, $2 \leq n < \aleph_0$). Assume that Y is a subcontinuum of X such that there are $n - 1$ mutually disjoint open subsets W_1 of $Y - (\overline{X - Y})$ and points $y_i \in W_i$, $1 \leq i \leq n - 1$, such that the sets

$$Y - \bigcup_{i=1}^{n-1} W_i, \quad Y - \{y_1, ..., y_{n-1}\}$$

are connected. Then, $Y = X$.

Proof. Let $W = \bigcup_{i=1}^{n-1} W_i$, and let $Z = Y - W$. Suppose that $Y \neq X$. We obtain a contradiction to 9.19 upon proving (1) and (2) below.
  (1) Z is a proper subcontinuum of X;
  (2) $X - Z$ has at least n distinct components.
We prove (1). By hypothesis, Z is connected. Also by hypothesis, each W_i is open in X. Hence, W is open in X. Thus, Z is compact. Suppose that $Z = \varnothing$. Then, since $W \subset Y$ (by hypothesis), $Y = W$. Hence, Y is open in X. Thus, since Y is also closed in X and since $Y \neq \varnothing$ and $Y \neq X$, we have a contradiction to the connectedness of X. Hence, $Z \neq \varnothing$. We have now proved that Z is a subcontinuum of X. Clearly, $Z \neq X$. Therefore, we have proved (1). Next, we prove (2). Observe that

$$X - Z = (X - Y) \cup W = (X - Y) \cup \left(\bigcup_{i=1}^{n-1} W_i \right)$$

where the sets $X - Y$, W_1, ..., W_{n-1} are nonempty, mutually disjoint, and open in $X - Z$. Therefore, by choosing a component of each of these sets, we see that (2) holds. Now, by (1) and (2), and since $n = D^s(X)$, we have a contradiction to 9.19. Therefore, $Y = X$ and we have proved 9.30.

The proof of 9.30 shows that the result can be generalized as follows: Given k sets W_i and k points y_i satisfying the conditions in 9.30 with $k < \aleph_0$, and letting m denote the number of nonempty components of $X - Y$ ($m < \aleph_0$ by 9.19 and $m = 0$ if and only if $X = Y$), then $k + m < D^s(X)$.

We have given a characterization of simple closed curves in 9.6. The following characterization is the one most often quoted. It is due to R. L.

Moore [9, p. 342]. The proof given below, which utilizes 9.24 to reduce the theorem to the setting of graphs, is different than other proofs in the literature (e.g., see [4, p. 55] or [7, p. 180]). We remark that some other characterizations are in 9.44 and [1].

9.31 Theorem. A continuum X is a simple closed curve if and only if X becomes disconnected upon the removal of any two points (i.e., $D^s(X) = 2$).

Proof. Assume that X is a continuum such that $D^s(X) = 2$. Then, X is nondegenerate and, by 9.24, X is a graph. Since a nondegenerate tree has at least two end points (by 6.6 and 9.27) and remains connected when they are both removed (by 6.27), and since $D^s(X) = 2$, X is not a tree. Hence, X contains a simple closed curve Y. Since X is a graph, there is by 9.10 a point $y_1 \in Y$ such that

$$\text{ord}(y_1, X) = 2.$$

Clearly, $y_1 \in Y - (\overline{X - Y})$. Thus, there is a connected open subset W_1 of Y such that

$$y_1 \in W_1 \quad \text{and} \quad W_1 \subset Y - (\overline{X - Y}).$$

Since Y is a simple closed curve, $Y - \{y_1\}$ and $Y - W_1$ are connected. Hence, the hypotheses of 9.30 are satisfied with $n = 2$. Therefore, by 9.30, $Y = X$ and, thus, X is a simple closed curve. This proves half of 9.31. Since the other half is trivial using the definition of a simple closed curve in 1.3, we have proved 9.31.

Therefore, there is exactly one continuum X such that $D^s(X) = 2$. In our final application of 9.24, we show that there are exactly five continua X such that $D^s(X) = 3$ (9.33). First, note the definitions of the following special continua.

9.32 Definitions of Noose, Figure Eight, Theta Curve, Dumbbell. A *noose* (*figure eight*, *theta curve*) is a space which is homeomorphic to the symbol representing the number six (the number eight, the greek letter theta, respectively). A *dumbbell* is a space which is homeomorphic to $C_1 \cup A \cup C_2$ where C_i is the simple closed curve in R^2 with center at $(i,0)$ and radius $= 1/4$, and A is the convex segment in R^2 from $(5/4,0)$ to $(7/4,0)$.

9.33 Theorem. Let X be a continuum. Then, $D^s(X) = 3$ if and only if X is one of the following five continua: an arc, a noose, a figure eight, a theta curve, a dumbbell.

Proof. Clearly, $D^s(X) = 3$ if X is any of the five continua listed above. We prove the converse. Assume that X is a continuum such that $D^s(X) = 3$. Then, X is nondegenerate and, by 9.24, X is a graph. We take three cases.

Case 1: X is a tree. Then, since $D^s(X) = 3$, we see by 9.27, 6.27, and 6.6 that X has exactly two non-cut points. Hence, by 9.29, X is an arc.

Case 2: X contains exactly one simple closed curve S. By 9.31, $S \neq X$. Hence, by 6.7, there is a non-cut point y_1 of X such that $y_1 \notin S$. Let A be an arc in X with end points y_1 and $v \in S$ such that $A \cap S = \{v\}$. Let $Y = A \cup S$. Clearly, Y is a noose. We shall show that $Y = X$ by using 9.30. Since S is the only simple closed curve in X, we have by 9.26 that y_1 is an end point of X. Hence, using (2) of 9.10, there is a connected open subset W_1 of X such that $y_1 \in W_1$ and $W_1 \subset A$. Note that $W_1 \subset Y - (\overline{X - Y})$. Again using (2) of 9.10, there exist $y_2 \in S$ and a connected open subset W_2 of $S - (\overline{X - S})$ such that $y_2 \in W_2$. Note that $W_2 \subset Y - (\overline{X - Y})$ and that $W_1 \cap W_2 = \varnothing$. Also, since Y is a noose, it follows from where y_i and W_i lie in Y and the connectedness of W_i that $Y - \{y_1, y_2\}$ and $Y - (W_1 \cup W_2)$ are connected. Hence, by 9.30 with $n = 3$, $Y = X$ and, thus, X is a noose.

Case 3: X contains two or more simple closed curves. Then, let S_1 and S_2 be two simple closed curves in X. If $|S_1 \cap S_2| = 1$, then $S_1 \cup S_2$ is a figure eight; if $|S_1 \cap S_2| \geq 2$, then $S_1 \cup S_2$ contains a theta curve; if $S_1 \cap S_2 = \varnothing$, then, letting A be an arc in X such that $A \cap S_i = \{a_i\}$ where a_i is an end point of A for each i, $S_1 \cup A \cup S_2$ is a dumbbell. Hence, $X \supset Y$ where Y is a figure eight, a theta curve, or a dumbbell. We shall show that $Y = X$ by using 9.30. Whichever of the three spaces Y may be, it is easy using (2) of 9.10 to find two disjoint, nonempty, open subsets W_1 and W_2 of $Y - (\overline{X - Y})$ and points $y_i \in W_i$, $i = 1$ and 2, such that $Y - (W_1 \cup W_2)$ and $Y - \{y_1, y_2\}$ are connected. Then, we can apply 9.30 with $n = 3$ to see that $Y = X$. Thus, X is a figure eight, a theta curve, or a dumbbell.

Since the situations covered by the three cases above take care of all possibilities, we have proved 9.33.

We have determined the continua X for which $D^s(X) = 2$ or 3 (9.31, 9.33). The case of $D^s(X) = 4$ is more complicated. At the present time it is known that there are at least 26 such continua, and that all such continua are graphs (9.24) which are embeddable in R^2 (8.2 of [10]). It is suspected that there are *exactly* 26 of them.

The following theorem gives a formula for computing $D^s(X)$ using only the nodes and edges of any given subdivision of a graph X [$\chi(X)$ is the Euler characteristic of X (9.47), and $X^{[1]}$ is the set of end points of X (6.25)].

9.34 Theorem. If X is a nondegenerate graph, then

$D^s(X) = 2 - \chi(X) + |X^{[1]}|$.

Proof. See 9.54.

Three comments are appropriate. First, by combining 9.34 and 9.24, we obtain a more substantive characterization than the one in 9.24. Second, for each integer $n \geq 2$, we see that there are only finitely many continua X such that $D^s(X) = n$ (the exact number is not known for any $n \geq 4$–see section 8 of [10]). Third, for graphs in the plane, the formula in 9.34 may be written as in the following corollary.

9.35 Corollary. If X is a nondegenerate graph in the plane R^2 and if $R(X;R^2)$ denotes the number of components (i.e., regions) of $R^2 - X$, then

$D^s(X) = R(X;R^2) + |X^{[1]}|$.

Proof. Euler's formula for a graph X in the plane [2, p. 281] states that $\chi(X) = 2 - R(X;R^2)$. Hence, the corollary follows from 9.34.

We remark that the notion of a disconnection number (9.14) and the material in 9.13-9.35 originated with this book (except, of course, the statements of 9.29 and 9.31). These and related results have been accepted for publication [10].

NOTE ADDED IN PROOF. I have recently discovered a conference proceedings paper by A. H. Stone entitled "Disconnectable spaces" (Top. Conf. Arizona State Univ., 1968, pp. 265-276) in which he gives a more general result than 9.24. His approach is different than the one used here (in particular, he does not obtain 9.13). Also, he does not obtain the formula in 9.34. The applications in this section and the other material about disconnection numbers in the exercises at the end of this chapter and in [10] are not in Stone's paper.

5. THE KURATOWSKI GRAPH THEOREM

We have devoted this chapter to characterizing graphs and giving some applications (more of both are in the exercises). Even though the following theorem from [6] is of a different character and shall not be proved, its elegance and far-reaching consequences demand its inclusion. We remark that it can be applied to a number of physical problems, e.g., the design of printed circuit boards (where one wishes to minimize wire crossings in order to be able to print as much of the circuit on a flat surface as possible).

9.36 Kuratowski Graph Theorem. Let X be a graph or, more generally, a Peano continuum which contains only finitely many simple closed curves. Then, X is embeddable in R^2 if and only if X does not contain either the complete graph K_5 on five vertices or the complete bipartite graph $K_{3,3}$ (see figures below).

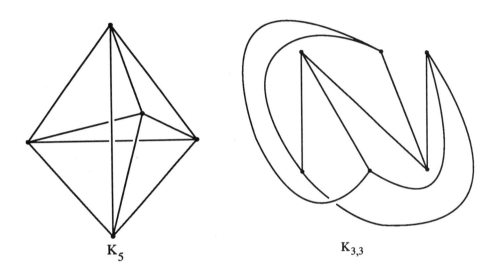

$$K_5 \qquad\qquad\qquad K_{3,3}$$

A generalization of 9.36 is in [3], a simpler proof of which is in [8].

EXERCISES

9.37 Exercise. Prove 9.2 (using only 9.1).

9.38 Exercise. In relation to 9.10, give an example of a continuum X such that $\operatorname{ord}(x,X) = 2$ for all but finitely many $x \in X$ and such that X is not a graph. Compare with (1) of [11, p. 182].

9.39 Exercise. To show the necessity of assuming X is a Peanian in the second part of 9.11 and in 9.13, give an example of a continuum X with infinitely many points of order ≥ 3 in X such that each subcontinuum of X has at most two end points.

9.40 Exercise. Using only 9.1, prove that any graph X contains only finitely many simple closed curves. [Hint: First prove that any simple closed curve $S \subset X$ is the union of some of the arcs in 9.1.]

9.41 Exercise. Prove that a continuum X is a graph if and only if each point of X has a (closed) neighborhood which is a simple n-od or an arc. [Hint: The "only if" part uses 9.4, 9.5, 9.9, and 9.10. The "if" part uses the existence of Lebesgue numbers of open covers (Corollary 4d of [7, p. 24])].

9.42 Exercise. Prove that a graph X is a tree if and only if any two points of X are separated in X by a third point (definition in 5.30). Compare with 10.2.

9.43 Exercise. We have characterized the arc as the only continuum with exactly two non-cut points (9.29). Prove that a continuum X is a simple triod if and only if X has exactly three non-cut points. [Hint: Use 9.28.] What are the continua with exactly four non-cut points?

9.44 Exercise. In 9.6 and 9.31, we gave two characterizations of simple closed curves. Work (a)–(c) below.
   (a) Prove that a nondegenerate continuum X is a simple closed curve if and only if $X - C$ is connected for all connected subsets C of X. [Hint: Easy using 9.31.]
   (b) Give an example to show that the "if" part of (a) would be false if we only required that $X - C$ be connected when C is a subcontinuum of X. See (c).
   (c) Prove that a nondegenerate Peano continuum X is a simple closed curve if and only if $X - C$ is connected for all subcontinua C of X. [Hint: If p, $q \in X$, $p \neq q$, such that $X - \{p,q\}$ is connected (comp., 9.31), then find disjoint open subsets U and V of X such that $p \in U$, $q \in V$, and $X - (U \cup V)$ is connected by applying 8.45 to $X/\{p,q\}$ (3.14).]
Some other characterizations are in [1].

9.45 Exercise. In 8.22, we proved that if a nondegenerate continuum Y is a monotone image of an arc, then Y is an arc. Prove (a) and (b) below.
   (a) If a nondegenerate continuum Y is a monotone image of a simple closed curve, then Y is a simple closed curve. [Hint: Use 9.31 and 8.46.]

(b) If a continuum Y is a monotone image of a graph, then Y is a
graph. [Hint: Use 9.12 to see that Y satisfies (1) and (2) of 9.10.]
Results for open images are in 9.46, and a better result is in 13.31 (also,
regarding (b), see 13.43).

9.46 Exercise. Let X and Y be continua, and let f be an open map
from X onto Y (defined in 2.18). Prove (a)–(e) below.
(a) If U is open in X, then $Bd(f[U]) \subset f[Bd(U)]$.
(b) For any $p \in X$, ord$(f(p),Y) \leq$ ord(p,X).
(c) If X is a graph, then Y is a graph.
(d) If X is an arc and Y is nondegenerate, then Y is an arc. [Hint:
Recall 9.5.]
(e) If X is a simple closed curve and Y is nondegenerate, then Y is a
simple closed curve or an arc. [Hint: Recall 9.5.]
Much more is known about the behavior of open maps on graphs–e.g.,
see [11, p. 182]. A generalization of (c)–(e) is in 13.31. Also, see 13.43
and the generalization of (d) in 12.15.

9.47 Exercise. We introduce the notions of subdivision and Euler
characteristic in the setting of graphs, and we give some preliminary facts
about them for use in later exercises. Let X be a graph. An *open free arc
in X* is any subset of X of the form $A - \{x,y\}$ where A is an arc in X with
end points x and y and $A - \{x,y\}$ is open in X; x and y are called *end
points of* $A - \{x,y\}$. A *subdivision of X* is an ordered pair

$$\Sigma = (N(X), E(X))$$

where $N(X)$ is a finite set of points of X called *nodes* (or *vertices*) and
$E(X)$ is a finite collection of open free arcs in X called *edges* such that the
two end points of each edge are in $N(X)$ and

$$N(X) \cup [\cup E(X)] = X.$$

If X is a graph and $\Sigma = (N(X),E(X))$ is a subdivision of X, then we define
$\chi(X;\Sigma)$ by

$$\chi(X;\Sigma) = |N(X)| - |E(X)|.$$

Prove (a)–(c) below.
(a) Every graph has a subdivision.
(b) If Σ_1 and Σ_2 are subdivisions of a graph X, then $\chi(X;\Sigma_1) =
\chi(X;\Sigma_2)$. [Hint: Let $\Sigma_i = (N_i(X), E_i(X))$ for each $i = 1$ and
2, define an appropriate subdivision $\Sigma = (N(X),E(X))$ where
$N(X) = N_1(X) \cup N_2(X)$, and show that $\chi(X;\Sigma_i) = \chi(X;\Sigma)$ for
each $i = 1$ and 2.]

By (a) and (b), we may associate with any graph X an integer $\chi(X;\Sigma)$ which, since it is independent Σ, will be denoted from now on by $\chi(X)$. It is called the *Euler characteristic of X*.

(c) If X and Y are homeomorphic graphs, then $\chi(X) = \chi(Y)$.
We remark that the ideas and results above may be generalized to, e.g., finite simplicial complexes [4, p. 241].

9.48 Exercise. Prove that if X is a graph and $\Sigma = (N(X),E(X))$ is a given subdivision of X, then there is a tree $T \subset X$ such that T has a subdivision $\Sigma' = (N(T),E(T))$ where

$$N(T) = N(X) \quad \text{and} \quad E(T) \subset E(X).$$

Such a tree is called a Σ-*spanning tree of X*. Its existence is extremely useful–some simple applications are in 9.49–9.51.

9.49 Exercise. Prove that a graph X is a tree if and only if $\chi(X) = 1$, where $\chi(X)$ is as defined in 9.47. [Hint: Prove the "only if" part first by inducting on $|N(X)|$ for trees X. Then, prove the "if" part by assuming X is not a tree and using 9.48 and the "only if" part.]

9.50 Exercise. Prove that $\chi(X) \leq 1$ for all graphs X. [Hint: Use 9.48 and 9.49.]

9.51 Exercise. Prove that if X is a graph, then $\chi(X) = 0$ if and only if X contains exactly one simple closed curve. [Hint: If X contains exactly one simple closed curve C, show X/C (3.14) is a tree, show $\chi(X/C) = \chi(X) + 1$ by finding a useful subdivision of X/C, and then apply 9.49. Conversely: If $\chi(X) = 0$, then, for T as in 9.48, show using 9.49 that $|E(X)| = |E(T)| + 1$, and use this to obtain the desired conclusion.]

9.52 Exercise. For each integer $n \leq -1$, give examples of graphs X and Y such that $\chi(X) = \chi(Y) = n$ and X and Y contain different numbers of simple closed curves. Compare with 9.49 and 9.51.

9.53 Exercise. Let X be a graph which is not a tree, and let $\Sigma = (N(X),E(X))$ be a given subdivision of X. We define $X^1(p_1,p_2)$ by the following simple surgery on X. Let $U \in E(X)$ such that U is contained in some simple closed curve in X. Let $p_1, p_2 \in U$ with $p_1 \neq p_2$, and let A be the unique arc in U with end points p_1 and p_2. Define $X^1(p_1,p_2)$ by

$$X^1(p_1,p_2) = X - (A - \{p_1,p_2\}).$$

For use in 9.54, prove the following: $X^1(p_1,p_2) = Y$ is a graph, $\chi(Y) = \chi(X) + 1$, $|Y^{[1]}| = |X^{[1]}| + 2$, and, by a useful choice of U, $D^s(Y) = D^s(X) + 1$. [Hint: Only the last assertion will cause difficulty. Let $n = D^s(X)$. There are $n - 1$ points $y_1, \ldots, y_{n-1}$ such that $X - \{y_1, \ldots, y_{n-1}\}$ is connected. Show at least one of them, y_i, lies in a simple closed curve C in X. Choose $U \in E(X)$ such that $U \subset C$ and $y_i \in \overline{U}$. Show that for $p_1 \neq p_2$ in this U and $Y = X^1(p_1,p_2)$, $D^s(Y) = D^s(X) + 1$.]

In the definition of $X^1(p_1,p_2)$, we require that $p_1,p_2 \in U \in E(X)$ where U is contained in a simple closed curve in X. The extra conditions imposed on U in the hint above were for the purpose of obtaining the equality

$$D^s(Y) = D^s(X) + 1, \quad \text{where } Y = X^1(p_1,p_2),$$

for some choice of U. This is used to do 9.54. Once 9.54 is done, 9.34 can easily be used to show that the equality above is valid for $Y = X^1(p_1,p_2)$ with no extra conditions on U.

9.54 Exercise. Prove 9.34. [Hint: Induct on the Euler characteristic k (consider *all* graphs X such that $\chi(X) = k$). You will use 6.27, 9.27, 9.49, 9.50, and 9.53.]

9.55 Exercise. Prove that if X is a nondegenerate graph, then

$$\Lambda(X) = \max\{D^s(Y): Y \text{ is a nondegenerate subcontinuum of } X\}$$

exists and is attained at a tree. We remark that a formula for $\Lambda(X)$ is in [10, 7.6].

9.56 Exercise. For each $n \geq 2$, compute $D^s(K_n)$ where K_n is the complete graph on n vertices, i.e., K_n is the graph obtained by joining each two of the points

$$e_i = (x_1, \ldots, x_i, \ldots, x_n), \quad x_i = 1 \text{ and } x_j = 0 \text{ for } j \neq i$$

by a convex arc in R^n. There is a surprising connection between the numbers $D^s(K_n)$ and a geometric result about R^2. Let $S(n)$ denote the maximum number of components into which R^2 can be divided by n convex lines (i.e., convex copies of R^1). Then, for each $n \geq 3$, $S(n - 2) = D^s(K_n)$. Lines giving $S(3)$ and $S(4)$ are drawn on the next page.

164

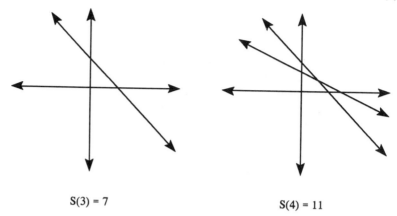

S(3) = 7   S(4) = 11

REFERENCES

1. R. H. Bing, Some characterizations of arcs and simple closed curves, Amer. J. Math., 70(1948), 497–506.

2. Richard A. Brualdi, *Introductory Combinatorics*, Elsevier North-Holland, Inc., New York, N.Y., 1977.

3. Schieffelin Claytor, Topological immersion of Peanian continua in a spherical surface, Ann. of Math., 35(1934), 809–835.

4. John G. Hocking and Gail S. Young, *Topology*, Addison-Wesley Publishing Company, Inc., Reading, Mass., 1961.

5. John L. Kelley, Hyperspaces of a continuum, Trans. Amer. Math. Soc., 52(1942), 22–36.

6. K. Kuratowski, Sur le problème des courbes gauches en topologie, Fund. Math, 15(1930), 271–283.

7. K. Kuratowski, *Topology*, Vol. II, Academic Press, New York, N.Y., 1968.

8. E. E. Moise, Remarks on the Claytor imbedding theorem, Duke Math. J., 19(1952), 199–202.

9. Robert L. Moore, Concerning simple continuous curves, Trans. Amer. Math. Soc., 21(1920), 333–347.

10. Sam B. Nadler, Jr., Continuum theory and graph theory: disconnection numbers, to appear in J. of London Math. Soc.

11. Gordon Thomas Whyburn, *Analytic Topology*, Amer. Math. Soc. Colloq. Publ., Vol. 28, Amer. Math. Soc., Providence, R.I., 1942.

X
Dendrites

We continue our study of special types of Peano continua by studying dendrites. In the previous chapter, we investigated graphs and, in the process, we obtained some results about trees. The definition in 10.1 shows that dendrites are defined in terms of Peano continua in the same way as trees are defined in terms of graphs (9.25). Moreover, the characterization of trees in 9.42 has a direct generalization for dendrites (10.2).

10.1 Definition of Dendrites. A *dendrite* is a Peano continuum which contains no simple closed curve.

We shall obtain numerous characterizations of dendrites. In addition, we shall determine many special basic properties of dendrites. As regards the latter, we show, e.g., that dendrites are regular (10.20), have the fixed point property (10.31), are tree-like (10.32), are embeddable in the plane (10.37), and contain only countably many points of order ≥ 3 (10.23). As regards the former, we give characterizations in terms of separation and/or cut points (10.2, 10.7, 10.8), unicoherence properties (10.10, 10.35), order (10.13), continuous selections (10.53(c)), a fixed point condition (10.55), and convergence (10.46, 10.56(b)).

In addition, we obtain a particularly well-structured and useful inverse limit representation for any dendrite in terms of trees in the dendrite

(10.33). We also determine two ways of writing any dendrite (10.28, 10.43).

We remark that our proofs do not use cyclic element theory, nor do they use the Cut Point Order Theorem [19, p. 49]. In this regard, they are different than the proofs in, e.g., [19].

1. SOME CHARACTERIZATIONS OF DENDRITES

We begin our study of dendrites with the following characterization which, as mentioned above, is a generalization of 9.42. Recall the definition in 5.30.

10.2 Theorem. A continuum X is a dendrite if and only if any two points of X are separated in X by a third point of X.

Proof. Assume that X is a dendrite. Let $p,q \in X$ such that $p \neq q$. By 8.23, there is an arc A in X from p to q. Let $r \in A - \{p,q\}$. Let U be the component of p in $X - \{r\}$. Suppose that $q \in U$. Then, since U is open in X [(ii) of 8.1], by 8.26 there is an arc B in U from p to q. Clearly, $A \cap B$ is not connected. Hence, it follows easily that $A \cup B$ contains a simple closed curve. This contradicts the assumption that X is a dendrite. Thus, $q \notin U$. Therefore, since U is open and closed in $X - \{r\}$, p and q are separated in X by r. Conversely, assume that any two points of X are separated in X by a third point of X. Then, by 5.30, X cannot contain a convergence continuum. Hence, by 5.12, X is a Peano continuum [see (iii) of 8.1]. Clearly, X cannot contain a simple closed curve (for if X did contain a simple closed curve S, then no two points of S could be separated in X by any point of X). Thus, X is a dendrite. This completes the proof of 10.2.

It seems appropriate at this point to clarify a situation which is obfuscated by the last part of the proof of 10.2. For this purpose, note the following definition and the corollary in 10.5.

10.3 Definition of Hereditarily Locally Connected Continuum. A continuum X is said to be *hereditarily locally connected*, written *hlc*, provided that every subcontinuum of X is a Peano continuum.

For example, every graph is hlc by 9.4.

It follows from 5.12 and (iii) of 8.1 that a continuum is hlc if it contains no convergence continuum. The following result shows that containing no convergence continuum characterizes hlc continua.

10.4 Theorem. A continuum X is hlc if and only if X contains no convergence continuum.

Proof. As remarked above, the "if" part of the theorem follows from 5.12 and (iii) of 8.1. To prove the other half, assume that X contains a convergence continuum A, i.e. (5.11), A is a nondegenerate subcontinuum of X and there is a sequence $\{A_i\}_{i=1}^{\infty}$ of subcontinua A_i of X such that (1)–(3) below hold:

(1) $A = \lim A_i$;

(2) $A \cap A_i = \varnothing$ for each i;

(3) $A_i \cap A_j = \varnothing$ for $i \neq j$ (5.23).

Let $p, q \in A$ such that $p \neq q$. Since we want to show X is not hlc, we may assume X itself is a Peano continuum. Hence, there is a connected open subset U of X such that $p \in U$ and $q \notin \bar{U}$. By (1), there exists N such that $A_i \cap U \neq \varnothing$ for all $i \geq N$. Now, let

$$Y = A \cup \bar{U} \cup \left[\bigcup_{i=N}^{\infty} A_i \right].$$

It follows easily that Y is a continuum. Suppose that Y is a Peano continuum. Then, since $q \in Y - \bar{U}$, there is a connected open subset V of Y such that $q \in V$ and $\bar{V} \cap \bar{U} = \varnothing$. Since $\bar{V} \subset Y$ and $\bar{V} \cap \bar{U} = \varnothing$, clearly

(4) $\bar{V} = [\bar{V} \cap A] \cup \left[\bigcup_{i=N}^{\infty} (\bar{V} \cap A_i) \right].$

Since $q \in V \cap A$, $V \cap A \neq \varnothing$. Thus, since V is open in Y, we see easily using (1) that $V \cap A_i \neq \varnothing$ for all but finitely many $i \geq N$. Hence, using (2)–(4), $\bar{V}$ is not σ-connected (5.15). However, since V is connected and nonempty, $\bar{V}$ is a continuum and, thus, we have a contradiction to 5.16. Hence, Y is not a Peano continuum. Therefore, X is not hlc. This completes the proof of 10.4.

10.5 Corollary. Every dendrite is hlc.

Proof. Recall 10.2 and the result in 5.30; then apply 10.4.

In fact, note the following stronger corollary.

10.6 Corollary. Every subcontinuum of a dendrite is a dendrite.

Proof. The result is immediate from 10.5 (and 10.1).

We remark that we shall prove in 10.20 that every dendrite is regular. Thus, since every regular continuum is hlc (10.16), 10.20 is a better result than 10.5.

We have seen that every end point of a continuum is a non-cut point [(b) of 6.25]. We now show that the converse characterizes dendrites.

10.7 Theorem. A nondegenerate continuum X is a dendrite if and only if each point of X is either a cut point of X or an end point of X.

Proof. Assume that X is a nondegenerate dendrite. Let p be a non-cut point of X. Let $\epsilon > 0$. Then, by 8.45, there is a connected open subset U of X such that

$$p \in U, \quad \text{diameter } (U) < \epsilon, \quad \text{and} \quad X - U \text{ is connected.}$$

Suppose that $|Bd(U)| \geq 2$. Then, by 8.25, we may apply (b) of 8.32 to obtain $q, r \in Bd(U)$ with $q \neq r$ such that q and r are each arcwise accessible from U. Thus, since U is arcwise connected (8.26), we see that there is an arc A in $U \cup \{q,r\}$ from q to r. Note that $q, r \in X - U$ and that, since $X - U$ is connected, $X - U$ is a Peano continuum (10.5). Hence, by 8.23, there is an arc B in $X - U$ from q to r. Clearly, $A \cup B$ is a simple closed curve. This contradicts our assumption that X is a dendrite. Hence, $|Bd(U)| \leq 1$. Therefore, since X is nondegenerate, it follows from what we have shown that p is an end point of X [use (f) of 5.17]. This proves half of 10.7. To prove the other half, assume that X is a continuum such that each point of X is either a cut point of X or an end point of X. Then, since clearly no end point of X can belong to a convergence continuum in X, it follows easily using (b) of 6.29 that X contains no convergence continuum. Hence, by 10.4, X is hlc. In particular, X is a Peano continuum. Suppose that X contains a simple closed curve Z. Clearly, no point of Z can be an end point of X. Hence, by our assumption, each point of Z is a cut point of X. Thus, by (d) of 6.29, there is a cut point of Z. However, this is impossible (since a simple closed curve has no cut point). Hence, X contains no simple closed curve. Therefore, X is a dendrite (by 10.1). This completes the proof of 10.7.

The following theorem, due to R. L. Moore [11], has an interesting variation for Hausdorff continua [16].

10.8 Theorem. A continuum X is a dendrite if and only if each nondegenerate subcontinuum of X contains uncountably many cut points of X.

Proof. Assume that X is a dendrite. Let Y be a nondegenerate subcontinuum of X. Let $p, q \in Y$ with $p \neq q$. Then, by 10.5 and 8.23, there is an arc A in Y with end points p and q. Let

$$r \in A - \{p, q\}.$$

Since $\text{ord}(r, A) = 2$, $\text{ord}(r, X) \geq 2$ and hence, by 10.7, r is a cut point of X. Therefore, we have proved that A, hence Y, contains uncountably many cut points of X. Conversely, assume that each nondegenerate subcontinuum of X contains uncountably many cut points of X. Then, by (b) of 6.29, X cannot contain a convergence continuum. Hence, by 10.4, X is hlc. In particular, X is a Peano continuum. Suppose that X contains a simple closed curve Z. Then, by assumption, Z contains uncountably many cut points of X. Thus, by (d) of 6.29, there is a cut point of Z. However, this is impossible. Thus, X does not contain a simple closed curve. Therefore, X is a dendrite (by 10.1). This completes the proof of 10.8.

The following result, which is of independent interest, will be used in the proof of our next characterization (10.10).

10.9 Proposition. Every connected subset of a dendrite is arcwise connected.

Proof. Let C be a nondegenerate connected subset of a dendrite X. Let $p, q \in C$ such that $p \neq q$. By 10.6, $\overline{C}$ is a dendrite. Hence, by 8.23, there is an arc A in $\overline{C}$ from p to q. We show that $A \subset C$. Since C is connected, each point of $\overline{C} - C$ is a non-cut point of $\overline{C}$ by Corollary 3(ii) of [9, p. 132]. Thus, since $\overline{C}$ is a dendrite, we have by 10.7 that each point of $\overline{C} - C$ is an end point of $\overline{C}$. Therefore, since the only points of A which can be end points of $\overline{C}$ are p and q, and since $p, q \in C$, we have that $A \cap (\overline{C} - C) = \varnothing$. Thus, since $A \subset \overline{C}$, $A \subset C$. This completes the proof of 10.9.

10.10 Theorem. A continuum X is a dendrite if and only if the intersection of any two connected subsets of X is connected.

Proof. Assume that X is a dendrite. Suppose that there are connected subsets C_1 and C_2 of X such that $C_1 \cap C_2$ is not connected. Let p and q be points of different components of $C_1 \cap C_2$. Then, by 10.9, there is an arc A_i in C_i from p to q for each $i = 1$ and 2. Clearly, $A_1 \cap A_2$ is not connected. Hence, $A_1 \cup A_2$ contains a simple closed curve. This contradicts our assumption that X is a dendrite. Therefore, we have proved half of 10.10. To prove the other half, assume that

(*) the intersection of any two connected subsets of X is connected. Suppose that X contains a convergence continuum K. Let $\mathfrak{C}$ be an uncountable collection of mutually disjoint closed cuttings of K [(f) of 6.29, which is easily done using a map to R^1]. By (b) of 6.29, there exists $C \in \mathfrak{C}$ such that $X - C$ is connected. Hence, by (*),

$K \cap (X - C)$ is connected.

However, this is not possible since $K \cap (X - C) = K - C$ and $C \in \mathfrak{C}$. Thus, X does not contain a convergence continuum. Hence, by 10.2, X is hlc. Clearly, by (*), X cannot contain a simple closed curve. Therefore, X is a dendrite (by 10.1). This completes the proof of 10.10.

The following simple definition facilitates the statement of our next characterization of dendrites (10.13).

10.11 Definition of Component Number. If S is a connected space and $p \in S$, then the *component number of p in S*, written $c(p,S)$, is the cardinality of the set of all components of $S - \{p\}$.

Recall the definition of $\text{ord}(p,X)$ in 9.3.

10.12 Lemma. Let X be a nondegenerate continuum and let $p \in X$. If $\text{ord}(p,X)$ is finite, then $c(p,X)$ is finite and, in fact, $c(p,X) \leq \text{ord}(p,X)$.

Proof. Assume that $\text{ord}(p,X) = n < \infty$. Suppose that $c(p,X) > n$, and let $Z_1, \ldots, Z_{n+1}$ be $n + 1$ components of $X - \{p\}$. Then, by 5.6, $p \in \bar{Z}_i$ for each i. Let $z_i \in Z_i$ for each i, and let

$\epsilon = \min\{d(p,z_i): 1 \leq i \leq n + 1\}$,

d being a metric for X. Let U be an open subset of X such that $p \in U$ and diameter $(U) < \epsilon$. Note that $U \cap Z_i \neq \varnothing$ and $Z_i - U \neq \varnothing$ for each i. Thus, since each Z_i is connected and U is open in X, we see that $Bd(U) \cap Z_i \neq \varnothing$ for each i (see (a) of 5.17). Therefore, since $Z_i \cap Z_j = \varnothing$ when $i \neq j$, we have that $|Bd(U)| \geq n + 1$. Thus, since U was chosen to be any open neighborhood of p having diameter $< \epsilon$, we have $\text{ord}(p,X) \geq n + 1$, a contradiction. Hence, $c(p,X) \leq n$. This proves 10.12.

10.13 Theorem. A nondegenerate continuum X is a dendrite if and only if $c(p,X) = \text{ord}(p,X)$ whenever either of these is finite.

Proof. If $c(p,X) = \text{ord}(p,X)$ whenever either of these is finite, then, clearly, every non-cut point of X must be an end point of X and, hence, by 10.7, X is a dendrite. Conversely, let X be a dendrite. Let $p \in X$. By

10.12, it suffices to prove $c(p,X) = \text{ord}(p,X)$ when $c(p,X)$ is finite. So assume that

$c(p,X) = n < \infty.$

Let $L_1, \ldots, L_n$ denote the n components of $X - \{p\}$. Then, by 5.9 and 10.6, each $L_i \cup \{p\}$ is a dendrite. Thus, since p is a non-cut point of each $L_i \cup \{p\}$, we have by 10.7 that p is an end point of each $L_i \cup \{p\}$. Therefore, it follows easily that $\text{ord}(p,X) = n$. This completes the proof of 10.13.

An application of 10.13 is in the proof of 10.23. Some other characterizations of dendrites are in 10.35, 10.46, 10.53(c), 10.55, and 10.56(b).

2. CARDINALITY PROPERTIES OF DENDRITES

We first introduce the notion of a regular continuum and prove every dendrite is regular (10.20). We then define a branch point and prove that a dendrite can have at most countably many branch points (10.23).

10.14 Definition of Regular Continuum. If X is a continuum and $p \in X$, then X is said to be *regular at p* provided that there is a local base $\mathcal{L}_p$ at p such that the boundary of each member of $\mathcal{L}_p$ is of finite cardinality [note: By (h) of 5.17, it makes no difference in our definition whether we consider the members of $\mathcal{L}_p$ to be open in X or simply to be neighborhoods of p in X]. A continuum X is said to be *regular* provided that X is regular at each of its points.

We note the following two basic results about regular continua.

10.15 Proposition. Every subcontinuum of a regular continuum is regular.

Proof. The result is an immediate consequence of 10.14.

10.16 Theorem. Every regular continuum is hlc.

Proof. It is clear from 10.14 that a regular continuum contains no convergence continuum. Therefore, the theorem follows from 10.4.

We remark that the converse of 10.16 is false (10.59).
We prove a general proposition in 10.18 which has a wide range of applications. We shall use it here to characterize regular continua (10.19)

and thereby deduce that dendrites are regular (10.20). First, note the following general definition.

10.17   Definition of Additive-Hereditary System.   Let $\mathbb{C}$ be a collection of closed subsets of a space S. We say that $\mathbb{C}$ is *additive* provided that $C_1 \cup C_2 \in \mathbb{C}$ whenever $C_1, C_2 \in \mathbb{C}$. We say that $\mathbb{C}$ is *hereditary* provided that if $C \in \mathbb{C}$ and A is a closed subset of C, then $A \in \mathbb{C}$. We say that $\mathbb{C}$ is an *additive-hereditary system* provided that $\mathbb{C}$ is additive and hereditary.

10.18   Proposition.   Let X be a continuum, and let $\mathbb{C}$ be an additive-hereditary system of closed subsets of X. Then, (1) and (2) below are equivalent:

(1)   Each $x \in X$ has a local base $\mathfrak{L}_x$ such that each $U \in \mathfrak{L}_x$ is open in X and $Bd(U) \in \mathbb{C}$;

(2)   Any two points of X are separated in X by some member of $\mathbb{C}$.

Proof.   Clearly, (1) implies (2). Assume (2). Let $p \in X$. Let W be an open subset of X such that $p \in W$. For each $x \in Bd(W)$, we have by (2) that for some $C_x \in \mathbb{C}$

$$X - C_x = G_x | H_x \qquad \text{with } p \in G_x \text{ and } x \in H_x \text{ (6.2)}.$$

Note that since $Bd(H_x)$ is a closed subset of C_x and $\mathbb{C}$ is hereditary,

(i) $Bd(H_x) \in \mathbb{C}$.

Since $Bd(W)$ is compact, there are finitely many of the sets H_x, say H_j for $1 \leq j \leq n$, such that

(ii) $Bd(W) \subset \bigcup_{j=1}^{n} H_j$.

Let $H = \bigcup_{j=1}^{n} H_j$. Clearly [use (a) of 5.17],

(iii) $Bd(H) \subset \bigcup_{j=1}^{n} Bd(H_j)$.

Since $\mathbb{C}$ is an additive-hereditary system, it follows from (i) and (iii) that

(iv) $Bd(H) \in \mathbb{C}$.

Now, let $V = W - \bar{H}$. Note:

$$\bar{V} - V \subset \bar{W} - (W - \bar{H}) = (\bar{W} - W) \cup (\bar{W} \cap \bar{H})$$

and, hence, by (ii),

$\bar{V} - V \subset H \cup (\bar{W} \cap \bar{H}) \subset \bar{H}.$

Also note that since H is open in X and $H \cap V = \varnothing$, $H \cap \bar{V} = \varnothing$ and, hence,

$(\bar{V} - V) \cap H = \varnothing.$

We now see that $\bar{V} - V \subset \bar{H} - H$. Thus, since V is open in X, we have by (a) of 5.17 that

$Bd(V) \subset Bd(H).$

Hence, $Bd(V)$ being a closed subset of $Bd(H)$ and $\mathfrak{C}$ being hereditary, we have by (iv) that $Bd(V) \in \mathfrak{C}$. Clearly, $p \in V \subset W$ and V is open in X. Therefore, it follows from what we have shown that $\mathcal{L}_p$ as required in (1) exists. This completes the proof of 10.18.

The following characterization of regular continua should be compared with the characterization of dendrites in 10.2, the characterization of rational continua in 10.57, and the characterization of being of dimension $\leq n$ mentioned at the end of 13.78.

10.19 Theorem. A continuum X is regular if and only if any two points of X are separated in X by some finite set.

Proof. Let $\mathfrak{C} = \{A \subset X : A \text{ is finite}\}$ and apply 10.18.

10.20 Theorem. Every dendrite is regular.

Proof. The theorem is immediate from 10.2 and 10.19.

10.20.1 Corollary. Each point of any dendrite D is of order $\leq \aleph_0$ in D.

Proof. The result is immediate from 10.20.

We have determined that the points of order $= 1$ (i.e., the end points (9.3)) in a nondegenerate dendrite D are precisely the non-cut points of D (10.7; comp., (b) of 6.25). Thus, it is natural, especially in view of 10.20.1, to inquire into the nature of the points of higher order in D. For reasons that will become clear, we focus our attention on the points of order ≥ 3 in D. These points may be dense, even continuumwise dense, in D (e.g., 10.37). However, as may be surprising, they are at most countable as we show in 10.23. This is in contrast to the end points of D since, as is perhaps also surprising, they may be uncountable (see 10.39 or 10.45). Regarding the points of order $= 2$, see 10.42.

10.21   Definition of Branch Point.   A point b of a dendrite X is called a *branch point of X* provided that ord$(b,X) > 2$.

The following general, easy-to-prove lemma will be used in the proof of 10.23. Recall the definition of a closed cutting in 6.29.

10.22   Lemma.   If X is a connected, separable metric space and $\mathbb{C}$ is an uncountable collection of mutually disjoint closed cuttings of X, then there exist $p,q \in X$ such that uncountably many members of $\mathbb{C}$ separate p and q in X.

Proof.   Let $D = \{x_i : i = 1, 2, \ldots\}$ be a countable dense subset of X. For each $i \neq j$, let

   $\mathbb{C}(i,j) = \{C \in \mathbb{C}: C \text{ separates } x_i \text{ and } x_j \text{ in } X\}$.

It is easy to see that each $C \in \mathbb{C}$ separates some two points of D in X. Hence,

   $\mathbb{C} = \cup \{\mathbb{C}(i,j): i \neq j\}$.

Thus, since $\mathbb{C}$ is uncountable, there exist k and ℓ such that $\mathbb{C}(k,\ell)$ is uncountable. Therefore, letting $p = x_k$ and $q = x_\ell$, we have proved 10.22.

The theorem below was discussed in the paragraph preceding 10.21.

10.23   Theorem.   The set of all branch points of a dendrite is countable.

Proof.   Let X be a dendrite. Suppose that there are uncountably many branch points of X. By 10.7, each branch point of X is a cut point of X. Hence, by 10.22, there exist $p,q \in X$ and an uncountable set B of branch points of X such that
   (1)   each $b \in B$ separates p and q in X.
By 8.23, there is an arc A in X from p to q. By (1),

   (2) $B \subset A - \{p,q\}$.

For each $b \in B$, we have: $A - \{b\}$ has exactly two components (by (2)) and $X - \{b\}$ has at least three components (by 10.13 and 10.21). Hence, for each $b \in B$, there must be a component W_b of $X - \{b\}$ such that $W_b \cap A = \varnothing$. By (2) and 5.7, $b \in \overline{W}_b \cap A$ for each $b \in B$. If, for some $b \in B$, there is a point $a \in \overline{W}_n \cap A$ such that $a \neq b$, then, since $W_b \cup \{a\}$ is a connected subset of $X - \{b\}$ [9, p. 132], we have that $a \in W_b$ which contradicts our choice of W_b. Hence,

(3) $\overline{W}_b \cap A = \{b\}$ for each $b \in B$.

Now, let $x, y \in B$ such that $x \neq y$. We show that $W_x \cap W_y = \emptyset$. First, note that by (3) we have

(4) $\overline{W}_x \cap (W_y \cup A) = (\overline{W}_x \cap W_y) \cup \{x\}$.

Observe that $\overline{W}_x$ is connected and, by (3), $W_y \cup A$ is connected [9, p. 132]. Hence, by 10.10, the left-hand side of (4) is connected. Thus, by (4),

(i) $\overline{W}_x \cap W_y = \emptyset$

or

(ii) $x \in (\overline{\overline{W}_x \cap W_y})$.

By (3), $\overline{W}_y \cap A = \{y\}$. Thus, since $x \neq y$ and $x \in A$, $x \notin \overline{W}_y$. Hence, (ii) must be false. Thus, (i) holds and, therefore, $W_x \cap W_y = \emptyset$. Hence, B being uncountable, $\{W_b : b \in B\}$ is an uncountable collection of mutually disjoint, open subsets of X, a contradiction. This proves 10.23.

3. APPROXIMATION BY TREES AND APPLICATIONS

One of the most important and productive mathematical principles is that it is usually best to examine complicated objects by approximating them with simpler (often related) objects, and then use the inherent control in the approximation and the properties of the simpler objects to draw appropriate conclusions. We have seen many examples of this principle at work in previous chapters. In the present situation, the complicated objects are dendrites and the simpler, related objects are trees. We show that any dendrite has a suitable, well-controlled approximation by trees in the dendrite (10.27–the all-important control is in (5)). We then use this result to obtain several different properties of dendrites. Perhaps the one whose proof best illustrates the principle mentioned above is 10.31.

The next two lemmas are devoted to defining and proving the continuity of a natural, useful retraction of a dendrite onto any of its subcontinua.

10.24 Lemma.   Let X be a dendrite, and let Y be a subcontinuum of X. For each $x \in X - Y$, there is a unique point $r(x) \in Y$ such that $r(x)$ is a point of any arc in X from x to any point of Y.

Proof. Fix $x \in X - Y$. By 8.23, there is an arc A in X from x to some point $y \in Y$. Considering A to be ordered by $\leq$ so that $x \leq y$, let $r(x)$ denote the first point of A in Y. By using 10.10, it follows easily that $r(x)$

is a point of any arc in X from x to any point of Y. From this fact, and since the subarc B of A from x to $r(x)$ is one such arc and satisfies $B \cap Y = \{r(x)\}$, the uniqueness of $r(x)$ is clear. This completes the proof of 10.24.

10.25 Lemma. Let X be a dendrite, and let Y be a subcontinuum of X. Define $r: X \to Y$ by letting $r(x)$ be as in 10.24 if $x \in X - Y$ and letting $r(x) = x$ if $x \in Y$. Then, the function r is continuous (and, hence, is a retraction of X onto Y).

Proof. By using 8.26 and 10.24, it follows easily that r is locally constant at points of $X - Y$ and that if W is a connected open subset of X such that $W \cap Y \neq \emptyset$, then $r(W) \subset W$. Clearly, these observations imply that r is continuous. This completes the proof of 10.25.

10.26 Terminology. The map $r: X \to Y$ defined in 10.25 is called the *first point map for Y*.

We are now ready to prove the following important theorem concerning the approximation of dendrites from within by trees.

10.27 Theorem. Let X be a nondegenerate dendrite. Then, there is a sequence $\{Y_i\}_{i=1}^{\infty}$ satisfying (1)–(5) below:

(1) each Y_i is a tree;
(2) $Y_{i+1} \supset Y_i$ for each i;
(3) $\lim Y_i = X$;
(4) $Y_1 = \{p_1\}$ and, for each i, $\overline{Y_{i+1} - Y_i}$ is an arc with an end point p_i such that $\overline{Y_{i+1} - Y_i} \cap Y_i = \{p_i\}$;

and, letting $r_i: X \to Y_i$ be the first point map for Y_i (10.26),

(5) $\{r_i\}_{i=1}^{\infty}$ converges uniformly to the identity map on X.

Proof. The theorem is easy to prove for the case when X itself is a tree (10.40). For the rest of the proof, assume that X is not a tree. Let $\{x_j: j = 1, 2, \ldots\}$ be a countable dense subset of X. Let $Y_1 = \{x_1\}$ and let $p_1 = x_1$. Assume inductively that we have defined a tree Y_k in X for some given positive integer k. Then, since $Y_k \neq X$,

$$m = \min\{j: x_j \notin Y_k\} \text{ exists.}$$

By 8.23, there is an arc A in X with end points x_m and some point p_k in Y_k such that $A \cap Y_k = \{p_k\}$. Let

$$Y_{k+1} = Y_k \cup A.$$

Now, by induction, we have defined Y_i and p_i for each $i = 1, 2, \ldots$. The inductive process shows clearly that (1), (2), and (4) of the lemma are satisfied, and so is (3) since, letting $A_i = \{x_1, \ldots, x_i\}$ for each i,

$Y_i \supset A_i$ for each i and $\lim A_i = X$.

To prove (5), let $\epsilon > 0$. Choose $\delta = \delta(\epsilon)$ as guaranteed by the uniform local arcwise connectedness of X (8.30). Then, by (3), there exists N such that, letting d be a metric for X and H_d be as specifically defined in 4.1,

$$H_d(X,Y_i) < \delta \quad \text{for all } i \geq N.$$

It now follows easily using the definition of r_i that

$$d(r_i(x),x) < \epsilon \quad \text{for each } x \in X \text{ and all } i \geq N.$$

Therefore, we have proved (5). This completes the proof of 10.27.

Note the following presentation theorem for dendrites. It is an easy consequence of 10.27. We remark that it is usually proved as a corollary to the Cyclic Chain Approximation Theorem [19, p. 89]. Recall the definition of $X^{[1]}$ in 6.25.

10.28 Corollary.   If X is any nondegenerate dendrite, then X can be written as follows:

$$X = X^{[1]} \cup \left(\bigcup_{i=1}^{\infty} A_i \right)$$

where each A_i is an arc with end points p_i and q_i such that

$$A_{i+1} \cap \left(\bigcup_{j=1}^{i} A_j \right) = \{p_{i+1}\} \quad \text{for each } i = 1, 2, \ldots$$

and diameter $(A_i) \to 0$ as $i \to \infty$.

Proof.   For each $i = 1, 2, \ldots$, let Y_i and p_i be as in 10.27, and let $A_i = \overline{Y_{i+1} - Y_i}$. Then, by 10.27, each A_i is an arc, p_i is an end point of A_i, and

$$A_{i+1} \cap \left(\bigcup_{j=1}^{i} A_j \right) = (\overline{Y_{i+2} - Y_{i+1}}) \cap Y_{i+1} = \{p_{n+1}\}.$$

It follows immediately from (5) of 10.27 that diameter $(A_i) \to 0$ as $i \to \infty$. Finally, the equation

$$(*) \quad X = X^{[1]} \cup \left(\bigcup_{i=1}^{\infty} A_i \right)$$

is readily proved as follows. First, since $\bigcup_{i=1}^{\infty} A_i = \bigcup_{i=1}^{\infty} Y_i$ and $\bigcup_{i=1}^{\infty} Y_i$

is a connected, dense set in X (10.27), $\bigcup_{i=1}^{\infty} A_i$ is a connected, dense set in X. Thus, since any connected, dense subset of any continuum Z must clearly contain all the cut points of Z, $\bigcup_{i=1}^{\infty} A_i$ must contain all the cut points of X. Thus by 10.7,

$$X - \bigcup_{i=1}^{\infty} A_i \subset X^{[1]}.$$

Hence, (*) is proved. This completes the proof of 10.28.

In our next application of 10.27, we show in 10.31 that dendrites have the fixed point property. It is convenient to prove a special case first (10.30). We use the following easy-to-prove general lemma about the "wedge" of two continua with the fixed point property.

10.29 Lemma. If X and Y are continua with the fixed point property such that $X \cap Y = \{p\}$, then $X \cup Y$ has the fixed point property.

Proof. Let $f: X \cup Y \to X \cup Y$ be continuous. Assume without loss of generality that $f(p) \in X$. Define $r: X \cup Y \to X$ by

$$r(z) = \begin{cases} z, & \text{if } z \in X \\ p, & \text{if } z \in Y \end{cases}$$

and note that r is continuous. Hence, $r \circ (f|X)$ is a continuous function from X into X. Thus, $r \circ (f|X)$ has a fixed point $q \in X$. It is easy to see that $f(q) = q$. This proves 10.29.

10.30 Lemma. Every tree has the fixed point property.

Proof. The proof is done by an easy induction on the number n of non-cut points of a tree (see 9.28). At the inductive step, 10.29 is used. The details are left to the reader.

10.31 Theorem. Every dendrite has the fixed point property.

Proof. Let X be a nondegenerate dendrite. By (1) and (5) of 10.27 and by 10.30, we may apply (g) of 1.19 to obtain that X has the fixed point property. This proves 10.31.

Three comments are appropriate. First, there are better theorems than 10.31. One is that every dendrite is an absolute retract ((iv) of [9, p. 339]–see Theorem 11 of [9, p. 343]), and another is in 10.54. Second, dendrites are the only one-dimensional Peano continua having the fixed point property (if X is a one-dimensional Peano continuum containing a simple closed curve S, use 13.54 to obtain a retraction from X onto S and then "rotate" S a little); in a similar vein, without considering dimension, see the characterization in 10.55. Third, every dendrite is tree-like, as our next result shows, but not every tree-like continuum has the fixed point property [1]. In connection with this, it is appropriate to remark that at the present time it is not known whether or not every tree-like continuum *in the plane* has the fixed point property (comp., [1, p. 12]).

10.32 Theorem.   Every dendrite is tree-like.

Proof.   Let X be a dendrite. Let $\epsilon > 0$. Then, by (1) and (5) of 10.27, there is a tree Y_ϵ in X such that the first point map $r_\epsilon\colon X \to Y_\epsilon$ is within ϵ of the identity map on X. Also, by 10.25, r_ϵ is continuous and $r_\epsilon(X) = Y_\epsilon$. It follows easily that r_ϵ is an ϵ-map (2.11). Therefore (by 2.12), we have proved that X is tree-like. This proves 10.32.

In connection with 10.31 and 10.32, see the comments near the end of 10.58.

The next result is stronger than 10.32 since it gives a particularly nice way to represent any given dendrite as an inverse limit of trees (comp., 2.13). It will be used in the verifications for 10.37.

10.33 Theorem.   Every nondegenerate dendrite satisfies the hypotheses of the Interior Approximation Theorem (2.29) in such a way that

$$\{Y_i\}_{i=1}^{\infty} \text{ in 2.29 satisfies 10.27}$$

and $g_i\colon Y_{i+1} \to Y_i$ in 2.29 is the identity on Y_i and, for p_i as in 10.27, $g_k^{-1} = Y_{i+1} - Y_i$.

Proof.   Let X be a nondegenerate dendrite. For each i, let Y_i and r_i be as in 10.27. To see that the hypotheses of 2.29 are satisfied, let

$$\varphi_i = r_i \quad \text{and} \quad g_i = r_i|Y_{i+1}\colon Y_{i+1} \to Y_i \qquad \text{for each } i.$$

Then: Use 10.25 for the continuity and ontoness of φ_i and g_i, use (2) of 10.27 to see that $g_i \circ \varphi_{i+1} = \varphi_i$ for each i, and note that, by (5) of 10.27, $\{\varphi_i\}_{i=1}^{\infty}$ converges uniformly to the identity map on X. Finally, the fact that each g_i has the other properties required in 10.33 follows immediately from the definition of r_i using (2) and (4) of 10.27. This completes the proof of 10.33.

4. A UNIVERSAL DENDRITE, PLANARITY OF DENDRITES

In general, for any class $\mathfrak{C}$ of spaces, it is of interest to know if there is a member of $\mathfrak{C}$, or a space which is minimal (in some sense), into which every member of $\mathfrak{C}$ can be embedded. Such a space is usually said to be *universal for* $\mathfrak{C}$. We have seen some examples of universal spaces (among others, 1.4, 1.11, and 9.36). In our next main result, which is actually in the form of an example, we show that there is a dendrite in R^2 which is universal for the class $\mathfrak{D}$ of all dendrites. Therefore, the example gives us two results at the same time: First, there is a universal dendrite (i.e., a member of $\mathfrak{D}$ which is universal for $\mathfrak{D}$) and, second, every dendrite is embeddable in R^2 (i.e., R^2 is universal for $\mathfrak{D}$). The example is in 10.37. Most of the results to be used for the verifications in 10.37 have already been proved. We need one more (10.36). First, note the following definition and theorem which will be used in the proof of 10.36.

10.34   Definition of Unicoherent.   A connected topological space S is said to be *unicoherent* provided that whenever A and B are closed, connected subsets of S such that $S = A \cup B$, then $A \cap B$ is connected. A connected topological space is said to be *hereditarily unicoherent* provided that each of its closed, connected subsets is unicoherent.

The following theorem is really just a corollary to 10.10.

10.35   Theorem.   A Peano continuum X is a dendrite if and only if X is hereditarily unicoherent.

Proof.   Clearly, by 10.10, every dendrite is hereditarily unicoherent. Conversely, since a simple closed curve is not unicoherent, every hereditarily unicoherent Peano continuum is a dendrite (by 10.1). This proves 10.35.

We mention that some other characterizations of dendrites within the class of Peano continua are in 10.46, 10.50, (c) of 10.53, and 10.55.

The following theorem is in [14] along with some more general results.

10.36   Theorem.   Any inverse limit of dendrites with monotone bonding maps (not necessarily onto) is a dendrite.

Proof.   Let $X_\infty = \varprojlim\{X_i, f_i\}_{i=1}^\infty$ where each X_i is a dendrite and each $f_i \colon X_{i+1} \to X_i$ is monotone (8.21). Let A and B be subcontinua of X_∞, and let $C = A \cap B$. For each i, let

$C_i = \pi_i(A) \cap \pi_i(B)$, $\pi_i \colon X_\infty \to X_i$ the ith projection map.

Then, by 2.19,

(1) $C = \varprojlim\{C_i, f_i | C_{i+1}\}_{i=1}^\infty.$

We shall show that C is connected. For this purpose, we may assume $C \neq \varnothing$. Then, clearly, each $C_i \neq \varnothing$. Hence, applying 10.35 to the two continua $\pi_i(A)$ and $\pi_i(B)$, we have that each C_i is a continuum. Thus, by (1) and 2.4, C is a continuum. Hence, it follows that we have proved

(2) X_∞ is hereditarily unicoherent.

Now, we show that X_∞ is a Peano continuum. (Note: If each f_i maps onto X_i, then X_∞ is a Peano continuum by 8.47.) By 2.6,

$X_\infty = \varprojlim\{\pi_i(X_\infty), f_i | \pi_{i+1}(X_\infty)\}_{i=1}^\infty$

and, using the hereditary unicoherence of each X_{i+1} (10.35), it is easy to verify that each $f_i | \pi_{i+1}(X_\infty)$ is a monotone map [onto $\pi_i(X_\infty)$ by 2.6]. Hence, by 8.47, X_∞ is a Peano continuum. Therefore, by (2) and 10.35, X_∞ is a dendrite. This proves 10.36.

We are now prepared to present the following interesting example from [17].

10.37 Wazewski's Universal Dendrite. We give an example of a dendrite D_∞ in R^2 which contains a topological copy of any dendrite whatsoever. Hence, we see in particular that every dendrite is embeddable in R^2. We construct D_∞ using inverse limits. We use 10.36 to know D_∞ is a dendrite, and we use 2.10 to know D_∞ is embeddable in R^2. We then show that D_∞ is universal (contains every dendrite). We break all this down into four steps.

Step 1: Construction of D_∞. Let $c \in R^2$. By a *star* with *center* c and *beams* B_j we mean a countable union $\bigcup_{j=1}^\infty B_j$ of convex arcs B_j in R^2, each having c as one end point, such that

$B_j \cap B_k = \{c\}$ when $j \neq k$ and diameter $(B_j) \to 0$ as $j \to \infty$.

Let D_1 be a star. Let m_j denote the midpoint of each beam B_j of D_1. For each j, form a star S_j with center m_j and otherwise disjoint from D_1, making sure that

$S_j \cap S_k = \varnothing$ when $j \neq k$ and diameter $(S_j) \to 0$ as $j \to \infty$.

Let $D_2 = D_1 \cup (\bigcup_{j=1}^\infty S_j)$. A picture of D_2 is drawn in Figure 10.37. Define D_3 in a similar manner putting a star at each midpoint of the two convex subarcs of each B_j determined by m_j and at the midpoint of each

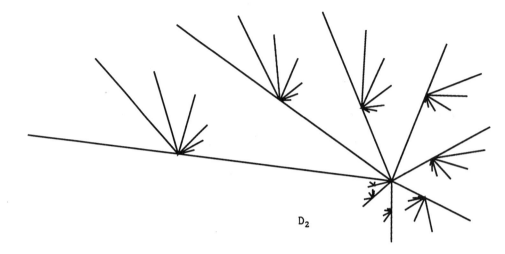

D_2

Figure 10.37

beam of each S_j as well (making sure that the new stars are disjoint from one another, intersect D_2 only in their centers, and get smaller and smaller). Continuing in this fashion, we obtain D_i for each $i = 1, 2, \ldots$. Note that each D_i is a dendrite and that $D_i \subset D_{i+1}$ for each i. Now, we define maps $f_i \colon D_{i+1} \rightarrow D_i$. The maps $f_i \colon D_{i+1} \rightarrow D_i$ are natural—f_i is the identity on D_i and sends each star added to D_i in obtaining D_{i+1} to its center [e.g., $f_1(x) = x$ if $x \in D_1$ and $f_1(x) = m_j$ if $x \in S_j$]. As is evident, each f_i is a monotone map. Let

$$D_\infty = \varprojlim\{D_i, f_i\}_{i=1}^\infty.$$

Step 2: D_∞ is a dendrite. This follows immediately from Step 1 and 10.36.

Step 3: D_∞ is embeddable in R^2. This is a matter of showing that the construction in Step 1 can be carried out with extra control so that (1) and (2) of 2.10 are satisfied. We leave these details to the reader (10.41).

Step 4: D_∞ is universal for dendrites. Assume Step 3 has been carried out as indicated (using 2.10). Then, since $D_i \subset D_{i+1}$ for each i, we may assume by 2.10 that

(1) $\quad D_\infty = \overline{\bigcup_{i=1}^\infty D_i}.$

Let $C_1 = \{c\}$ where c is the center of the star D_1 and, for each $i = 2$, $3, \ldots$, let C_i denote the set of all the centers of stars in $\bigcup_{j=i}^{\infty} D_j$ which are not centers of any star in D_{i-1}. Let $C = C_1 \cup C_2$. It is clear that C is the set of all the branch points of D_∞ and that (2) and (3) below hold:

    (2) $\mathrm{ord}(z, D_\infty) = \aleph_0$ for all $z \in C$ (recall 10.20.1 to know that $\mathrm{ord}(z, D_\infty) \leq \aleph_0$ for all $z \in C$);

    (3) For each $i \geq 2$, $C_i \cap (A - \{x, y\})$ is a countable, dense subset of any arc A in D_∞ with end points x and y.

We shall also use the following notation, where f_i is as in Step 1:

$$f_{i,j} = f_i \circ \cdots \circ f_{j-1}\colon D_j \to D_i \text{ if } j > i + 1 \quad \text{and} \quad f_{i\,i+1} = f_i.$$

Now, we prove D_∞ is universal. The reader should remember that $D_i \subset D_{i+1}$ for each i and that (1) holds since these facts will be used implicitly in the proof. Let X be a nondegenerate dendrite. We show that X is embeddable in D_∞ by using 2.22. Let

$$X = \varprojlim \{Y_i, g_i\}_{i=1}^{\infty}$$

as guaranteed by 10.33. Let B denote the set of all the branch points of X. By 10.23,

    (4) $|B| \leq \aleph_0$.

Recall from (4) of 10.27 that $Y_1 = \{p_1\}$. Let $n(1) = 1$. By (3), we may let $h_1\colon Y_1 \to D_{n(1)}$ be such that $h_1(p_1) \in C_2$. Now, by (4) of 10.27, Y_2 is an arc one of whose end points is p_1. Let $Y_2' = Y_2 - \{p_1, q_1\}$ where q_1 is the end point of Y_2 different from p_1. Even though $B \cap Y_2'$ may not be dense in Y_2', (4) allows us to obtain a countable, dense subset E_2 of Y_2' such that $E_2 \supset B \cap Y_2'$. Then, since $h_1(p_1) \in D_{n(1)} \cap C_2$, (3) allows us to apply (e) of 6.22 to obtain an embedding h_2 of Y_2 in $D_{n(2)}$ for some $n(2) > n(1)$ such that

$$h_2(B \cap Y_2') \subset C_{n(2)+1}, \quad h_2(p_2) \in C_{n(2)+1},$$
$$h_2(p_1) = h_1(p_1), \quad \text{and} \quad h_2(Y_2 - Y_1) \subset D_{n(2)} - D_{n(1)}.$$

Note that the following diagram is commutative (use 10.33):

$$
\begin{array}{ccc}
 & g_1 & \\
Y_1 & \longleftarrow & Y_2 \\
h_1 \downarrow & \circlearrowright & \downarrow h_2 \\
D_{n(1)} & \longleftarrow & D_{n(2)} \\
 & f_{n(1)n(2)} &
\end{array}
$$

Now, by (4) of 10.27, $\overline{Y_3 - Y_2}$ is an arc A_3 one of whose end points is p_2. Let $A_3' = A_3 - \{p_2, q_2\}$ where q_2 is the end point of A_3 different from p_2. Then, making use of (4) as we did before, we obtain a countable, dense subset E_3 of A_3' such that $E_3 \supset B \cap A_3'$. Then, since $h_2(p_2) \in D_{n(2)} \cap C_{n(2)+1}$, (3) allows us to apply (e) of 6.22 to obtain an embedding h_3 of Y_3 in $D_{n(3)}$ for some $n(3) > n(2)$ such that

$$h_3(B \cap A_3') \subset C_{n(3)+1}, \quad h_3(p_3) \in C_{n(3)+1},$$
$$h_3|Y_2 = h_2, \quad \text{and} \quad h_3(Y_3 - Y_2) \subset D_{n(3)} - D_{n(2)}.$$

Note that the following diagram is commutative (use 10.33):

$$
\begin{array}{ccc}
 & g_2 & \\
Y_2 & \longleftarrow & Y_3 \\
h_2 \downarrow & \circlearrowright & \downarrow h_3 \\
D_{n(2)} & \longleftarrow & D_{n(3)} \\
 & f_{n(2)n(3)} &
\end{array}
$$

Continuing as above, but using (2) in conjunction with (3) to define h_i for each $i \leq 4$, we obtain the diagram below where each rectangle is commutative, each h_i is an embedding of Y_i in $D_{n(i)}$, and $n(i + 1) > n(i)$ for each i:

$$
\begin{array}{ccccccccccc}
 & g_1 & & & & & & g_i & & & \\
Y_1 & \longleftarrow & Y_2 & \longleftarrow \cdots \longleftarrow & Y_i & \longleftarrow & Y_{i+1} & \longleftarrow \cdots & X \\
h_1 \downarrow & \circlearrowright & \downarrow h_2 & & h_i \downarrow & \circlearrowright & \downarrow h_{i+1} & & \\
D_{n(1)} & \longleftarrow & D_{n(2)} & \longleftarrow \cdots \longleftarrow & D_{n(i)} & \longleftarrow & D_{n(i+1)} & \longleftarrow \cdots & Z \\
 & f_{n(1)n(2)} & & & & f_{n(i)n(i+1)} & & &
\end{array}
$$

Hence, the induced map $h_\infty : X \to Z$ as defined in 2.22 exists. By (a)–(c) of 2.22, h_∞ is an embedding of X in Z. By 2.37, Z is homeomorphic to D_∞. Thus, X is embeddable in D_∞. Therefore, we have proved that D_∞ is universal for dendrites. This completes 10.37.

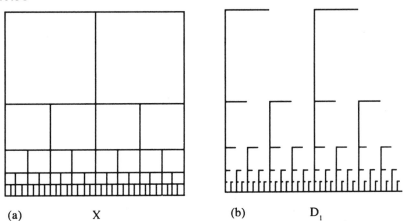

(a)   X   (b)   D₁

Figure 10.38 (From Ref. 9, with permission of Academic Press and Panstwowe Wydawnictwo Naukowe, Warsaw.)

Some dendrites which are universal within certain classes of dendrites are discussed in 10.49. We also remark that a universal arc-like continuum is constructed in 12.22.

Finally, we note that any dendrite D whose set of branch points is continuumwise dense in D and each of whose branch points is of infinite order in D is homeomorphic to the dendrite D_∞ in 10.37 [17, pp. 123–124].

5. TWO EXAMPLES

We conclude our discussion of dendrites by giving two examples. The first one concerns unions of dendrites.

It is clear that a continuum which is the union of two dendrites need not be a dendrite (e.g., S^1 is the union of two arcs). The following example shows that a continuum which is the union of two dendrites need not even be hlc (see 10.5). We remark that a continuum which is the union of two Peano continua must, however, be a Peano continuum (8.51).

10.38 Example. We give an example of a non-hlc continuum X which is the union of two dendrites. The continuum X is drawn in Figure 10.38(a), and is defined analytically as follows. Let

$$A_0 = \{(x,0) \in R^2: 0 \leq x \leq 1\};$$

for each $n = 1, 2, \ldots$, let

$A_n = \{(x, 2^{-n+1}) \in R^2 \colon 0 \leq x \leq 1\};$

for each $n = 0, 1, 2, \ldots$ and each m such that $0 \leq m \leq 2^{n+1}$, let

$B_{n,m} = \{(m \cdot 2^{-n-1}, y) \in R^2 \colon 0 \leq y \leq 2^{-n}\};$

finally, let

$$X = \left[\bigcup_{n=0}^{\infty} A_n \right] \cup \left[\bigcup_{n=0}^{\infty} \left(\bigcup_{m=0}^{2^{n+1}} B_{n,m} \right) \right].$$

It is clear that X is a non-hlc continuum; also, X is the union of two dendrites D_1 and D_2, symmetric with respect to the line $x = 1/2$, where D_1 is as drawn in Figure 10.38(b). We remark that the example also shows that the union of two regular continua need not be regular or even hlc (see 10.16 and 10.59).

We have seen that the set of all branch points of a dendrite is countable (10.23). The same is not always true for the set of all end points. Actually, Wazewski's universal dendrite D_∞ (10.37) has uncountably many end points (10.45). A somewhat clearer example illustrating this phenomenon is given below:

10.39 Example. Let C denote the Cantor Middle-Third set (7.5), and let G denote the dendrite drawn below in Figure 10.39. Briefly, G is the closure of the union of countably many line segments of slope ± 1, each of which joins a suitably chosen point above the x-axis to an appropriate so-called "end point" of C (meaning an end point of a maximal subinterval of one of the sets C_i in 7.5). It is evident that G is a dendrite and that $G^{[1]} = C$, so $G^{[1]}$ is uncountable.

In the process of studying dendrites, we have been led to a few results and examples involving hlc continua and regular continua. These continua are one-dimensional (by 13.57), and one-dimensional continua are called *curves*. In this book, we discuss in depth only three classes of curves: graphs, dendrites, and arc-like continua. By necessity, any kind of thorough treatment of curve theory would consume several lengthy volumes. We refer the interested reader to [9, pp. 274–307] and [19, pp. 89–99] for more information about curves. We shall briefly mention two other types of curves in 10.57 and 10.58.

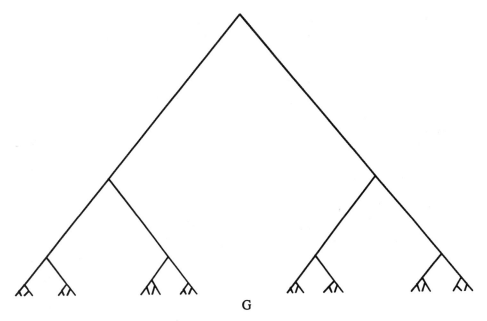

G

Figure 10.39

EXERCISES

10.40 Exercise. Verify the statement in the first line of the proof of 10.27 by proving 10.27 for the case when X itself is a nondegenerate tree. [Hint: X has an end point e (why?), and e has a neighborhood which is an arc (why?).]

10.41 Exercise. Do the details for Step 3 of 10.37.

10.42 Exercise. Let X be a nondegenerate dendrite. The set of all cut points of X is continuumwise dense in X (by 10.8), and each point of order 2 in X is a cut point of X (by 10.7). Improve the first of these facts by proving the following result: If X is any nondegenerate dendrite, then the set of all points of order 2 in X is continuumwise dense in X. [Hint: Use 10.5, 8.23, and 10.23.] Give an example of a dendrite X and a dense subset Z of X such that Z is not continuumwise dense in X. On the other hand, note that any connected, dense subset of any dendrite X is continuumwise dense in X (why?).

10.43 Exercise. Prove that if X is a dendrite, then, for each $\epsilon > 0$, there are finitely many dendrites $X_1, \ldots, X_n$ such that (1)–(4) hold:
(1) $X = \bigcup_{i=1}^{n} X_i$;
(2) diameter $(X_i) < \epsilon$;
(3) $|X_i \cap X_j| \leq 1$ when $i \neq j$;
(4) $X_i \cap X_j \cap X_k = \varnothing$ when $i \neq j$, $i \neq k$, and $j \neq k$.
[Hint: Using 10.42, find a finite set F of points of order 2 in X such that each component U_i of $X - F$ has diameter $< \epsilon$. There are only finitely many such U_i (why?), say $U_1, \ldots, U_n$ with $U_i \neq U_j$ for $i \neq j$. Let $X_i = \bar{U}_i$ for each i (recall 10.6). Use 10.35 to prove (3). Use that ord$(x,X) = 2$ for each $x \in F$ to prove (4).]

10.44 Exercise. Prove that if X is a nondegenerate dendrite and $p \in X$, then $p \in X^{[1]}$ (6.25) if and only if p is an end point of every arc in X to which p belongs. [Hint: Use 10.7, 10.6, and 6.3.]
We remark that the equivalence is true for any nondegenerate Peano continuum X (Theorem 15 of [9, p. 320]). A point p of an arcwise connected continuum X which is an end point of every arc in X to which p belongs is called an *end point of X in the classical sense*. The set of such points has been examined for dendroids in [10].

10.45 Exercise. Verify the statement above 10.39 that Wazewski's universal dendrite D_∞ (10.37) has uncountably many end points.

10.46 Exercise. In 10.35, we gave a characterization of dendrites within the class of Peano continua. We give some more such characterizations here, all of which are somewhat similar to one another. Prove that a Peano continuum X is a dendrite if and only if any one of (1)–(3) holds:
(1) Every convergent sequence of arcs in X converges to an arc or a one-point set; (2) If $\{A_i\}_{i=1}^{\infty}$ is a sequence of arcs in X such that $A_i \subset A_{i+1}$ for each i, then $\bigcup_{i=1}^{\infty} A_i$ is contained in an arc; (3) If f is any one-to-one continuous function from $[0,1)$ into X, then $\overline{f([0,1))}$ is an arc. [Hint: In proving that (1) implies X is a dendrite, 8.30 will be useful as will the fact that any nondegenerate subcontinuum of a dendrite which is not an arc must contain a simple triod (why?).]

10.47 Exercise. Prove that every point p of a nondegenerate dendrite X lies in a maximal arc M in X (i.e., $p \in M$ and the arc M is not properly contained in any arc in X). [Hint: Use 10.46 and 4.34.]

10.48 Exercise. A general characterization of trees is in 9.28. Prove that for a dendrite X, (1)–(4) are equivalent: (1) X is a tree; (2) X is a graph; (3) X has only finitely many end points; (4) X is the union of finitely many arcs. [Hint: All the work has been done–it is simply a matter of collecting appropriate facts from Chapter IX and this chapter.]

10.49 Exercise. For each cardinal number n, $3 \le n \le \aleph_0$, let $\mathfrak{C}(n)$ denote the following collection of dendrites:

$$\mathfrak{C}(n) = \{D: D \text{ is a dendrite and } \operatorname{ord}(x,D) \le n \text{ for all } x \in D\}.$$

By a *universal dendrite for* $\mathfrak{C}(n)$, $3 \le n \le \aleph_0$, we mean a dendrite $U_n \in \mathfrak{C}(n)$ such that any dendrite $D \in \mathfrak{C}(n)$ can be topologically embedded in U_n. Since $\mathfrak{C}(\aleph_0)$ consists of all dendrites (10.20.1), we constructed a universal dendrite for $\mathfrak{C}(\aleph_0)$ in 10.37. By a procedure analogous to the one used in 10.37, we may construct a universal dendrite for $\mathfrak{C}(n)$ for each integer $n \ge 3$: Start with a simple n-od (9.8), and then use simple $(n-2)$-ods (as defined in 9.8 when $n - 2 \ge 3$; here, we take a simple 1-od to be a convex arc and a simple 2-od to be an arc shaped like the letter V). Carry out the details. We mention that some interesting results about these types of universal dendrites are in [3]–see, especially, the Corollary in [3, p. 493].

10.50 Exercise. Prove the following result, half of which is 10.32: A Peano continuum is tree-like if and only if it is a dendrite. [Hint: It suffices (why?) to prove S^1 is not tree-like. Prove this using only 2.12 and the fact that trees are unicoherent (10.35).]

10.51 Exercise. Prove that every dendrite is contractible. [Hint: Given what we have done in the chapter, there are two approaches. First: 10.27 can be used as follows. Let X, Y_i, p_i, and r_i be as in 10.27. Given a homeomorphism φ_i from $[0,1]$ onto $\overline{Y_{i+1} - Y_i}$ with $\varphi_i(1) = p_i$, define α_i on $Y_{i+1} \times [0,1]$ by

$$\alpha_i(x,t) = \begin{cases} x, & x \in Y_i \text{ and } 0 \le t \le 1 \\ x, & x = \varphi_i(s) \text{ where } t \le s \le 1 \\ \varphi_i(t), & x = \varphi_i(s) \text{ where } 0 \le s \le t \end{cases}$$

Then, a desired contraction of X to p_1 can be obtained by pasting the maps $\alpha_i(r_{i+1},\cdot)$ together appropriately. The second approach is more geometric. Let c denote the center of the star D_1 in 10.37. Show that Step 3 of 10.37 can be done so that, in addition to (1) of Step 4, D_∞ has the following property: If $x \in D_\infty$, $x \ne c$, and A is the unique arc in D_∞ from c to x, then, for any $y \in A - \{c,x\}$,

$$\|c - y\| < \|c - x\|.$$

Then, it is easy to see that D_∞ is contractible. Thus, since the map r in 10.26 is a retraction, every subcontinuum of D_∞ is contractible (prove that any retract of a contractible space is contractible). Therefore, since D_∞ is universal (10.37), every dendrite is contractible.]

We mention that the result in this exercise is a simple consequence of the fact that every dendrite is an absolute retract ((iv) of [9, p. 339]).

10.52 Exercise. Prove that if X is a dendrite and f is a monotone map from X onto a continuum Y, then Y is a dendrite. [Hint: Use 8.46 and 10.10.] The analogous result for open maps will be proved in 13.41. Also, see 13.35 ($r(X)$ is defined in 13.32).

10.53 Exercise. Let X be a continuum. If $\mathfrak{C} \subset 2^X$, then a *selection for* $\mathfrak{C}$ is a function $\sigma\colon \mathfrak{C} \to X$ such that $\sigma(A) \in A$ for all $A \in \mathfrak{C}$. The Axiom Of Choice guarantees the existence of a selection for any $\mathfrak{C} \subset 2^X$. However, the existence of a continuous selection is quite another matter. Prove (a)–(f) below; a characterization of dendrites is in (c), and a characterization of the arc is in (f).

(a) There does not exist a continuous selection for $C(S^1)$. [Hint: Use 4.29 and the fact that S^1 is not a retract of B^2 (which follows from 12.39 since, if r is a retraction of B^2 onto S^1, $-r\colon B^2 \to B^2$ would have no fixed point).]

(b) If X is a dendrite, then there is a continuous selection for $C(X)$. [Hint: Fix a point $p \in X$. For each $K \in C(X)$ such that $p \notin K$, let $\sigma(K) = r_K(p)$ where r_K is the first point map for K (10.26); if $p \in K \in C(X)$, let $\sigma(K) = p$. Prove that σ is continuous.] See (c).

(c) A Peano continuum X is a dendrite if and only if there is a continuous selection for $C(X)$. [Hint: Use (a) and (b)—you will also need 4.7.] See (d).

(d) There is a continuous selection for $C(X)$ when X is the cone over $\{0\} \cup \{1/n\colon n = 1, 2, \ldots\}$. Thus, (c) would be false without the assumption that X is a Peano continuum.

(e) There is a continuous selection for 2^I where $I = [0,1]$.

(f) If X is a nondegenerate continuum, then there is a continuous selection for $F_2(X) = \{A \in 2^X\colon |A| \le 2\}$ if and only if X is an arc. [Hint: If σ is a continuous selection for $F_2(X)$, then, for $x, y \in X$, define $x <_\sigma y$ to mean $x \ne y$ and $\sigma(\{x,y\}) = x$; show that $<_\sigma$ is a simple ordering (6.14) and that the order topology (6.14) equals the given topology on X—then, apply 6.16 and 6.17.]

For more on continuous selections in relation to (a)–(f), see Chapter V of [13] and the references contained therein. Also, see (b) of 10.58.

10.54 Exercise. Prove that if X is a dendrite and $F: X \to 2^X$ is continuous, then there is a point $p \in X$ such that $p \in F(p)$. [Hint: First, prove the result for trees by an easy induction on the number of non-cut points (9.28)–at the inductive step, "chase points along an appropriately chosen arc" (an analogue of the hint for (e) of 1.19—the dog now chases a bunch of rabbits). Then, use 10.27 together with the observation that the maps

$$r_i^*: 2^X \to 2^{Y_i} \, (4.27), \qquad r_i: X \to Y_i \text{ as in } 10.27,$$

converge uniformly to the identity map on 2^X.]

The theorem is due to Plunkett [15], and is obviously stronger than 10.31. A converse is in 10.55, and a similar result for arc-like continua is in 12.56. Finally, we remark that the proof we have suggested above is substantially simpler than the one given in [15].

10.55 Exercise. Prove that if X is a Peano continuum, then (1)–(3) are equivalent: (1) X is a dendrite; (2) every continuous function $F: X \to 2^X$ has a fixed point in the sense of 10.54 (i.e., a point p such that $p \in F(p)$); (3) same as (2) but with 2^X replaced by $C(X)$. [Hint: (1) implies (2) by 10.54. Clearly, (2) implies (3). To prove (3) implies (1), assume that X is a Peano continuum which is not a dendrite, and use 4.29 together with the fact that a 2-cell is an absolute extensor (defined following the proof of 13.49).]

The theorem is due to Plunkett [15, Theorem 3]. There are continua satisfying (2) which are not Peano continua–see 12.56.

10.56 Exercise. We define the notion of 0-regular convergence and characterize dendrites in terms of it. Let X be a continuum, and let $\{Y_i\}_{i=1}^{\infty}$ be a sequence of subcontinua of X. Then: $\{Y_i\}_{i=1}^{\infty}$ is said to *converge 0-regularly to* Y, written

$$Y_i \to Y \quad \text{0-reg.,}$$

provided that $\lim Y_i = Y$ (4.12) and, for each $\epsilon > 0$, there exist $\delta(\epsilon)$ and $N(\epsilon)$ such that for each $i \geq N(\epsilon)$, any two points of Y_i less than $\delta(\epsilon)$ apart lie together in a connected subset of Y_i of diameter $< \epsilon$. Prove (a) and (b) below.

(a) If $\{Y_i\}_{i=1}^{\infty}$ is a sequence of subcontinua of a continuum X and $Y_i \to Y$ 0-reg., then Y is a Peano continuum. [Hint: Use (a) of 8.42].

(b) A continuum X is a dendrite if and only if convergence (4.12) is equivalent to 0-regular convergence for sequences of subcontinua of X. [Hint: For the "if" part, show X satisfies 10.1 by using (a) in a trivial way to see that X is a Peano continuum and by finding a convergent sequence of arcs in S^1 which does not converge 0-regularly. For the "only if" part, make use of 8.30 and 10.35.] The notion of 0-regular convergence is a special case of the notion of r-regular convergence due to Whyburn [20]. The theorem in (b) is essentially due to White [18, 7.32]. For more information, see [13, pp. 518–522] and [19, pp. 174–181]. To gain a little more familiarity with the notion of 0-regular convergence, work (c) and (d) below.

(c) Give an example of a sequence $\{A_i\}_{i=1}^{\infty}$ of arcs in a continuum X such that $\lim A_i$ (4.12) is an arc A but $A_i \nrightarrow A$ 0-reg..

(d) Prove that if $\{A_i\}_{i=1}^{\infty}$ is a sequence of arcs in a continuum X such that $A_i \rightarrow A$ 0-reg., then A is an arc or a one-point set. [Hint: First show that if $A_{i(j)} = B_{i(j)} \cup C_{i(j)}$ where $B_{i(j)}$ and $C_{i(j)}$ are arcs such that $B_{i(j)} \cap C_{i(j)} = \{a_{i(j)}\}$ and $\lim B_{i(j)}$, $\lim C_{i(j)}$, and $\lim\{a_{i(j)}\}$ exist, then

$$[\lim B_{i(j)}] \cap [\lim C_{i(j)}] = \lim\{a_{i(j)}\}.$$

Show how to use this fact to apply 9.29.]

10.57 Exercise. A continuum X is said to be *rational* provided that each point p of X has a local base $\mathcal{L}_p$ such that the boundary of each member of $\mathcal{L}_p$ is at most countable. Clearly, regular continua are rational. Prove the following analogue of 10.19: A continuum X is rational if and only if any two points of X are separated in X by some countable set. We remark that hlc continua are rational; a proof is in [19, p. 94].

10.58 Exercise. A *dendroid* is an arcwise connected, hereditarily unicoherent continuum. Thus, by 8.23 and 10.35, a Peano continuum is a dendroid if and only if it is a dendrite. Prove (a) and (b) below.

(a) Every subcontinuum of a dendroid is a dendroid. See 11.54.

(b) If X is a continuum such that there is a continuous selection for $C(X)$, then X is a dendroid. [Hint: Use (d) of 5.25 and 8.28 to get that every subcontinuum of X is arcwise connected. Use this and (a) of 10.53 to show that X is hereditarily unicoherent.]

The converse of (b) is false [12, p. 372]–we note that the attempted simplification of this done in 5.10 of [13, p. 258] contains a basic error (see [5]).

We have shown that dendrites have the fixed point property and are tree-like (10.31, 10.32). The same is also true of dendroids ([2], [4]). However, it has been an open question [6, p. 261] if for any dendroid X and any $\epsilon > 0$, there is a tree T_ϵ in X and a retraction r_ϵ from X onto T_ϵ such that r_ϵ is within ϵ of the identity map on X (comp., 10.27). Recently, Robert Cauty has announced an affirmative answer.

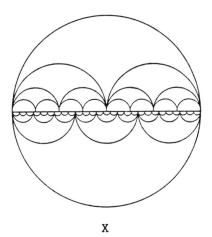

X

Figure 10.59 (From Ref. 9, with permission of Academic Press and Panstwowe Wydawnictwo Naukowe, Warsaw.)

In connection with 10.37, we remark that there is no universal dendroid [7].

For results about special types of maps on dendroids, see 13.40 and 13.41.

10.59 Exercise. We give an example of an hlc continuum X which is not regular (thus showing that the converse of 10.16 is false). The continuum X is drawn in Figure 10.59, and is defined analytically as follows. Let

$$A = \{(x,0) \in R^2: 0 \leq x \leq 1\};$$

for each $n = 1, 2, \ldots$ and each $k = 1, \ldots 2^{n-1}$, let

$$B_{n,k} = \left\{(x,y) \in R^2: \left(x - \frac{2k-1}{2^n}\right)^2 + y^2 = 4^{-n} \text{ and } y \geq 0\right\};$$

for each $n = 0, 1, 2, \ldots$ and each $k = 1, \ldots, 3^n$, let

$$C_{n,k} = \left\{(x,y) \in R^2: \left(x - \frac{2k-1}{2 \cdot 3^n}\right)^2 + y^2 = \frac{1}{4} \cdot 9^{-n} \text{ and } y \leq 0\right\};$$

let

$$X = A \cup \left[\bigcup_{n=1}^{\infty} \left(\bigcup_{k=1}^{2^{n-1}} B_{n,k}\right)\right] \cup \left[\bigcup_{n=0}^{\infty} \left(\bigcup_{k=1}^{3^n} C_{n,k}\right)\right].$$

Verify that X has the desired properties. In addition, show that X is the union of two regular continua. Thus, an hlc continuum can be the union of two regular continua and, yet, fail to be regular. In this sense, the example here is stronger than 10.38. In another sense (the continua D_1 and D_2 in 10.38 being dendrites), 10.38 is a stronger example than this one (see 10.20).

REFERENCES

1. David P. Bellamy, A tree-like continuum without the fixed-point property, Houston J. Math, 6(1980), 1–13.

2. K. Borsuk, A theorem on fixed points, Bull. Pol. Acad. Sci., 2(1954), 17–20.

3. J. J. Charatonik, Open mappings of universal dendrites, Bull. Pol. Acad. Sci., 28(1980), 489–494.

4. H. Cook, Tree-likeness of dendroids and λ-dendroids, Fund. Math., 68(1970), 19–22.

5. S. T. Czuba, On dendroids with Kelley's property, Proc. Amer. Math. Soc., 102(1988), 728–730.

6. J. B. Fugate, Small retractions of smooth dendroids onto trees, Fund. Math., 71(1971), 255–262.

7. J. Krasinkiewicz and P. Minc, Nonexistence of universal continua for certain classes of curves, Bull. Acad. Polon. Sci. Ser. Sci. Math. Astronom. Phys., 24(1976), 733–741.

8. K. Kuratowski, *Topology*, Vol. I, Academic Press, New York, N.Y., 1966.

9. K. Kuratowski, *Topology*, Vol. II, Academic Press, New York, N.Y., 1968.

10. A. Lelek, On plane dendroids and their end points in the classical sense, Fund. Math., 49(1961), 301–319.

11. R. L. Moore, Concerning the cutpoints of continuous curves and other closed and connected point-sets, Proc. Nat. Acad. Sci., 9(1923), 101–106.

12. Sam B. Nadler, Jr. and L. E. Ward, Jr., Concerning continuous selections, Proc. Amer. Math. Soc., 25(1970), 369–374.

13. Sam B. Nadler, Jr., *Hyperspaces of Sets*, Monographs and Text-

books in Pure and Applied Math., vol. 49, Marcel Dekker, Inc., New York, N.Y., 1978.

14. Sam B. Nadler, Jr., Multicoherence techniques applied to inverse limits, Trans. Amer. Math. Soc., 157(1971), 227–234.

15. Robert L. Plunkett, A fixed point theorem for continuous multi-valued transformations, Proc. Amer. Math. Soc., 7(1956), 160–163.

16. L. E. Ward, Jr., A new characterization of trees, Proc. Amer. Math. Soc., 104(1988), 1252–1255.

17. Tadé Wazewski, Sur les courbes de Jordan ne renfermant aucune courbe simple fermée de Jordan, Polskie Tow. Mat. Annales Rocznik, 2(1923), 49–170.

18. Paul A. White, r-regular convergence spaces, Amer. J. Math., 66(1944), 69–96.

19. Gordon Thomas Whyburn, *Analytic Topology*, Amer. Math. Soc. Colloq. Publ., vol. 28, Amer. Math. Soc., Providence, R.I., 1942.

20. Gordan Thomas Whyburn, On sequences and limiting sets, Fund. Math., 25(1935), 408–426.

XI

Irreducible Continua

Recall from 4.35 that a continuum X is said to be *irreducible* provided that there exist p, $q \in X$ such that no proper subcontinuum of X contains $\{p,q\}$, in which case, when we wish to specify p and q, we say X is *irreducible between p and q*. We see from (b) of 4.35 that any nondegenerate continuum contains many nondegenerate irreducible continua. In view of this abundance of irreducible continua, it is clear that knowledge about them is of interest and would be useful in investigating continua in general. In connection with this, we mention that members of some important classes of continua are irreducible (e.g., indecomposable continua, 11.18, and arc-like continua, 12.5); furthermore, their irreducibility plays a significant role in the study of their other properties. Also, we have seen examples of the use of irreducible continua in investigating properties of general continua (e.g., in 6.27).

We ask the reader to recall the definition of a composant $\kappa(p)$ in 5.20. Clearly,

$$\kappa(p) = \{x \in X: X \text{ is not irreducible between } p \text{ and } x\}.$$

Throughout this chapter, the symbol $\kappa(p)$ will always denote the composant of p in the given continuum under consideration. The general relationship between irreducibility and composants is evident and is stated in 11.3 (in more precise fashion, in 11.2). Thus, the study of irreducible continua necessitates the study of composants, and conversely.

In this chapter, we begin with a few of the more elementary properties of irreducible continua (11.2–11.9). We then show that a nondegenerate continuum has either one, three, or uncountably many composants (11.16; also, see 11.13 and 11.15). In the process, we obtain the result that indecomposable continua are irreducible (11.15.1). This result is improved in 11.18, which we then use to obtain two characterizations of indecomposability in terms of irreducibility (11.19 and 11.20). The second of these characterizations is especially significant in relation to the procedure used in constructing the continuum in 1.10. In the remainder of the chapter, we obtain two characterizations of irreducible continua (11.21 and 11.35). The first of these is a complete characterization. The second is a characterization for unicoherent continua. We remark that we shall apply the nontrivial half of 11.35, namely 11.34, in Chapter XII to prove that arc-like continua are irreducible (12.5).

1. SOME ELEMENTARY PROPERTIES OF IRREDUCIBLE CONTINUA

Let us note the following definition (recall that we have already defined what it means for a continuum to be irreducible between p and q).

11.1 Definition of Point of Irreducibility. If X is a continuum and $p \in X$, then p is called a *point of irreducibility of* X provided that X is irreducible between p and some point $q \in X$.

We note the following simple theorem and corollary.

11.2 Theorem. If X is a nondegenerate continuum and $p \in X$, then p is a point of irreducibility of X if and only if the composant $\kappa(p) \neq X$.

Proof. Clear from definitions in 4.35, 5.20, and 11.1.

11.3 Corollary. A nondegenerate continuum X is irreducible if and only if some composant of X is a proper subset of X.

Proof. Immediate from 11.2.

From the definition in 5.20, it is clear that any composant is connected [comp., (a) of 5.20]. We now show that the complement of any composant is connected.

11.4 Theorem. If X is a continuum and $p \in X$, then $X - \kappa(p)$ is connected (possibly empty), i.e.,

$$\{x \in X: X \text{ is irreducible between } p \text{ and } x\}$$

is connected.

Proof. If $\kappa(p) = X$, then $X - \kappa(p) = \varnothing$ and, thus, is connected. Hence, by 11.2, we may assume for the purpose of proof that X is irreducible between p and some point $q \in X$. Suppose that $X - \kappa(p)$ is not connected, i.e. (6.2),

(1) $X - \kappa(p) = M | N, \quad q \in M$

(without loss of generality). Since M and N are mutually separated, there is an open subset U of X such that $M \subset U$ and $\overline{U} \cap N = \varnothing$ [7, p. 130]. Hence,

(2) $(\overline{U} - U) \cap (M \cup N) = \varnothing$.

Since $q \in M \subset U$, the component Q of U containing q exists. Also, note that $U \neq X$ (since $\overline{U} \cap N = \varnothing$ and $N \neq \varnothing$). Thus, by 5.7, $\overline{Q} - U \neq \varnothing$. Let $r \in \overline{Q} - U$. Then, since $\overline{Q} \subset \overline{U}$, $r \in \overline{U} - U$. Hence, by (1) and (2), $r \in \kappa(p)$. Therefore, there is a proper subcontinuum A of X such that $p, r \in A$. Thus, since $r \in \overline{Q}$ and $\overline{Q}$ is a continuum, $A \cup \overline{Q}$ is a continuum. Thus, since $p, q \in A \cup \overline{Q}$ and X is irreducible between p and q, we have that

(3) $A \cup \overline{Q} = X$.

Since $\overline{U} \cap N = \varnothing$ and $\overline{Q} \subset \overline{U}$, $\overline{Q} \cap N = \varnothing$. Thus, by (3), $N \subset A$. Hence, noting that $A \subset \kappa(p)$, we have a contradiction to (1) since $N \neq \varnothing$ (6.2). Therefore, $X - \kappa(p)$ is connected. This proves 11.4.

In relation to 11.4, we remark that $X - \kappa(p)$ may not be a continuum even when $X - \kappa(p) \neq \varnothing$ and X is a decomposable continuum (see (b) of 11.40).

The results in 11.6–11.8 are related to one another and to 11.4 in that they are all concerned with connectedness properties of complements. We note the following simple lemma (whose converse will be proved in 11.21).

11.5 Lemma. Let X be an irreducible continuum, and let p be a point of irreducibility of X. Then, X is not the union of two proper subcontinua both of which contain p.

Proof. Assuming that X is irreducible between p and q and that X is such a union, one easily obtains a contradiction.

11.6 Theorem. Let X be a nondegenerate irreducible continuum, and let p be a point of irreducibility of X. Then, for all subcontinua A of X such that $p \in A$, $X - A$ is connected.

Proof. Let C be a subcontinuum of X such that $X - C$ is not connected, i.e. (6.2), $X - C = P|Q$. Then, letting $Y = P \cup C$ and $Z = Q \cup C$, we have by 6.3 that

(1) Y and Z are continua.

Clearly,

(2) $X = Y \cup Z$, $Y \cap Z = C$, and $Y \neq X \neq Z$.

By (1), (2), and 11.5, $p \notin C$. This completes the proof of 11.6.

Some further results relating directly to 11.6 are in 11.41 and 11.42.

11.7 Theorem. Let X be a nondegenerate continuum such that X is irreducible between p and q. If C is a subcontinuum of X such that $X - C$ is not connected, then $X - C$ has exactly two components, each open in X, one containing p and the other containing q.

Proof. Beginning as in the proof of 11.6, $X - C = P|Q$, $Y = P \cup C$, $Z = Q \cup C$, and (1) and (2) of the proof of 11.6 hold. Thus, since p and q are each a point of irreducibility of X, $p \notin C$ and $q \notin C$ (by 11.5). Hence, we may assume without loss of generality that $p \in P$ and $q \in Q$. Then, $p \in Y$ and $q \in Z$; thus, since Y and Z are subcontinua of X [by (1) of the proof of 11.6], we have by 11.6 that $X - Y$ and $X - Z$ are connected. Therefore, noting that

$P = X - Z$ and $Q = X - Y$,

11.7 now follows.

11.8 Theorem. Let X be a nondegenerate continuum such that X is irreducible between p and q. If A and B are subcontinua of X such that $p \in A$ and $q \in B$, then $X - (A \cup B)$ is connected.

Proof. If $A \cap B \neq \varnothing$, then, since $p, q \in A \cup B$, $X = A \cup B$ and, so, $X - (A \cup B) = \varnothing$ and, thus, is connected. Hence, for the purpose of proof, assume that

(1) $A \cap B = \varnothing$.

Suppose that $X - (A \cup B)$ is not connected, i.e.,

(2) $X - (A \cup B) = P|Q$ (6.2).

Let $S = X - A$. Note that S is connected by 11.6, $B \subset S$ (by (1)), B is connected, and, by (2),

(3) $S - B = X - (A \cup B) = P|Q$.

Hence, by 6.3, $P \cup B$ and $Q \cup B$ are connected. Therefore,

(4) $\bar{P} \cup B$ and $\bar{Q} \cup B$ are continua.

Next, note that $X = A \cup [B \cup (\overline{S - B})]$. Thus, since X is connected and $A \cap B = \varnothing$ (by (1)), we must have that $A \cap (\overline{S - B}) \neq \varnothing$. Therefore, since

$$\overline{S - B} = \bar{P} \cup \bar{Q} \quad \text{(by (3))},$$

$A \cap \bar{P} \neq \varnothing$ or $A \cap \bar{Q} \neq \varnothing$, say $A \cap \bar{P} \neq \varnothing$. Then, by (4), $A \cup \bar{P} \cup B$ is a continuum. Thus, since $p, q \in A \cup \bar{P} \cup B$,

$$A \cup \bar{P} \cup B = X, \quad \text{i.e.,} \quad X - (A \cup B) \subset \bar{P}.$$

But, this contradicts (2). Therefore, we have proved 11.8.

As an application of 11.6–11.8, we prove the following result.

11.9 Corollary. If X is a continuum which is irreducible between p and q, then, for any subcontinuum Y of X, the interior Y° of Y in X is connected.

Proof. Since the corollary is trivial if $Y = X$, we may assume for the purpose of proof that $Y \neq X$. Hence, X is nondegenerate and $p \in X - Y$ or $q \in X - Y$, say $p \in X - Y$. Recall that

(1) $Y^\circ = X - (\overline{X - Y})$.

We take two cases. First, assume that $X - Y$ is connected. Then, $\overline{X - Y}$ is a subcontinuum of X. Thus, since $p \in \overline{X - Y}$, we have by 11.6 and (1) that Y° is connected. Second, assume that $X - Y$ is not connected. Then, by 11.7, $X - Y$ has exactly two components, P and Q, where, without loss of generality, $p \in P$ and $q \in Q$. Hence, by 11.8,

$$X - (\bar{P} \cup \bar{Q}) \text{ is connected.}$$

Thus, since $\bar{P} \cup \bar{Q} = \overline{X - Y}$, we again have by (1) that Y° is connected. This completes the proof of 11.9.

We remark that the notion of irreducibility can be generalized as follows. A connected topological space X is said to be irreducible provided that there exist p, $q \in X$ such that no proper, closed, connected subset of X contains $\{p,q\}$. With this more general definition, the results in 11.5–11.9 remain valid on replacing continuum with connected space and subcontinuum with closed connected subset (the proofs are the same). Note, however, that 11.4 does not remain valid in this more general setting (example?), and neither do most of the rest of the results in this chapter (except, e.g., 11.41–11.43).

2. COUNTING COMPOSANTS

We now turn our attention to determining the number of composants a continuum may have. Decomposable continua have either one or three composants (11.13). Nondegenerate indecomposable continua have uncountably many composants (11.15) and they are mutually disjoint (11.17). We also give some consequences of these results, e.g., indecomposable continua are irreducible (11.15.1) and two characterizations of indecomposability (11.19 and 11.20).

11.10 Lemma. If X is a decomposable continuum, then X is a composant of X.

Proof. Let A and B be proper subcontinua of X such that $X = A \cup B$. Note that $A \cap B \neq \varnothing$ since X is connected, and let $z \in A \cap B$. Then, noting that $A \subset \kappa(z)$ and $B \subset \kappa(z)$, we see that $X = \kappa(z)$. This proves 11.10.

The following two lemmas will be used in the proof of 11.13.

11.11 Lemma. If X is a decomposable continuum such that X is irreducible between p and q, then X, $\kappa(p)$, and $\kappa(q)$ are three different composants of X.

Proof. By 11.10, X is a composant of X. Clearly, $p \in \kappa(p)$ and $p \notin \kappa(q)$. Hence, $\kappa(p) \neq \kappa(q)$ and $\kappa(q) \neq X$. By 11.2, $\kappa(p) \neq X$. This completes the proof of 11.11.

11.12 Lemma. Let X be a decomposable continuum such that X is irreducible between p and q. If K is a composant of X, then $p \in K$ or $q \in K$; furthermore, if p, $q \in K$, $K = X$.

Proof. Let A and B be proper subcontinua of X such that $X = A \cup B$. Clearly, p and q do not both belong to A and they do not both belong to B. Hence, we may assume without loss of generality that $p \in A$ and $q \in B$. Let $r \in X$ such that $K = \kappa(r)$. If $r \in A$, then $A \subset K$ so $p \in K$; if $r \in B$, then $B \subset K$ so $q \in K$. Thus, since $r \in A$ or $r \in B$, $p \in K$ or $q \in K$. This proves the first part of 11.12. To prove the second part, assume that

$$p, q \in K, \quad \text{i.e.,} \quad p, q \in \kappa(r).$$

Then, there are proper subcontinua E and F of X such that p, $r \in E$ and q, $r \in F$. Since $r \in E$ and E is a proper subcontinuum of X, $E \subset K$. Similarly, $F \subset K$. Hence, $E \cup F \subset K$. Since E and F are continua and $r \in E \cap F$, $E \cup F$ is a continuum. Thus, since p, $q \in E \cup F$, $E \cup F = X$. Therefore, since $E \cup F \subset K$, $K = X$. This completes the proof of 11.12.

11.13   Theorem.   Let X be a decomposable continuum. Then, X has either exactly one or exactly three composants. More precisely: If X is not irreducible, X has exactly one composant (namely, X itself), and, if X is irreducible, X has exactly three composants (namely, the ones in 11.11).

Proof. If X is not irreducible, then clearly (11.2) $\kappa(x) = X$ for each $x \in X$. Hence, assume for the rest of the proof that X is irreducible between, say, p and q. Then, letting K be a composant of X, it suffices by 11.11 to prove that

$$K = X, \quad \kappa(p), \quad \text{or} \quad \kappa(q).$$

To prove this, assume that $K \neq X$. Then, by 11.12, p and q do not both belong to K. Hence, we may assume that $p \in K$ and $q \notin K$. We show that $K = \kappa(p)$. Let $r \in X$ such that $K = \kappa(r)$. Now, let $x \in K$. Since $x, p \in \kappa(r)$, there are proper subcontinua A and B of X such that x, $r \in A$ and p, $r \in B$. Since $r \in A$ and A is a proper subcontinuum of X, $A \subset K$. Similarly, $B \subset K$. Hence, $A \cup B \subset K$. Thus, since $q \notin K$, $A \cup B \neq X$. Hence, noting that $A \cup B$ is a continuum (since $A \cap B \neq \emptyset$) and that $p, x \in A \cup B$, we see that $x \in \kappa(p)$. Therefore, $K \subset \kappa(p)$. Next, let $y \in \kappa(p)$. Then, there is a proper subcontinuum C of X such that y, $p \in C$. Since $p \in K = \kappa(r)$, there is a proper subcontinuum D of X such that p, $r \in D$. Since X is irreducible between p and q and since $p \in C$ and $p \in D$, we have that $q \notin C$ and $q \notin D$. Thus, $C \cup D \neq X$. Hence, noting that $C \cup D$ is a continuum (since $C \cap D \neq \emptyset$) and that y, $r \in C \cup D$, we see that $y \in \kappa(r) = K$. Therefore, $\kappa(p) \subset K$. Thus, we have proved that $K = \kappa(p)$. This completes the proof of 11.13.

Next, we determine the number of composants of an indecomposable continuum (11.15). This is done with the aid of the following general proposition which is of independent interest.

11.14 Proposition. If X is any nondegenerate continuum, then the composant $\kappa(p)$ of any point $p \in X$ is a union of countably many proper subcontinua of X each of which contains p.

Proof. Let $\{U_i: i = 1, 2, ...\}$ be a countable open base for $X - \{p\}$ such that each $U_i \neq \varnothing$. For each i, let C_i be the component of $X - U_i$ containing p. Since each $U_i \neq \varnothing$, $C_i \neq X$ for any i. Thus, since each U_i is open in X,

(1)   Each C_i is a proper subcontinuum of X.

Since $p \in C_i$ for each i, we have by (1) that

(2) $\displaystyle\bigcup_{i=1}^{\infty} C_i \subset \kappa(p)$.

Now, let $x \in \kappa(p)$. Then, by definition of $\kappa(p)$ (5.20), there is a proper subcontinuum A of X such that $p, x \in A$. Clearly, for some U_k, $A \cap U_k = \varnothing$, i.e., $A \subset X - U_k$. Hence, by the definition of C_k, $A \subset C_k$. Thus, $x \in C_k$. Hence, we have proved that

(3) $\kappa(p) \subset \displaystyle\bigcup_{i=1}^{\infty} C_i$.

By (1)–(3), we have proved 11.14.

11.15 Theorem. If X is a nondegenerate indecomposable continuum, then X has uncountably many composants.

Proof. By 11.14 and 6.19, each composant of X is the countable union of sets which are nowhere dense in X. Therefore, since X is the union of its composants, X must have uncountably many composants by the Baire Theorem (Theorem 6 of [8, p. 9]). This proves 11.15.

The following corollary will be significantly strengthened in 11.18.

11.15.1 Corollary. Every indecomposable continuum is irreducible.

Proof. The result is immediate from 11.15 and 11.3.

We note that 11.15.1 implies that nondegenerate arcwise connected continua are decomposable (11.54).

For a number of years the question remained open as to whether or not 11.15 is true for Hausdorff continua (its proof used 6.19 and 11.14—even though 6.19 is true for Hausdorff continua, the proof of 11.14 used a countable base). It is now known that there are nondegenerate indecomposable Hausdorff continua with only one composant [1]. Hence, even 11.15.1 is false for Hausdorff continua.

Let us note that we have now determined the number of composants that a continuum may have.

11.16 Theorem. A nondegenerate continuum has either exactly one, exactly three, or uncountably many composants.

Proof. The corollary follows immediately from 11.13 and 11.15.

Next, we prove the following important theorem which we will apply in the proof of 11.18.

11.17 Theorem. If X is a nondegenerate indecomposable continuum, then the composants of X are mutually disjoint.

Proof. Let K and L be composants of X, $K = \kappa(p)$ and $L = \kappa(q)$. Assume that $K \cap L \neq \varnothing$, and let $z \in K \cap L$. We show that $K = L$. Let $x \in K$. Then, there are proper subcontinua A, B, and C of X such that

$$p, x \in A, \quad p, z \in B, \quad q, z \in C.$$

Let $D = A \cup B \cup C$. Clearly, D is a continuum. Also, by 6.19 and the Baire Theorem (Theorem 6 of [8, p. 9]), $D \neq X$. Thus, since $x, q \in D$, $x \in L$. Hence, we have proved that $K \subset L$. Similarly, $L \subset K$. Thus, $K = L$. Therefore, we have proved 11.17.

The following result is a significant improvement of the corollary in 11.15.1.

11.18 Theorem. If X is a nondegenerate indecomposable continuum, then each point of X is a point of irreducibility of X. In fact, there is an uncountable subset J of X such that X is irreducible between any two points of J.

Proof. The first part is immediate from 11.15, 11.17, and 11.2. To prove the second part, use 11.17 and the Axiom of Choice [3, p. 1] to obtain a subset J of X such that

$$|J \cap K| = 1 \quad \text{for each composant } K \text{ of } X.$$

Then, by 11.15 and 11.17, J is uncountable and, by 11.17, X is irreducible between any two points of J. This completes the proof of 11.18.

The first part of 11.18 is reversible and, thus, we have the following characterization of indecomposable continua.

11.19 Corollary. A continuum X is indecomposable if and only if each point of X is a point of irreducibility of X.

Proof. Half is in 11.18, and the other half is immediate from 11.2 and 11.10.

Recall that in 1.10 we started with three points a, b, $c \in R^2$ and used simple chains to construct a continuum X as a nested intersection. We concluded that X was indecomposable by claiming that the construction could be used to show that X was irreducible between each two of the points a, b, c (from which the proof of the indecomposability of X is easy). We now show that the indecomposability of any nondegenerate continuum is actually equivalent to its irreducibility between each two of some triple of points. Thus, in this sense, the construction in 1.10 was conceptually natural.

11.20 Corollary. A nondegenerate continuum X is indecomposable if and only if there are three points of X such that X is irreducible between each two of these three points.

Proof. If X is indecomposable, then, by the second part of 11.18, there certainly are three such points. If X is decomposable, then, clearly (by the pigeonhole principle), there cannot be three such points. This completes the proof of 11.20.

3. KURATOWSKI'S CHARACTERIZATION

Kuratowski wrote two of the most important and fundamental papers concerning the theory of irreducible continua ([5] and [6]). In this section, we prove his characterization theorem 11.21 which is Theorem XIX of [6, p. 270]. We note that it shows that the condition in the conclusion of

11.5 actually characterizes irreducibility. We also note that it contains 11.15.1 as an immediate corollary (so does 11.34–however, we point out that 11.34, unlike 11.21, uses 11.15.1 in its proof).

11.21 Kuratowski's Theorem. Let X be a continuum and let $p \in X$. Then, p is point of irreducibility of X if and only if

(#) X is not the union of two proper subcontinua both of which contain p.

Hence, X is irreducible if and only if there exists $p \in X$ such that (#) holds.

Proof. To prove the half which is not 11.5, assume that (#) holds for p. Then, an easy induction using (#) shows that:

(1) If $C_1, \ldots, C_k$ are finitely many proper subcontinua of X such that $p \in C_i$ for each i, then $X \neq \bigcup_{i=1}^{k} C_i$.

Now, suppose that p is not a point of irreducibility of X, i.e. (11.2), $\kappa(p) = X$. Then, by 11.14, there are proper subcontinua A_i of X, $i = 1, 2, \ldots$, such that

(2) $X = \bigcup_{i=1}^{\infty} A_i$

and

(3) $p \in A_i$ for each i.

Furthermore, by using (1) to replace the continua A_i with appropriate finite unions $\bigcup_{i=1}^{\ell} A_i$ if necessary, we may assume without loss of generality that

(4) $A_i \subset A_{i+1} \neq A_i$ for each i.

By (4), there exists $x_i \in A_{i+1} - A_i$ for each i. Let x denote the limit of a convergent subsequence of $\{x_i\}_{i=1}^{\infty}$. Using (4), we see that, for any given i, $x_j \in X - A_i$ for each $j \geq i$. Thus, $x \in \overline{X - A_i}$ for all i. By (2), $x \in A_m$ for some m. Hence,

(5) $(\overline{X - A_i}) \cap A_m \neq \varnothing$ for all i.

By (2), $X - A_m = \bigcup_{i=1}^{\infty} (A_i - A_m)$. Each A_i is closed in X and each $A_i - A_m$ is open in A_i. Hence, for each i,

$$A_i - A_m = \bigcup_{j=1}^{\infty} B_{i,j} \quad \text{where each } B_{i,j} \text{ is closed in } X.$$

Thus, $X - A_m = \bigcup_{i=1}^{\infty} (\bigcup_{j=1}^{\infty} B_{i,j})$. Therefore, since $X - A_m$ is a nonempty open subset of X, we have by the Baire Theorem [8, p. 9] that, for some n and ℓ, $B_{n,\ell}$ has nonempty interior in X. Hence, $A_n - A_m$ has nonempty interior in X, i.e.,

$$\overline{X - (A_n - A_m)} \neq X.$$

Thus, since $X - (A_n - A_m) = (X - A_n) \cup A_m$,

(6) $(\overline{X - A_n}) \cup A_m \neq X$.

We show that $(\overline{X - A_n}) \cup A_m$ is a continuum, (10) below, by first showing that $\overline{X - A_n}$ is a continuum (and then using (5)). It suffices to show that $\overline{X - A_n}$ is connected. Suppose that $\overline{X - A_n}$ is not connected, i.e.,

(*) $\overline{X - A_n} = E|F$ (6.2).

Then, (7) and (8) below hold for the sets $A_n \cup E$ and $A_n \cup F$:

(7) $(A_n \cup E) \cup (A_n \cup F) = X$;

(8) $(A_n \cup E) \cap (A_n \cup F) = A_n$.

By (*), E and F are compact. Thus, $A_n \cup E$ and $A_n \cup F$ are compact. Hence, by (7) and (8), we may apply (a) of 6.18 to see that $A_n \cup E$ and $A_n \cup F$ are continua. Thus, by (3), (7), and our initial assumption that (#) holds for p, we have that

$A_n \cup E = X$ or $A_n \cup F = X$.

This contradicts (*). Hence, $\overline{X - A_n}$ is connected and, thus, is a continuum. Therefore, by (5),

(9) $(\overline{X - A_n}) \cup A_m$ is a continuum.

Now, note that

(10) $X = A_n \cup [(\overline{X - A_n}) \cup A_m]$.

Recall that A_n is a proper subcontinuum of X containing p (by (3)); also, by (3), (6) and (9), $(\overline{X - A_n}) \cup A_m$ is a proper subcontinuum of X containing p. Hence, (10) contradicts our initial assumption that (#) holds for p. Therefore, p is a point of irreducibility of X. This completes the proof of 11.21.

4. TRIODS AND WEAK TRIODS

Our next main result about irreducible continua does not occur until 11.34. It involves the following notion which is a straightforward generalization of the definition of a simple triod (2.21).

11.22 Definition of Triod. A continuum X is called a *triod* provided that there is a subcontinuum Z of X such that $X - Z$ is the union of three nonempty sets each two of which are mutually separated in X.

For example, any n-cell ($n \geq 2$), the Hilbert cube, and any n-sphere ($n \geq 2$) is a triod. On the other hand, no arc-like continuum is a triod (12.4) and, clearly, no indecomposable continuum is a triod.

It is evident that a triod cannot be irreducible (use 6.3). Also, a continuum may not be a triod and yet still fail to be irreducible (e.g., S^1). We shall prove that for unicoherent continua, not being a triod is equivalent to being irreducible (11.35). The substantial half of this equivalence is called Sorgenfrey's Theorem (11.34). This important theorem has many applications (one of which is in 12.5).

There are other general types of triods besides the type defined in 11.22 (e.g., see [11]). Various types of triods play an important role in continuum theory—the theorem in 11.34 and its application in 12.5 are specific illustrations of this. We do not have the time to investigate the general theory of various types of triods and its ramifications. Instead, we shall confine our attention generally to those facts which are pertinent to the proof of 11.34.

Specifically, we shall be concerned only with triods as defined in 11.22 and what we call weak triods (11.24). We determine their equivalence for unicoherent continua in 11.26.

The notation introduced in the following definition will be used often throughout the rest of the chapter.

11.23 Definition of Essential Sum. A continuum X is said to be the *essential sum* of the subcontinua X_i ($1 \leq i \leq n$), written

$$X = X_1 \oplus X_2 \oplus \cdots \oplus X_n,$$

provided that

$$X = \bigcup_{i=1}^{n} X_i \quad \text{and} \quad X_k \not\subset \bigcup_{i \neq k} X_i \text{ for each } k = 1, \ldots, n.$$

We emphasize that when we write $X = X_1 \oplus X_2 \oplus \cdots \oplus X_n$, we tacitly assume that X and each X_i are continua. In particular, note that writing

$$X = Y \oplus Z$$

says that X is a decomposable continuum and that Y and Z are proper subcontinua of X whose union is X.

In the proof of Sorgenfrey's Theorem (11.34), there will be three separate occasions in which we will want to contradict the assumption that X is not a triod. It will be easier to obtain these contradictions by using the following notion which, for unicoherent continua, is equivalent to being a triod (11.26).

11.24 Definition of Weak Triod. A continuum X is called a *weak triod* provided that $X = X_1 \oplus X_2 \oplus X_3$ (11.23) where $\cap_{i=1}^{3} X_i \neq \varnothing$.

A noose (9.32) is an example of a weak triod which is not a triod. In fact, any nondegenerate Peano continuum which is not an arc or a simple closed curve is a weak triod (11.50).

It follows easily using 11.5 that no weak triod can be irreducible.

We remark that in [11, pp. 440–441] Sorgenfrey defines eight different types of triods. What we have called a weak triod in 11.24 is what is called a type one triod in [11, p. 440]. The following proposition justifies our terminology.

11.25 Proposition. Every triod is a weak triod.

Proof. Let X be a triod. Then (11.22), there is a subcontinuum Z of X such that $X - Z = \cup_{i=1}^{3} U_i$ where each U_i is nonempty and each two of the sets U_1, U_2, U_3 are mutually separated in X (6.2). Hence,

$$X - Z = U_i | (U_j \cup U_k) \quad \text{when } i \neq j, i \neq k, \text{ and } j \neq k \text{ (6.2)}.$$

Thus, by 6.3, $U_i \cup Z$ is a continuum for each $i = 1, 2, 3$. Hence, letting $X_i = U_i \cup Z$ for each $i = 1, 2, 3$, it follows easily that

$$X = X_1 \oplus X_2 \oplus X_3 \quad \text{and} \quad \bigcap_{i=1}^{3} X_i = Z \neq \varnothing.$$

Therefore, X is a weak triod (11.24). This proves 11.25.

For reasons already commented on, the following theorem is important to the proof of 11.34.

11.26 Theorem. If X is a unicoherent continuum, then X is a triod if and only if X is a weak triod.

Proof. Half of the result is in 11.25. To prove the other half, assume that X is a unicoherent weak triod. Let (11.24).

$$X = X_1 \oplus X_2 \oplus X_3 \quad \text{where } \bigcap_{i=1}^{3} X_i \neq \emptyset.$$

Recall from 11.23 that each X_i is a continuum and, thus, since $\bigcap_{i=1}^{3} X_i \neq \emptyset$, we see that $X_1 \cup X_2$, $X_1 \cup X_3$, and $X_2 \cup X_3$ are also continua. Thus, since $X = \bigcup_{i=1}^{3} X_i$ (11.23) and X is unicoherent,

$X_j \cap (X_k \cup X_\ell)$ is a continuum Z_j whenever $j \neq k$, $j \neq \ell$, and $k \neq \ell$. Let $Z = \bigcup_{j=1}^{3} Z_j$. Since each Z_j is a continuum containing $\bigcap_{i=1}^{3} X_i$ and $\bigcap_{i=1}^{3} X_i \neq \emptyset$, Z is a continuum. Observe that being a point of Z is equivalent to being a point of at least two of the sets X_1, X_2, X_3. Thus, since $X = \bigcup_{i=1}^{3} X_i$, it follows easily that

(*) $X - Z = [X - (X_1 \cup X_2)] \cup [X - (X_1 \cup X_3)] \cup [X - (X_2 \cup X_3)]$.

Clearly, each of the three sets in square brackets on the right-hand side of (*) is open in X and, since $X = \bigcup_{i=1}^{3} X_i$, any two of them are disjoint. Furthermore, each of them is nonempty as is immediate from the second condition in 11.23. Therefore, X is a triod (by 11.22). This completes the proof of 11.26.

For the most part, the results in 11.27–11.33 are concerned with properties of decomposable continua which are not weak triods. Some of them are technical in nature and are included mainly for use in the proof of 11.34. Others are of independent interest.

The following proposition should be compared with 11.6. Though the proposition is of some independent interest, it is of such interest only for non-unicoherent continua in view of 11.34 (recall 11.6 and 11.25).

11.27 Proposition.   If $X = Y \oplus Z$ and X is not a weak triod, then $X - Y$ and $X - Z$ are connected.

Proof.   Suppose that $X - Y = U|V$ (6.2). We first show that U is connected. Suppose that $U = W_1|W_2$. Then, noting that

$X - Y = W_i|(W_j \cup V) \quad$ when $i \neq j$,

we have by 6.3 that $W_1 \cup Y$, $W_2 \cup Y$, and (since $X - Y = U|V$) $V \cup Y$ are continua. Clearly, X is the essential sum (11.23) of these three continua and their total intersection is Y. Hence, X is a weak triod—a contradiction. Therefore, U is connected. Similarly, V is connected. Hence,

(1) $\overline{U}$ and $\overline{V}$ are continua.

Next, we show (2) below. Suppose there is a subcontinuum K of $Y \cap Z$

such that $K \cap \overline{U} \neq \varnothing$ and $K \cap \overline{V} \neq \varnothing$. Then, by (1), $K \cup \overline{U}$ and $K \cup \overline{V}$ are continua. Thus, since $U - \overline{V}$, $V - \overline{U}$, and $Y - Z$ are each nonempty, we see that

$$X = (K \cup \overline{U}) \oplus (K \cup \overline{V}) \oplus Y.$$

Thus, since $(K \cup \overline{U}) \cap (K \cup \overline{V}) \cap Y \supset K \neq \varnothing$, X is a weak triod—a contradiction. Hence:

(2)   No subcontinuum of $Y \cap Z$ intersects both $\overline{U}$ and $\overline{V}$.

Note that since $X - Y = U|V$, $\overline{U} \cap \overline{V} \subset Y$. Hence, as is trivial by (2),

(3) $\overline{U} \cap \overline{V} = \varnothing$.

By (2), we may apply 5.2 to obtain disjoint closed subsets M and N of $Y \cap Z$ such that (4) and (5) hold:

(4) $Y \cap Z = M \cup N$;

(5) $\overline{U} \cap (Y \cap Z) \subset M$ and $\overline{V} \cap (Y \cap Z) \subset N$.

We now show using $M \cup \overline{U}$ and $N \cup \overline{V}$ that Z is not connected (a contradiction, which establishes the proposition). Since $Z - Y = X - Y$, $Z - Y = U \cup V$ and $\overline{U} \cup \overline{V} \subset Z$. Hence, we see using (4) that

(6) $Z = (M \cup \overline{U}) \cup (N \cup \overline{V})$.

Since $M \cap N = \varnothing$, we see using (3)–(5) that

(7) $(M \cup \overline{U}) \cap (N \cup \overline{V}) = \varnothing$.

Since $U \neq \varnothing$ and $V \neq \varnothing$, clearly

(8) $M \cup \overline{U} \neq \varnothing$ and $N \cup \overline{V} \neq \varnothing$.

Since $M \cup \overline{U}$ and $N \cup \overline{V}$ are closed sets, (6)–(8) show that Z is not connected, a contradiction. Therefore, $X - Y$ is connected. This completes the proof of 11.27.

We include the following corollary for use later.

11.28   Corollary.   Let $X = Y \oplus Z$ and assume that X is not a weak triod. Let $U = X - Z$. If H is a subcontinuum of $\overline{U}$ such that $H \supset \overline{U} \cap Z$, then $U - H$ is connected.

Proof.   By the connectedness of X, $\overline{U} \cap Z \neq \varnothing$. Hence, $H \cap Z \neq \varnothing$ and thus, $H \cup Z$ is a continuum. By 11.27, $\overline{U}$ is a continuum. Since $Z - Y \neq \varnothing$ (11.23) and $\overline{U} \subset Y$, $Z \cup H \not\subset \overline{U}$. Finally, assuming for the purpose of proof that $U - H \neq \varnothing$, we see that $\overline{U} \not\subset Z \cup H$. Therefore, (by 11.23),

$$X = \overline{U} \oplus (Z \cup H).$$

Hence, by 11.27, $X - (Z \cup H)$ is connected. Thus, since $X - (Z \cup H) = U - H$, we have proved 11.28.

The reader should be careful to note the difference between the following general definition and the notion of being irreducible *about* $A \cup B$ (4.35).

11.29   **Definition of Irreducible from A to B.**   Let X be a continuum, and let A and B be nonempty compact subsets of X. We say that a subcontinuum C of X is *irreducible from A to B*, written

$$C = \text{irr}(A,B),$$

provided that $C \cap A \neq \varnothing$, $C \cap B \neq \varnothing$, and no proper subcontinuum of C intersects both A and B. If A or B is a singleton set, we omit the set notation; thus, e.g., if $A = \{p\}$ and C is as above, we say that C is *irreducible from p to B* and write

$$C = \text{irr}(p,B).$$

Clearly, C is irreducible from p to q (i.e., $C = \text{irr}(p,q)$) if and only if C is irreducible between p and q (in the sense of 4.35). See 11.37.

We note the following general existence result.

11.30   **Proposition.**   If X is a continuum and A and B are nonempty compact subsets of X, then there is a subcontinuum C of X such that $C = \text{irr}(A,B)$.

Proof.   The easy proof is left as an exercise (11.38).

The following lemma will be used in the proof of 11.34, in a contrapositive fashion, to guarantee that the induction near the beginning of the proof can be done and results in sets with the desired properties.

11.31   **Lemma.**   Let $X = Y \oplus Z$ and assume that X is not a weak triod. Let $U = X - Z$. Assume that A and B are subcontinua of $\overline{U}$ and $p \in U - A$ such that

$$A \supset \overline{U} \cap Z, \quad B = \text{irr}(p,A), \quad \text{and} \quad A \cup B \supset U.$$

Then ($\overline{U}$ is a continuum by 11.27), $\overline{U} = \text{irr}(p, \overline{U} \cap Z)$.

Proof. Let L be a subcontinuum of $\overline{U}$ such that $p \in L$ and $L \cap Z \neq \emptyset$. We show that $L \supset \overline{U}$. We first prove (*):

(*) $B \subset L$.

To prove (*), note that $p \in L - A$ and let K be the component of $L - A$ containing p. Since $K \subset \overline{U} - A$ and $A \cup B \supset \overline{U}$, $K \subset B$. Hence, $\overline{K} \subset B$. By 5.7, $\overline{K} \cap A \neq \emptyset$. Thus, since $\overline{K}$ is a subcontinuum of B and since $p \in \overline{K}$ and $B = \text{irr}(p,A)$, we have that $\overline{K} = B$. Therefore, since $\overline{K} \subset L$, $B \subset L$. This proves (*). By (*) and our assumption that $A \cup B \supset U = X - Z$, we see that

(1) $X = A \cup L \cup Z$ (where A, L, and Z are continua).

Now, suppose that $L \not\supset \overline{U}$. Then, $L \not\supset U$ and, hence, there is a point $x \in U - L$. By (*), $x \notin B$. Thus, since $A \cup B \supset U$ and $x \in U - L$, we have that $x \in A - L$. Thus, since $x \in U = X - Z$, $x \in A - (L - Z)$. Hence,

(2) $A \not\subset L \cup Z$.

Recall that $p \in L - A$ and, since $p \in U$, $p \notin Z$. Hence,

(3) $L \not\subset A \cup Z$.

Let $q \in Z - Y$. Since $Z - Y$ is open in X and $(Z - Y) \cap U = \emptyset$, $(Z - Y) \cap \overline{U} = \emptyset$. Hence, $q \notin \overline{U}$. Thus, $q \notin A \cup L$. Hence,

(4) $Z \not\subset A \cup L$.

Since $L \subset \overline{U}$ and since $\overline{U} \cap Z \subset A$ (by hypothesis), $L \cap Z \subset A$. Thus, since $L \cap Z \neq \emptyset$,

(5) $L \cap Z \cap A \neq \emptyset$.

By (1)–(5), X is a weak triod. However, by assumption, X is not a weak triod. Therefore, $L \supset \overline{U}$. It follows that we have proved 11.31.

We now state and prove the last two lemmas to be used in the proof of Sorgenfrey's Theorem.

11.32 Lemma. Let $X = Y \oplus Z$ be a unicoherent continuum which is not a weak triod. Let $U = X - Z$, and let A be a proper subcontinuum of $\overline{U}$ such that $A \supset \overline{U} \cap Z$. Let $W = U - A$. Then ($\overline{W}$ is a continuum by 11.28): If $\overline{W} = E \oplus F$, either $E \cap A = \emptyset$ (and $F \cap A \neq \emptyset$) or $F \cap A = \emptyset$ (and $E \cap A \neq \emptyset$).

Proof. We begin the proof by observing that since $U = X - Z$ and $W = U - A$, clearly

(1) $W = X - (Z \cup A)$.

Note that since X is connected, $\bar{U} \cap Z \neq \emptyset$. Thus, since $A \supset \bar{U} \cap Z$, $A \cap Z \neq \emptyset$. Hence,

(2) $Z \cup A$ is a continuum.

Next, we prove (*) below [Note: The equivalence in (*) is due to the fact that $E \cap F \cap Z \subset \bar{U} \cap Z \subset A$]:

(*) $E \cap F \cap (Z \cup A) = \emptyset$

(equivalently, $E \cap F \cap A = \emptyset$). To prove (*), first note that by (1), and since $W \subset E \cup F$,

(3) $X = E \cup F \cup (Z \cup A)$.

Since $F \neq \bar{W}$, $F \not\supset$ W. Hence, there exists $x \in W - F$. Thus, noting that $x \in E$, we see from (1) that

(4) $E \not\subset F \cup (Z \cup A)$.

An argument similar to the one just used to prove (4) shows that

(5) $F \not\subset E \cup (Z \cup A)$.

Since $(Z - Y) \cap U = \emptyset$, $(Z - Y) \cap W = \emptyset$. Hence, $Z - Y$ being open in X, $(Z - Y) \cap \bar{W} = \emptyset$. Thus, since $Z - Y \neq \emptyset$ and $\bar{W} = E \cup F$,

(6) $(Z \cup A) \not\subset E \cup F$.

Since $W = E \oplus F$, E and F are continua (11.23). Also, $Z \cup A$ is a continuum (by (2)). Therefore: If (*) were false, then, by (3)–(6), X would be a weak triod. Hence, (*) is true. Now, to complete the proof of the lemma, recall that

(7) $X = (Z \cup A) \cup \bar{W}$ (by (1))

where $Z \cup A$ is a continuum (by (2)) and $\bar{W}$ is a continuum (as noted in the hypotheses of the lemma). Also, note that

(8) $(Z \cup A) \cap \bar{W} = (Z \cup A) \cap (E \cup F) = [(Z \cup A) \cap E] \cup [(Z \cup A) \cap F]$

where, by (*),

(9) $[(Z \cup A) \cap E] \cap [(Z \cup A) \cap F] = \emptyset$.

Thus, since X is unicoherent, we see from (7)–(9) that $(Z \cup A) \cap E = \emptyset$ or $(Z \cup A) \cap F = \emptyset$. Therefore, clearly

(10) $E \cap A = \emptyset$ or $F \cap A = \emptyset$.

Now, suppose that $E \cap A = \emptyset$ and $F \cap A = \emptyset$. Then, since $\overline{W} = E \cup F$,

(11) $\overline{W} \cap A = \emptyset$.

If there exists $x \in \overline{W} \cap Z$, then, since $W = U - A$, we have that $x \in \overline{U} \cap Z$; thus, since $\overline{U} \cap Z \subset A$, $x \in A$ and, hence, $x \in \overline{W} \cap A$, a contradiction to (11). Thus, $\overline{W} \cap Z = \emptyset$. Hence, by (11),

$$\overline{W} \cap (Z \cup A) = \emptyset$$

which, by (1), contradicts the connectedness of X. Thus, it is false that $E \cap A = \emptyset$ and $F \cap A = \emptyset$. Therefore, by (10), we have proved 11.32.

11.33 Lemma. Let $X = Y \oplus Z$ be a unicoherent continuum which is not a weak triod. Let $U = X - Z$ and $V = X - Y$ (note: $\overline{U}$ and $\overline{V}$ are continua by 11.27). If there exist $p \in U$ and $q \in V$ such that

$$\overline{U} = \text{irr}(p, \overline{U} \cap Z) \quad \text{and} \quad \overline{V} = \text{irr}(q, \overline{V} \cap Y),$$

then $X = \text{irr}(p, q)$.

Proof. Suppose there is a proper subcontinuum K of X such that $p, q \in K$. We first prove (*):

(*) $K \supset U \cup V$, i.e., $K \supset X - (Y \cap Z)$.

To prove (*), let $L = (K \cup Z) \cap \overline{U}$. Note that $X = (K \cup Z) \cup \overline{U}$ and that $K \cup Z$ and $\overline{U}$ are continua. Thus, since X is unicoherent, L is a continuum. Since, $p, q \in K \cup Z$ and since $p \in U$ and $q \in X - \overline{U}$, the connectedness of $K \cup Z$ gives us that

$$(K \cup Z) \cap (\overline{U} - U) \neq \emptyset.$$

Thus, since $\overline{U} - U \subset Z$, we see that

$$L \cap (\overline{U} \cap Z) = \emptyset.$$

Therefore: Since $p \in L$ and L is a subcontinuum of $\overline{U} = \text{irr}(p, \overline{U} \cap Z)$, we have that $L \supset \overline{U}$. Hence, $L \supset U$. Thus, from the definitions of U and L, we have that $K \supset U$. A similar argument shows that $K \supset V$. Therefore, we have proved (*). We now show that X is a weak triod (which is a contradiction). Note that V is connected (11.27), $V \cap \overline{U} = \emptyset$ (since V is open in X and $V \cap U = \emptyset$), and $V \subset K$ (by (*)). Hence, the component J of $K - \overline{U}$ containing V exists. Clearly,

(1) $X = \overline{J} \cup \overline{U} \cup (Y \cap Z)$.

Any point of V is in $\bar{J}$ and not in $\bar{U} \cup (Y \cap Z)$. No point of U is in $\bar{J} \cup (Y \cap Z)$. Since $K \neq X$ (by assumption), there is an $x \in X - K$; $x \in Y \cap Z$ (by (*)) and, since $\bar{J} \subset K$ and, by (*), $\bar{U} \subset K$, $x \notin \bar{J} \cup \bar{U}$. Finally, note that $\bar{J}$, $\bar{U}$, and, by the unicoherence of X, $Y \cap Z$ are continua. All these facts together with (1) show that

(2) $X = \bar{J} \oplus \bar{U} \oplus (Y \cap Z)$.

By 5.7, $\bar{J} \cap \bar{U} \neq \varnothing$. Since $J \cap \bar{U} = \varnothing$, $J \cap U = \varnothing$. Thus, since U is open in X, $\bar{J} \cap U = \varnothing$. Hence, $\bar{J} \cap \bar{U} \subset Y \cap Z$. Thus, since $\bar{J} \cap \bar{U} \neq \varnothing$, we have that

(3) $\bar{J} \cap \bar{U} \cap (Y \cap Z) \neq \varnothing$.

By (2) and (3), X is a weak triod. This contradiction proves 11.33.

5. SORGENFREY'S THEOREM

In 1944, Sorgenfrey published a theorem which has proved to be extremely useful for determining that various types of continua are irreducible. In this section, we prove this theorem (11.34, which is 3.2 of [11, p. 456]). We have devoted a large portion of the previous section to obtaining some results for use in the proof. However, there is still much to do and, hence, the proof is fairly long. We note that the theorem actually characterizes those unicoherent continua which are irreducible (11.35).

11.34 Sorgenfrey's Theorem. Every nondegenerate unicoherent continuum X which is not a triod is irreducible.

Proof. By 11.15.1, we may assume for the purpose of proof that X is decomposable, i.e. (11.23), $X = Y \oplus Z$. By 11.26,

(1) X is not a weak triod.

Let $U = X - Z$ and $V = X - Y$. By (1) and 11.27,

(2) $\bar{U}$ and $\bar{V}$ are continua.

Since X is unicoherent and satisfies (1), it suffices by 11.33 to prove (a) and (b):

(a) $\bar{U} = \text{irr}(p, \bar{U} \cap Z)$ for some $p \in U$;

(b) $\bar{V} = \text{irr}(q, \bar{V} \cap Y)$ for some $q \in V$.

Suppose that (a) is false. Using this supposition, we show that we can use induction to define points $p_{n(i)}$, continua K_i, and open sets U_i for each $i = 1, 2, \ldots$ whose properties will yield a contradiction.

For use in the first step of the induction, note that since X is unicoherent and $X = \overline{U} \cup Z$ where $\overline{U}$ (by (2)) and Z are continua,

(3) $\overline{U} \cap Z$ is a continuum.

In particular,

(4) $\overline{U} \cap Z \neq \varnothing$.

Now, let $D = \{p_i: 1, 2, \ldots\}$ be a countable dense subset of U. Let $p_{n(1)} = p_1$. By (2) and (4), we may apply 11.30 to obtain a subcontinuum C_1 of $\overline{U}$ such that

(5) $C_i = \mathrm{irr}(p_{n(1)}, \overline{U} \cap Z)$.

Let $K_1 = C_1 \cup (\overline{U} \cap Z)$ and let $U_1 = U - K_1$. By (3) and (5), K_1 is a subcontinuum of $\overline{U}$. Again using (3) and (5), and recalling our supposition that (a) is false, we see from 11.31 (with $A = \overline{U} \cap Z$, $B = C_1$, and $p = p_{n(1)}$) that $K_1 \not\supset U$, i.e., $U_1 \neq \varnothing$. Now, assume inductively that, for some $j \geq 1$, we have defined $p_{n(j)}$ with $n(j) \geq j$, a proper subcontinuum K_j of $\overline{U}$ such that

$$K_j \supset \overline{U} \cap Z \quad \text{and} \quad K_j \cap D = \{p_1, \ldots, p_{n(j)}\},$$

and $U_j = U - K_j$. Then, noting that $U_j \cap D \neq \varnothing$, let $n(j + 1)$ denote the smallest i such that $p_i \in U_j$. Note that

$$\overline{U} = \overline{(U - K_j) \cup K_j} = \overline{(U - K_j)} \cup K_j = \overline{U_j} \cup K_j.$$

Hence, by (2), $\overline{U_j} \cap K_j \neq \varnothing$. Also, by 11.28, $\overline{U_j}$ is a continuum. Thus, by 11.30, there is a subcontinuum C_{j+1} of $\overline{U_j}$ such that

(6) $C_{j+1} = \mathrm{irr}(p_{n(j+1)}, \overline{U_j} \cap K_j)$.

Let $K_{j+1} = C_{j+1} \cup K_j$ and let $U_{j+1} = U - K_{j+1}$. Since K_j and C_{j+1} are subcontinua of $\overline{U}$ and $C_{j+1} \cap K_j \neq \varnothing$ (by (6)), K_{j+1} is a subcontinuum of $\overline{U}$. Since $C_{j+1} \subset \overline{U_j}$, we see from (6) that

$$C_{j+1} = \mathrm{irr}(p_{n(j+1)}, K_j).$$

Therefore, recalling our supposition that (a) is false, we see from 11.31 (with $A = K_j$, $B = C_{j+1}$, and $p = p_{n(j+1)}$) that $K_{j+1} \not\supset U$, i.e., $U_{j+1} \neq \varnothing$.

We summarize the facts which are evident from the induction above as follows:

(7)   For each $i = 1, 2, \ldots$, K_i is a proper subcontinuum of $\overline{U}$, $K_i \supset \overline{U} \cap Z \neq \varnothing$, $K_{i+1} \supset K_i$, $K_i \cap D = \{p_1, \ldots, p_{n(i)}\}$ where $n(i) \geq i$, $U_i = U - K_i \neq \varnothing$ is open in X, and $\overline{U_i}$ is a continuum.

By (7), $\overline{U_i} - U \subset \overline{U} \cap Z \subset K_i$ and $U - U_i \subset K_i$. Hence:

(8)   For each $i = 1, 2, \ldots, \overline{U}_i - U_i \subset K_i$.

Recall that $U = X - Z$. Hence, by (7), $X = \overline{U}_i \cup (Z \cup K_i)$ where $\overline{U}_i$ and $Z \cup K_i$ are continua and $\overline{U}_i \cap (Z \cup K_i) = \overline{U}_i \cap K_i$. Thus, since X is unicoherent,

(9) $\overline{U}_i \cap K_i$ is a continuum for each $i = 1, 2, \ldots$.

We now show that each $\overline{U}_i$ is a decomposable continuum. By (7), each $\overline{U}_i$ is a nondegenerate continuum. Suppose that $\overline{U}_m$ is indecomposable for some m. By (9) and (7), $\overline{U}_m \cap K_m$ is a proper subcontinuum of $\overline{U}_m$. Hence, $\overline{U}_m \cap K_m$ is contained in a composant M of $\overline{U}_m$. By 11.15 and 11.17, $\overline{U}_m - M \neq \varnothing$. Let $r \in \overline{U}_m - M$. By our supposition that (a) is false (and by (2)), there is a proper subcontinuum L of $\overline{U}$ such that $r \in L$ and $L \cap Z \neq \varnothing$. Let

$$W = L \cap U_m.$$

Since $r \in \overline{U}_m - M$ and $\overline{U}_m \cap K_m \subset M$, $r \in \overline{U}_m - K_m$. Hence, by (8), $r \in U_m$. Thus, since $r \in L$, $r \in W$. Therefore, the component C of W containing r exists. Since $L \cap Z \neq \varnothing$ and, by (7), $U_m \subset U$, $W \neq L$. Thus, since W is open in L, we may apply 5.7 to obtain that

$$\overline{C} \cap (L - W) \neq \varnothing, \quad \text{i.e. (since } W = L \cap U_m), \quad \overline{C} \cap (L - U_m) \neq \varnothing.$$

Hence, there exists $x \in \overline{C} \cap (L - U_m)$. Since $C \subset W$, $\overline{C} \subset \overline{U}_m$. Thus, $x \in \overline{U}_m - U_m$. Hence, by (8), $x \in K_m$ and, therefore, $x \in \overline{U}_m \cap K_m$. Thus, $x \in M$. Therefore, since $x \in \overline{C}$, $\overline{C} \cap M \neq \varnothing$. Also, since $r \in C$ and $r \in \overline{U}_m - M$, $\overline{C} \cap (\overline{U}_m - M) \neq \varnothing$. Therefore, $\overline{C}$ being a subcontinuum of $\overline{U}_m$ (since $C \subset W$) and $\overline{U}_m$ being indecomposable, we have by 11.17 that $\overline{C} = \overline{U}_m$. Thus, since $\overline{C} \subset L$, we have proved that

(10) $\overline{L} \supset \overline{U}_m$.

By (10) and (7), $X = Z \cup L \cup K_m$. Also: $Z \not\subset L \cup K_m$ [use V and recall that $L \cup K_m \subset \overline{U}$], $L \not\subset Z \cup K_m$ [use U_m, (10), and (7)], and $K_m \not\subset Z \cup L$ [if $K_m - Z \subset L$, i.e., $K_m \cap U \subset L$, then, by (10) and (7), $U \subset L$; hence, $\overline{U} = L$, a contradiction]. Thus (11.23),

(11) $X = Z \oplus L \oplus K_m$.

Recall that $L \cap Z \neq \varnothing$, and let $y \in L \cap Z$. Since $L \subset \overline{U}$, $y \in \overline{U} \cap Z$ and, hence, by (7), $y \in K_m$. Thus, $L \cap Z \cap K_m \neq \varnothing$. Therefore, by (11), X is a weak triod. This contradicts (1). Hence, our supposition that $\overline{U}_m$ is indecomposable for some m is false. Therefore, we have proved that

(12) $\overline{U}_i$ is a decomposable continuum for each $i = 1, 2, \ldots$.

Next, we prove (*):

(*)   For each positive integer ℓ, $K_\ell \cap \overline{U}_n = \varnothing$ for some $n > \ell$.

Suppose that (*) is false, i.e.: For some ℓ, $K_\ell \cap \overline{U}_i \neq \varnothing$ for all $i > \ell$. By (12), there are continua E and F such that $\overline{U}_\ell = E \oplus F$ (11.23). Hence, by 11.32 [with $A = K_\ell$ and $W = U_\ell$ (see (7))], we may assume without loss of generality that

(13) $E \cap K_\ell \neq \varnothing$

(14) $F \cap K_\ell = \varnothing$.

Since $F \subset \overline{U}_\ell$, $F \subset U_\ell$ by (8) and (14). Also, since $\overline{U}_\ell = E \oplus F$, $F - E$ is nonempty and open in $\overline{U}_\ell$. Thus, since $F - E \subset U_\ell$ and, by (7), U_ℓ is open in $\overline{U}$, we have that

(15) $F - E$ is a nonempty open subset of $\overline{U}$.

Recall that D was chosen so as to be dense in U. Hence, by (7) and (15), $F \cap K_j \neq \varnothing$ for some $j > \ell$. By (7), $K_\ell \subset K_j$. Hence, by (14), $K_\ell \subset K_j - F$. Therefore, since K_ℓ is connected (by (7)), the component J of $K_j - F$ containing K_ℓ exists. Since $U \subset K_\ell \cup U_\ell$ (by (7)) and $U = X - Z$, clearly

$X = Z \cup K_\ell \cup U_\ell$.

Therefore, since $J \supset K_\ell$ and $E \cup F \supset U_\ell$,

(16) $X = (Z \cup \overline{J}) \cup (E \cup \overline{J}) \cup (F \cup \overline{J})$.

We use the way of writing X in (16) to contradict (1), thereby proving (*). For this purpose, let

$A = Z \cup \overline{J}, B = E \cup \overline{J}$,   and $C = F \cup \overline{J}$.

By (16), $X = A \cup B \cup C$ and, since $J = \varnothing$, $A \cap B \cap C \neq \varnothing$. In the next two paragraphs, we show that A, B, and C are continua and that no one of A, B, or C is contained in the union of the other two.

Recall that E, F, Z, and $\overline{J}$ are continua. Since $J \supset K_\ell$ and, by (7), $K_\ell \supset \overline{U} \cap Z \neq \varnothing$, $J \cap Z \neq \varnothing$. Hence, A is a continuum. Since $J \supset K_\ell$, we see using (13) that B is a continuum. Recall that K_j was chosen so that $F \cap K_j \neq \varnothing$. Hence, $K_j - F$ is a proper (open) subset of the continuum K_j. Thus, by 5.7,

$\overline{J} \cap [K_j - (K_j - F)] \neq \varnothing$.

Hence, $\overline{J} \cap F \neq \varnothing$. Therefore, C is a continuum.

By (7), $B \cup C \subset \overline{U}$. Hence, using any point of V, we see that $A \not\subset B \cup C$. Next, we show that $C \not\subset A \cup B$. By (15), there exists $y \in F - E$.

Clearly, $y \in C$. Since $F \subset \overline{U}_\ell$, $F \subset U_\ell$ by (8) and (14). Thus, since $U_\ell \subset U$ (by (7)), $F \subset U$. Hence, $y \notin Z$. Recall that $J \subset K_j - F$. Thus, in particular, $(F - E) \cap J = \varnothing$ and, by (7), $\overline{J} \subset \overline{U}$. Hence, by (15),

(17')...

$(F - E) \cap \overline{J} = \varnothing.$

Thus, $y \notin \overline{J}$. Since we have shown $y \in C - (Z \cup E \cup \overline{J})$, we have proved that $C \not\subset A \cup B$. Finally, we show that $B \not\subset A \cup C$. Recall from (7) that

(17) $\quad \overline{U} \cap Z \subset K_j.$

Since $\overline{U}_\ell = E \cup F$ and, by (7), $\overline{U}_\ell \subset \overline{U}$, note that

(18) $\quad E - F \subset \overline{U}.$

Since $j > \ell$, we see from (7) that $U_j \subset U_\ell$; thus, since $\overline{U}_\ell = E \cup F$,

(19) $\quad U_j \subset E \cup F.$

Now, suppose that $B \subset A \cup C$. Then, using only the definitions of A, B, and C, we have that

$E - F \subset Z \cup \overline{J}.$

Thus, since $\overline{J} \subset K_j$, $E - F \subset Z \cup K_j$. Hence, by (17) and (18), $E - F \subset K_j$. Thus, since $U - K_j = U_j$ (by (7)),

$U - (E - F) \supset U_j.$

Therefore, by (19), $U_j \subset F$. Thus, $\overline{U}_j \subset F$. Hence, by (14), $\overline{U}_j \cap K_\ell = \varnothing$ which, since $j > \ell$, contradicts our supposition that (*) is false for ℓ. Therefore, $B \not\subset A \cup C$.

From what we have shown in the last two paragraphs and from (16), we now have (by 11.23) that

$X = A \oplus B \oplus C.$

Thus, since $A \cap B \cap C \supset \overline{J} \neq \varnothing$, X is a weak triod (by 11.24). This contradicts (1). Therefore, we have proved (*).

We now use (*) to contradict (1) again and thus obtain contradiction to our initial supposition that (a) is false.

By (7), $\overline{U}_i \supset \overline{U}_{i+1}$ and $\overline{U}_i \neq \varnothing$ for each $i = 1, 2, \ldots$. Hence, by the second part of 1.7, $\cap_{i=1}^\infty \overline{U}_i \neq \varnothing$. Fix $t \in \cap_{i=1}^\infty \overline{U}_i$. Then, by (*),

(20) $\quad t \notin \bigcup_{i=1}^\infty K_i.$

Clearly, $t \in \overline{U}$ (by (7)). Hence, by our supposition that (a) is false (and by (2)), there is a proper subcontinuum Q of $\overline{U}$ such that $t \in Q$ and $Q \cap Z \neq$

$\varnothing$. Now, suppose that $Q \supset K_i - Z$ for infinitely many i (hence, for all i by (7)). Then, since $D \subset U$, we see from (7) that $Q \supset D$. Thus, since $\bar{D} = \bar{U}$, $Q = \bar{U}$, a contradiction. Hence:

(21)   There exists ℓ such that $Q \not\supset K_i - Z$ for any $i \geq \ell$.

For ℓ in (21), we have by (*) that

(22) $K_\ell \cap \bar{U}_n = \varnothing$ for some $n > \ell$.

Since $U = X - Z$, we have by (7) that $X = Z \cup K_n \cup U_n$. Thus, clearly,

(23) $X = Z \cup K_n \cup (Q \cup \bar{U}_n)$.

We use the way of writing X in (23) to contradict (1). First, Z and, by (7), K_n are continua. Also, since $t \in Q \cap \bar{U}_n$ and since Q and, by (7), $\bar{U}_n$ are continua, $Q \cup \bar{U}_n$ is a continuum. Next, we prove (i)–(iii) below:

   (i) $Z \not\subset K_n \cup (Q \cup \bar{U}_n)$;

   (ii) $K_n \not\subset Z \cup (Q \cup \bar{U}_n)$;

   (iii) $Q \cup \bar{U}_n \not\subset Z \cup K_n$.

To prove (i), simply recall that $K_n \cup (Q \cup \bar{U}_n) \subset \bar{U}$ and that $V \cap \bar{U} = \varnothing$. To prove (ii), first use (21) to obtain a point x such that

$x \in (K_\ell - Z) - Q$.

Then (since $x \in K_\ell$), by (22), $x \notin \bar{U}_n$ and, by (7), $x \in K_n$ (since $n > \ell$). Now, we prove (iii). Since $t \in Q$, $t \in Q \cup \bar{U}_n$. If $t \in Z$, then, since $t \in Q \subset \bar{U}$, we have that $t \in \bar{U} \cap Z$; hence, by (7), $t \in K_i$ for all i which clearly contradicts (20). Thus, $t \notin Z$. Also, by (20), $t \notin K_n$. This proves (iii). Now: By (23) and what we have subsequently shown, we have that

(24) $X = Z \oplus K_n \oplus (Q \cup \bar{U}_n)$.

Since $Q \subset \bar{U}$, $Q \cap Z \subset \bar{U} \cap Z$. Hence, by (7), $Q \cap Z \subset K_n$. Thus, since Q was chosen so that $Q \cap Z \neq \varnothing$, we have that

(25) $Z \cap K_n \cap (Q \cup \bar{U}_n) \neq \varnothing$.

By (24) and (25), X is a weak triod. This contradicts (1), and we have finally run out of contradictions! Therefore, it must be that (a) is true. By symmetry, (b) is also true. Therefore, by 11.33, we have proved 11.34.

11.35   Corollary.   A nondegenerate unicoherent continuum X is irreducible if and only if X is not a triod.

Proof. Half is 11.34, and the other half is trivial.

It is clear that if X is a triod (or, more generally, a weak triod), then, for any point $p \in X$, X is the union of two proper subcontinua A and B such that $p \in A \cap B$. The converse is obviously false (e.g., $X = S^1$). However, for unicoherent continua, the converse is true (Theorem 3.3 of [11]):

11.36 Corollary. Let X be a unicoherent continuum. If, for each point $p \in X$, X is the union of two proper subcontinua each of which contains p, then X is a triod (and conversely).

Proof. Clearly, X is not irreducible (see 11.5). Hence, by 11.34, X is a triod. This proves 11.36.

We remark that the hypotheses in 11.34 are considerably stronger than they may appear to be at first glance. In particular, assuming X is not a triod in the presence of unicoherence is really assuming X is not a weak triod (by 11.26). However, the conclusion to 11.34 is compensatingly strong and, thus, 11.34 has many important applications.

Let us conclude this chapter by mentioning without proof the following general characterization theorem due to Sorgenfrey [10]: *A continuum X is irreducible about some n points, $2 \le n < \infty$, if and only if whenever*

$$X = X_1 \oplus X_2 \oplus \cdots \oplus X_{n+1},$$

the union of some n of the continua X_i *is not connected.* An easy argument shows that this theorem implies 11.34 (see 11.51).

EXERCISES

11.37 Exercise. Let X be a continuum and let C be a subcontinuum of X. Recall from 11.29 that $C = \text{irr}(p,q)$ if and only if C is irreducible between p and q (in the sense of 4.35). Generalize half of this tautology by proving (a):

(a) If $C = \text{irr}(A,B)$, then $C = \text{irr}(p,q)$ for some $p \in A$ and some $q \in B$.

Easy examples show that the converse of (a) is false. In fact, prove (b):

(b) If X is any nondegenerate continuum, then there exist nonempty compact subsets A and B of X and a subcontinuum C of X such that $C = \text{irr}(p,q)$ for some $p \in A$ and $q \in B$, but $C \ne \text{irr}(A,B)$. [Hint: First, obtain C using (b) of 4.35. Then, obtain A and B in different ways depending on whether or not C is indecomposable.]

11.38 Exercise. Prove 11.30 using 4.34 (and 4.18).

11.39 Exercise. We used the Baire Theorem in the proof of 11.17. Give a direct proof of 11.17 which does not use the Baire Theorem.

11.40 Exercise. This exercise concerns complements of composants which we showed in 11.4 are connected. Work (a)–(d) below.
(a) If X is any nondegenerate indecomposable continuum, then no composant of X has compact complement. [Hint: Recall some results in this chapter and (a) of 5.20.]
(b) Verify the statement following the proof of 11.4 by giving an example of a composant K of a decomposable continuum X such that $X - K$ is not compact. [Hint: Start with two nondegenerate disjoint continua, one of which is indecomposable. Use 3.20.] See (c).
(c) If X is an hereditarily decomposable continuum, then the complement of any composant K of X is compact; hence, if $K \neq X$, then $X - K$ is a continuum. [Hint: Use 11.4.] See (d).
(d) This is a generalization of (c): If $X = \text{irr}(p,q)$ is a continuum such that every indecomposable subcontinuum of X has empty interior in X, then $X - \kappa(p)$ is a continuum. [Hint: Let $J = X - \kappa(p)$. By 11.2 and 11.4, it suffices to prove J is closed in X. Suppose there exists $x \in \bar{J} - J$. Then, there is a proper subcontinuum A of X with $p, x \in A$. Let $y \in J$ and, by (b) of 4.35, let

$B = \text{irr}(x,y).$

Show B is decomposable, hence $B = B_x \oplus B_y$ (11.23) with $x \in B_x$ and $y \in B_y$, and show $B_x \cap J \neq \varnothing$. Obtain a contradiction by considering $A \cup B_x$.]
The condition in (d) that every indecomposable subcontinuum of X has empty interior is important to the study of usc decompositions. In particular, it characterizes those irreducible continua X for which there is a so-called finest monotone map f of X onto an arc with each $f^{-1}(t)$ having empty interior (see [6, p. 259] and Theorems 3, 6, and 10 of [12], and recall our 3.21). For the significance of this result and more information, see [8, pp. 199–200] and [12]. We also remark that (d), which is a special case of Theorem 7 of [12], is obtained in [2] from a general set of equivalences–see Theorem 1.22 of [2].

11.41 Exercise. Let X be a continuum which is irreducible between p and q. Prove that if A is a subcontinuum of X such that $p \in Bd(A)$, then A is nowhere dense in X. [Hint: Follows easily using only 11.5 and 11.6.]

11.42   Exercise.   Let X be a continuum which is irreducible between p and q, and let A be a proper subcontinuum of X such that $p \in A$. Note that $\overline{X - A}$ is a continuum by 11.6. Prove that $\overline{X - A}$ is irreducible between q and any point of $Bd(A)$; hence, in particular, if A is nowhere dense in X, X is irreducible between q and any point of A.

11.43   Exercise.   Prove that if X is a continuum such that $X = \mathrm{irr}(p,q)$ and $X = \mathrm{irr}(r,s)$, then $X = \mathrm{irr}(p,s)$ or $X = \mathrm{irr}(q,s)$. We remark that this result and the ones in 11.41 and 11.42 hold for arbitrary irreducible spaces as discussed in the paragraph following the proof of 11.9.

11.44   Exercise.   Let X and Y be continua, and let f be a continuous function from X onto Y. Prove (a) and (b) below.
  (a)   If Y is irreducible, then X contains an irreducible subcontinuum A such that $f(A) = Y$. [Hint: Use (b) of 4.36.]
  (b)   If $X = \mathrm{irr}(p,q)$ and f is monotone, then $Y = \mathrm{irr}(f(p),f(q))$. [Hint: Use 8.46.]

11.45   Exercise.   Prove that a nondegenerate continuum X is decomposable if and only if for some (or, equivalently, for each) point $p \in X$, p is not a limit point of the set of all $x \in X$ such that $X = \mathrm{irr}(p,x)$.

11.46   Exercise.   Prove that if X is an indecomposable continuum and A_i is a subcontinuum of X for each $i = 1, 2, \ldots$, then $X - \bigcup_{i=1}^{\infty} A_i$ is connected. [Hint: Recall (a) of 5.20.]

11.47   Exercise.   Prove that a cartesian product of two nondegenerate continua is always a decomposable continuum.

11.48   Exercise.   We use results about composants to obtain two hyperspace characterizations of indecomposable continua ((c) and (d) below). A follow-up result is in 11.49. Prove (a)–(d) below.
  (a)   If X is a continuum and $\mathcal{C}$ is a subcontinuum of $C(X)$, then $\bigcup \mathcal{C} \in C(X)$. Hence, $\bigcup$ may be considered as being a function from $C(C(X))$ onto $C(X)$.
  (b)   The function $\bigcup : C(C(X)) \to C(X)$ in (a) is continuous.
  (c)   A continuum X is indecomposable if and only if whenever $\mathcal{Q}$ is an arc in $C(X)$ such that $\bigcup \mathcal{Q} = X$, then $X \in \mathcal{Q}$. [Hint: Assume X is indecomposable. Let $\mathcal{Q}_{AB}$ be an arc in $C(X)$ with end points A and B such that $\bigcup \mathcal{Q}_{AB} = X$. Let $<$ be a simple ordering on $\mathcal{Q}_{AB}$ such that the order topology induced by $<$ is the topology on $\mathcal{Q}_{AB}$ (6.14) and $A < B$. Use (a) and (b) to obtain the first $F \in \mathcal{Q}_{AB}$ such that $\bigcup$

$\mathcal{C}_{AF} = X$. Then, show $F = X$ using (a), (b), and facts about composants (consider $\cup \mathcal{C}_{AE}$ and $\cup \mathcal{C}_{EF}$ for $A < E < F$). Conversely, assume X is decomposable, $X = Y \oplus Z$ (11.23). Make use of the fact that $C(Y)$ and $C(Z)$ are arcwise connected (5.25).]

(d) A nondegenerate continuum X is indecomposable if and only if $C(X) - \{X\}$ is not arcwise connected. [Hint: Assume X is indecomposable. Then, $X = irr(p,q)$ by 11.15.1. Let $\mathcal{C}$ be an arc in $C(X)$ from $\{p\}$ to $\{q\}$. Use (a), then (c). Conversely, assume X is decomposable. Then, by (c), there is an arc $\mathcal{C}$ in $C(X)$ such that $\cup \mathcal{C} = X$ and $X \notin \mathcal{C}$. Let $B \in C(X) - \{X\}$. Then, $B \cap A \neq \varnothing$ for some $A \in \mathcal{C}$. Make use of the fact that $C(A)$ and $C(B)$ are arcwise connected (5.25).]

The results in (a)–(d) are in section 1 and section 8 of [4], except that only half of (c) is explicitly stated (8.1 of [4]). Characterizations of those A in $C(X)$, or 2^X, which arcwise disconnect $C(X)$, or 2^X, for *any* continuum X are in section 4 of [9]. In particular, we note that (d) remains true with $C(X)$ replaced by 2^X (4.3 of [9]).

11.49 Exercise. This exercise is a natural continuation of 11.48. Recall that a space S is said to be *uniquely arcwise connected* provided that given p, $q \in S$ with $p \neq q$, there is one and only one arc in S with end points p and q. Prove the following theorem (8.4 and [4]) which gives an insightful picture of what it means to be hereditarily indecomposable:

THEOREM. A continuum X is hereditarily indecomposable if and only if $C(X)$ is uniquely arcwise connected.

[Hint: Assume (#) X is hereditarily indecomposable. By (d) of 5.25, $C(X)$ is arcwise connected. Let $\mathcal{C}$ be an arc in $C(X)$. Use (#) together with (a) and (c) of 11.48 to see that if $\mathcal{C}'$ is any subarc of $\mathcal{C}$, then $\cup \mathcal{C}' \in \mathcal{C}'$. Using this and recalling the definition of an order arc following (e) of 5.25, show that either $\mathcal{C}$ is an order arc or $\mathcal{C} = \mathcal{B} \cup \mathcal{C}$ where $\mathcal{B}$ and $\mathcal{C}$ are order arcs, $\mathcal{B} \cap \mathcal{C} = \{\cup \mathcal{C}\}$, the end points of $\mathcal{B}$ are $\cup \mathcal{C}$ and one of the end points of $\mathcal{C}$, and the end points of $\mathcal{C}$ are $\cup \mathcal{C}$ and the other end point of $\mathcal{C}$. Finally, prove using (#) that two order arcs in $C(X)$ with the same end points are equal; this follows once (#) is used to see that if A, $B \in C(X)$ such that $A \cap B \neq \varnothing$ and $\mu(A) = \mu(B)$ where μ is a size function (4.33), then $A = B$. Conversely, assume that X contains a decomposable subcontinuum Y, $Y = Y_1 \oplus Y_2$ (11.23). Let $p \in Y_1 \cap Y_2$. Then, using the Order Arc Theorem below (e) of 5.25, find two different arcs in $C(Y)$ from $\{p\}$ to Y.]

11.50 Exercise. Prove the statement in the paragraph following 11.24 that any nondegenerate Peano continuum X which is not an arc or a simple closed curve is a weak triod. [Hint: By (b) of 8.40, X contains a simple triod T. Let p_1, p_2, and p_3 denote the three end points of T, and let q denote the branch point of T. For each i, let A_i denote the arc in T from q to p_i and let U_i be a connected open subset of X such that $p_i \in U_i$ and $\overline{U}_i \cap (\overline{U}_j \cup A_j) = \varnothing$ for $i \neq j$. For each i, let

$$\mathbb{C}_i = \left\{ K: K \text{ is a component of } X - \bigcup_{i=1}^{3} \overline{U}_i \text{ and } \overline{K} \cap \overline{U}_i \neq \varnothing \right\}.$$

Let $X_i = \overline{U}_i \cup (\overline{\bigcup \mathbb{C}_i})$ for each $i = 1, 2, 3$. Show that X_1, X_2, X_3 satisfy 11.24 (you will use 5.7).] The result is in [11, p. 446].

11.51 Exercise. Show how to obtain 11.34 from the theorem in [10] stated in the last paragraph of this chapter. [Hint: Assume X satisfies the hypotheses of 11.34 but that X is not irreducible. Then, by the theorem in [10], $X = X_1 \oplus X_2 \oplus X_3$ where $X_i \cap X_j \neq \varnothing$ for any i and j. Use unicoherence to show that $\bigcap_{i=1}^{3} X_i \neq \varnothing$. Define Z as in the proof of 11.26, and use Z as in the proof of 11.26.] This proof of 11.34 is in [2, pp. 21–22].

11.52 Exercise. The existence of subcontinua which are irreducible or which are irr(A,B) is useful in proving fundamental results which would seem to have nothing to do with irreducibility. We give two illustrations of this in (a) and (b) below, which we ask you to prove.

(a) If X is a continuum and if A and B are mutually disjoint subcontinua of X, then there is a component K of $X - (A \cup B)$ such that $\overline{K} \cap A \neq \varnothing$ and $\overline{K} \cap B \neq \varnothing$. [Hint: Let C be as in 11.30. Show that $C - (A \cup B)$ is connected by using the usc decomposition $\mathcal{D}$ of $C \cup A \cup B$ obtained by shrinking A and B to different points and then applying 11.8 to $\mathcal{D}$.]

(b) If X is an hereditarily unicoherent continuum and if $\mathbb{C}$ is any collection of subcontinua of X, then $\bigcap \mathbb{C}$ is connected and, hence, if $\bigcap \mathbb{C} \neq \varnothing$, $\bigcap \mathbb{C}$ is a continuum. [Hint: Suppose that $\bigcap \mathbb{C} = P|Q$ (6.2). Let $p \in P$, $q \in Q$, and consider $K = $ irr(p,q).]

11.53 Exercise. Prove that if X is a decomposable, irreducible continuum, then $X = Y \oplus Z$ (11.23) where

$$Y = \overline{X - Z} \quad \text{and} \quad Z = \overline{X - Y}.$$

[Hint: Write $X = A \oplus B$, let $Y = \overline{X - B}$, let $Z = \overline{X - Y}$, and use 11.6.]

11.54 Exercise. Prove that every nondegenerate arcwise connected continuum is decomposable. [Hint: Use 11.15.1.] Hence, by (a) of 10.58, every nondegenerate dendroid is hereditarily decomposable. We note that this implies nondegenerate dendroids are one-dimensional by 13.57.

11.55 Exercise. A continuum Y is said to be *uniquely irreducible* provided that Y is irreducible about a unique pair of points, i.e., $Y = \text{irr}(p,q)$ for some $p, q \in Y$ and, if $Y = \text{irr}(p',q')$, $\{p',q'\} = \{p,q\}$. Prove that a nondegenerate continuum X is an arc if and only if each subcontinuum of X (as well as X itself) is uniquely irreducible. [Hint: Assuming the condition in the "if" part, there are unique $p, q \in X$ such that $X = \text{irr}(p,q)$. Let $r \in X - \{p,q\}$. By (b) of 4.35 (or, by 11.30), there exist K and L such that

$$K = \text{irr}(p,r) \quad \text{and} \quad L = \text{irr}(q,r).$$

Show that $K \cap L = \{r\}$ and then use this to apply 9.29.]

REFERENCES

1. David P. Bellamy, Indecomposable continua with one and two composants, Fund. Math., 101(1978), 129–134.
2. D. E. Bennett and J. B. Fugate, Continua and their non-separating subcontinua, Dissertationes Math., 149(1977).
3. Thomas J. Jech, *The Axiom of Choice*, North-Holland Publ. Co., Amsterdam, Holland, 1973.
4. J. L. Kelley, Hyperspaces of a continuum, Trans. Amer. Math. Soc., 52(1942), 22–36.
5. K. Kuratowski, Théorie des continus irréductibles entre deux points I, Fund. Math., 3(1922), 200–231.
6. K. Kuratowski, Théorie des continus irréductibles entre deux points II, Fund. Math., 10(1927), 225–275.
7. K. Kuratowski, *Topology*, Vol. I, Academic Press, New York, N.Y., 1966.
8. K. Kuratowski, *Topology*, Vol. II, Academic Press, New York, N.Y., 1968.
9. Sam B. Nadler, Jr., Arcwise accessibility in hyperspaces, Dissertationes Math., 138(1976).
10. R. H. Sorgenfrey, Concerning continua irreducible about *n* points, Amer. J. of Math., 68(1946), 667–671.
11. R. H. Sorgenfrey, Concerning triodic continua, Amer. J. of Math., 66(1944), 439–460.
12. E. S. Thomas, Jr., Monotone decompositions of irreducible continua, Dissertationes Math., 50(1966).

XII
Arc-Like Continua

Recall the definition of $\mathcal{P}$-like in 2.12. We see that $\mathcal{P}$-like continua can be changed into members of $\mathcal{P}$ by means of arbitrarily small usc decompositions, i.e., usc decompositions each of whose elements has as small a diameter as we wish (use 3.21). Thus, $\mathcal{P}$-like continua should share, or have properties which are "close to," some of the properties that all the members of $\mathcal{P}$ have in common. When $\mathcal{P}$ consists of just one continuum P, we may think of a $\mathcal{P}$-like continuum X as resembling P in the sense that there are arbitrarily small usc decompositions of X homeomorphic to P. In this sense, arc-like continua resemble an arc. Since the simplest nondegenerate continuum is an arc, the most fundamental $\mathcal{P}$-like continua must be the arc-like continua. This is one of the reasons we have chosen to devote a chapter to them.

We shall see that arc-like continua have many properties in common with an arc. However, arc-like continua may be very complicated and, thus, are a source of numerous examples and counterexamples. In connection with this, we note that, e.g., the continua in 1.10, (c) of 1.23, 2.9, and 2.27 are all arc-like. It is without a doubt that the most famous example of an arc-like continuum aside from the arc and the $\sin(1/x)$-continuum (1.5) is the pseudo-arc which we constructed in 1.23. We give the following brief history. In 1920, Knaster and Kuratowski [32] asked if a nondegenerate homogeneous continuum in R^2 must be a simple closed curve. In 1921, Mazurkiewicz [44] asked if every continuum in R^2 which is homeomorphic to each of its nondegenerate subcontinua must be an arc. In 1922, Knaster [31] described the first example of a nondegenerate hereditarily indecomposable continuum. In 1948, Bing [1] answered the question in [32] negatively with a continuum we denote by B and Moise [48] answered the question in [44] negatively with a continuum we denote by M. Because of the fundamental property Moise's continuum had in common with an arc (namely, being homeomorphic with each of its non-degenerate subcontinua), Moise called M a pseudo-arc. We remark that Moise showed M is hereditarily indecomposable (Theorem 6 of [48]) and that a later result of Henderson [21] showed this was necessary for a negative answer to the question in [44]. Bing knew about the pseudo-arc M from a lecture Moise gave [1, p. 730], and Bing modified Moise's construction of M in describing his continuum B. On showing that B gave a negative answer to the question asked in [32], Bing refuted some previously published results of Waraszkiewicz and Choquet (see [1, p. 729]). Three years later, Bing showed there is only one hereditarily indecomposable arc-like continuum (Theorem 1 of [2] and 12.11). This result showed that B, M, and Knaster's continuum K constructed in 1922 are all homeomorphic! Also, it clearly implies Moise's result [48] (use 12.1). In addition, Bing showed that "most" continua in R^n ($n \geq 2$) or Hilbert space are pseudo-arcs (Theorem 2 of [2]–see 12.70). This chain of events and results is undoubtedly responsible for the continuing interest in and development of the theory of arc-like continua.

We have constructed the pseudo-arc and indicated how to prove it is hereditarily indecomposable in 1.23. A systematic treatment of the pseudo-arc, even if it were limited to the results mentioned above, would require space we do not have. Instead, we shall devote this chapter to some of the general theory of arc-like continua.

Because of the length of this chapter, we take this opportunity to summarize what is done. We start by proving that arc-like continua are hereditarily unicoherent (12.2) and a-triodic (12.4); hence, by 11.34, they are hereditarily irreducible (12.5). We obtain two characterizations of arc-like continua, one in terms of chains (12.11) and the other in terms of inverse limits of arcs (12.19). The first of these characterizations is used to prove the invariance of being arc-like under monotone or open maps (12.14 and 12.15). The second is used to prove that arc-like continua are planar (12.20) and that there is a universal arc-like continuum (12.22). We introduce universal maps in 12.23 (i.e., maps which have a coincidence with every other map). After proving a few general results about universal maps, we show that every map of a continuum onto an arc-like continuum is universal (12.29). Thus, arc-like continua have the fixed point property (12.30). Also, 12.29 leads to some notions of span which we define in 12.31 and briefly discuss (more discussion is in the exercises in 12.57–12.60). Finally, after a short digression into some basic homotopy theory (Section 5), we prove a general result about the hyperspace $C(Y)$ of any arc-like continuum Y (12.44). This result is then used to obtain two different kinds of results. The first is that $C(Y)$ has the fixed point property (12.45), and the second is that maps from a continuum onto an arc-like continuum have a ''continuum covering'' property (12.46). More results about arc-like continua are in the exercises at the end of the chapter.

1. IRREDUCIBILITY AND RELATED PROPERTIES

A continuum is said to be *hereditarily irreducible* provided that each of its subcontinua is irreducible. Our main result in this section is that arc-like continua are hereditarily irreducible (12.5). We note that 12.1 and 12.2 give additional information. Also, 12.6 is of special interest.

12.1 Theorem. Every nondegenerate subcontinuum of an arc-like continuum is arc-like.

Proof. Let X be an arc-like continuum, and let Y be a nondegenerate subcontinuum of X. Let $\epsilon > 0$ such that $\epsilon <$ diameter (Y). Let f be an ϵ-map from X onto an arc A. Let $g = f|Y$. Then, since

$$g^{-1}(g(y)) = f^{-1}(g(y)) \cap Y,$$

g is an ϵ-map. Thus, since $\epsilon <$ diameter (Y), $g(Y)$ must be nondegenerate and hence, $g(Y)$ is an arc. We have now shown that there is an ϵ-map from Y onto an arc for each $\epsilon > 0$. This proves 12.1.

12.2 Theorem. Every arc-like continuum is hereditarily unicoherent.

Proof. By 12.1, it suffices to show that every arc-like continuum is unicoherent. Let X be an arc-like continuum. Suppose that X is not unicoherent. Then, there are subcontinua A and B of X such that $X = A \cup B$ and

(1) $A \cap B = P|Q$ (6.2).

By (1), there are open subsets U and V of X such that $P \subset U$, $Q \subset V$, and $\bar{U} \cap \bar{V} = \varnothing$. Let d denote a given metric for X and, for any two nonempty subsets K and L of X, let

$d(K,L) = \text{glb}\{d(y,z): y \in K \text{ and } z \in L\}.$

By (1) and since $P \subset U$ and $Q \subset V$, $A \cap U \neq \emptyset$ and $A \cap V \neq \emptyset$. Thus, since A is connected, we see that $A \not\subset U \cup V$. Similarly, $B \not\subset U \cup V$. Hence, letting

(2) $\epsilon = \min\{d(\overline{U},\overline{V}), d(A - [U \cup V],B), d(A,B - [U \cup V])\},$

we see that ϵ exists. Note that, by (1),

(3) $A \cap B \subset U \cup V.$

Since U and V are open in X and $\overline{U} \cap \overline{V} = \emptyset$, it follows using (3) and compactness that $\epsilon > 0$. Thus, since X is arc-like, there is an ϵ-map f from X onto $[0,1]$. Let

(4) $J = f(A) \cap f(B).$

Since J is the intersection of two subintervals of $[0,1]$,

(5) J is connected.

We shall contradict (5) by using sets G and H defined as follows:

$G = \{t \in J : f^{-1}(t) \subset U\}, \quad H = \{t \in J : f^{-1}(t) \subset V\}.$

By (a) and (b) of 7.15, G and H are each open in J. Since $U \cap V = \emptyset$, $G \cap H = \emptyset$. Recall that $P \subset U$ and, by (1), $P \neq \emptyset$ and $P \subset A \cap B$; thus, since f is an ϵ-map, it follows using (2) that $f(P) \subset G$ and, hence, $G \neq \emptyset$. Similarly (using Q), $H \neq \emptyset$. Next, suppose that $J \neq G \cup H$. Then, there exists $t \in J$ such that $t \notin G \cup H$. Thus (since f is an ϵ-map), by (2), $f^{-1}(t) \not\subset U \cup V$. Hence, there exists $x \in f^{-1}(t)$ such that $x \notin U \cup V$, say

$x \in A - [U \cup V]$

(without loss of generality). By (4), there exists $b \in B$ such that $f(b) = t$. Thus, since $x, b \in f^{-1}(t)$, $d(x,b) < \epsilon$. This contradicts (2). Hence, $J = G \cup H$. From what we have shown about G and H, we have a contradiction to (5). Therefore, X must be unicoherent. This completes the proof of 12.2.

12.3   Definition of A-Triodic.   A continuum X is said to be *a-triodic* provided that X does not contain a triod (11.22).

12.4   Theorem.   Every arc-like continuum is *a*-triodic.

Proof.   By 12.1, it suffices to show that no arc-like continuum is itself a triod. Let X be an arc-like continuum. Suppose that X is a triod. Then

(11.22), there is a subcontinuum Z of X such that $X - Z = \bigcup_{i=1}^{3} U_i$ where each U_i is nonempty and each two of U_1, U_2, U_3 and mutually separated in X (6.2). Hence,

$$X - Z = U_i|(U_j \cup U_k) \quad \text{when } i \neq j, i \neq k, \text{ and } j \neq k \quad (6.2).$$

Thus, by 6.3,

(1) $U_i \cup Z$ is a continuum for each $i = 1, 2, 3$.

Fix $p_i \in U_i$ for each i. Let d denote a metric for X, and let

$$d(p_i, X - U_i) = \text{glb}\{d(p_i,y): y \in X - U_i\} \quad \text{for each } i.$$

Now, let

(2) $\epsilon = \min\{d(p_i, X - U_i): i = 1, 2, 3\}$

and note that $\epsilon > 0$. Thus, since X is arc-like, there is an ϵ-map f from X onto $[0,1]$. Let

$$J_i = f(U_i \cup Z) \quad \text{for each } i = 1, 2, 3.$$

By (1), each J_i is a closed subinterval of $[0,1]$. Hence, noting that

$$[0,1] = \bigcup_{i=1}^{3} J_i \quad \text{and} \quad \bigcap_{i=1}^{3} J_i \supset f(Z) = \varnothing,$$

it is clear that one of J_1, J_2, J_3 is contained in the union of the other two, say $J_1 \subset J_2 \cup J_3$. Hence,

$$f(p_1) = f(q) \quad \text{for some } q \in Z \cup U_2 \cup U_3.$$

This gives a contradiction since, on the one hand, f is an ϵ-map, so $d(p_1,q) < \epsilon$, and, on the other hand, $q \in X - U_1$, so, by (2), $d(p_1,q) \geq \epsilon$. Therefore, X is not a triod. This completes the proof of 12.4.

We have shown that arc-like continua are a-triodic (12.4) and hereditarily unicoherent (12.2). Theorem 11 of [4] says that for hereditarily decomposable continua, being arc-like is equivalent to being a-triodic and hereditarily unicoherent (see [4] for the proof). It is easy to see that without some restriction, this equivalence would fail (e.g., see 12.52). In connection with this, we remark that even though tree-like continua are hereditarily unicoherent, there are a-triodic tree-like continua which are not arc-like ([28]; arc-like continua have span zero (12.32)). For some related results, see, e.g., Corollary 2 of [19].

The following result is of fundamental importance.

12.5 Theorem. Every arc-like continuum is irreducible and, in fact, hereditarily irreducible.

Proof. By 12.2 and 12.4, we may apply Sorgenfrey's Theorem in 11.34 to any arc-like continuum and to any of its subcontinua. This proves 12.5.

Even without using 12.5, it is easy to prove that a locally connected arc-like continuum must be an arc (see 8.41). By virtue of 12.5, the following stronger result is easy to prove.

12.6 Corollary. The only arcwise connected arc-like continuum is an arc.

Proof. Use 12.5.

12.7 Corollary. An arc-like continuum does not contain a weak triod.

Proof. Weak triods are not irreducible (as is noted following 11.24). Therefore, the corollary follows from the second part of 12.5 (see the comment below).

The proof we gave for 12.7 used 12.5 which depended essentially on 11.34. We note that a proof of 12.7 which does not depend in any way on 11.34 is easily done using 12.4, 12.2, and 11.26.

2. CHAINABILITY AND MONOTONE, OPEN IMAGES

We constructed the continuum in 1.10 with what we called simple chains, and we used these chains to verify that the resulting continuum is indecomposable. Many interesting and complicated continua can be constructed by using nested intersections of (the unions of) such chains—roughly speaking, the "pattern" the $n + 1$st chain makes in the nth chain determines the topological properties of the resulting continuum. When a continuum can be covered by chains with arbitrarily small links, the continuum is said to be chainable (12.8). The chains give the continuum a sort of linear flavor which facilitates the study of its properties.

We prove that a continuum is arc-like if and only if it is chainable (12.11). Then, we use chains to prove two mapping theorems (12.14 and 12.15). As we shall see from the proofs of these two theorems, the conditions being considered are probably best examined with chains rather than, e.g., with ϵ-maps or inverse limits (12.19).

The reader may recognize that much of what we do in this section is part of the general procedure of using a defining sequence of covers and their nerves to determine properties of the original space.

Note that the following definition is an abstraction of what was used in 1.10.

12.8 Definition of Chainable.   Let (X,d) be a continuum. A *chain in X* is a nonempty, finite, indexed collection $\mathcal{C} = \{U_1, ..., U_n\}$ of open subsets U_i of X such that

$$U_i \cap U_j \neq \varnothing \text{ if and only if } |i - j| \leq 1.$$

If $\mathcal{C} = \{U_1, ..., U_n\}$ is a chain in X, then the members of $\mathcal{C}$ are called *links of* $\mathcal{C}$, U_i being called the *ith link of* $\mathcal{C}$; the *mesh of* $\mathcal{C}$, written *mesh*($\mathcal{C}$), is defined by

$$\text{mesh}(\mathcal{C}) = \max\{\text{diameter } (U_i): 1 \leq i \leq n\}.$$

If $\mathcal{C}$ is a chain in X and mesh($\mathcal{C}$) $< \epsilon$, then we say that $\mathcal{C}$ is an ϵ-*chain in X*. We say that the continuum X is *chainable* provided that there is an ϵ-chain in X covering X for each $\epsilon > 0$.

We remark that chainable continua are sometimes called *linearly chainable* or *snake-like*. Also, they are sometimes called arc-like as is justified by virtue of the theorem in 12.11.

Note the examples of chains drawn below. The one on the left shows that even though $U_i \cap U_{i+2} = \varnothing$, $\overline{U}_i$ may intersect $\overline{U}_{i+2}$ (see 12.9 and 12.10). The one on the right shows that one link may be contained in another link (however, as is easy to prove, this can happen only with the first and last links).

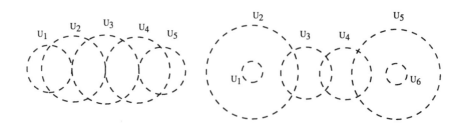

12.9 Lemma. If X is a continuum and $\mathcal{L} = \{V_1, ..., V_k\}$ is a chain in X covering X, then

$$\overline{V}_i \cap \overline{V}_j = \varnothing \qquad \text{whenever } |i - j| \geq 3.$$

Proof. Assume that $\overline{V}_i \cap \overline{V}_j \neq \varnothing$ for some i and j. Then, since $\cup \, \mathcal{L} = X$, there exists $V_\ell \in \mathcal{L}$ such that

$$V_\ell \cap (\overline{V}_i \cap \overline{V}_j) \neq \varnothing.$$

Hence, in particular, $V_\ell \cap \overline{V}_i \neq \varnothing$ and $V_\ell \cap \overline{V}_j \neq \varnothing$. Thus, since V_ℓ is open in X,

(1) $V_\ell \cap V_i \neq \varnothing$

(2) $V_\ell \cap V_j \neq \varnothing$.

By (1), $|\ell - i| \leq 1$ and, by (2), $|\ell - j| \leq 1$. Hence, $|i - j| \leq 2$. Therefore, we have proved 12.9.

12.10 Lemma. If X is a chainable continuum and $\epsilon > 0$, then there is an ϵ-chain $\mathcal{C} = \{U_1, ..., U_n\}$ in X covering X such that

$$\overline{U}_i \cap \overline{U}_j = \varnothing \qquad \text{whenever } |i - j| \geq 2.$$

Proof. Let $\mathcal{L} = \{V_1, ..., V_k\}$ be an $(\epsilon/2)$-chain in X covering X. Clearly, we may assume that $k \geq 2$. Let

$$m = \max\{i : 2i \leq k\}.$$

Note that $m \geq 1$ (since $k \geq 2$), and let

$$U_i = V_{2i-1} \cup V_{2i} \qquad \text{for each } i \leq m.$$

If k is even, let $\mathcal{C} = \{U_1, ..., U_m\}$; if k is odd, let $U_{m+1} = V_k$ and let $\mathcal{C} = \{U_1, ..., U_{m+1}\}$. In either case, it is easy to see that $\mathcal{C}$ has the desired properties (using that 12.9 holds for $\mathcal{L}$, we see that $\overline{U}_i \cap \overline{U}_j = \varnothing$ whenever $|i - j| \geq 2$). This completes the proof of 12.10.

12.11 Theorem. A nondegenerate continuum X is arc-like if and only if it is chainable.

Proof. Assume that X is a nondegenerate chainable continuum. Let $\epsilon > 0$. By 12.10, there is an $(\epsilon/2)$-chain $\mathcal{C} = \{U_1, ..., U_n\}$ in X covering X such that

(1) $\overline{U}_i \cap \overline{U}_j = \varnothing \qquad \text{whenever } |i - j| \geq 2.$

Since a chain in X of sufficiently small mesh must have at least three links, we may assume that $n \geq 3$. Let

$$m = \max\{i: 2i + 1 \leq n\}.$$

Note that $m \geq 1$ (since $n \geq 3$), and define M_k and I_k for each $k = 1, \ldots, m$ as follows:

$$M_k = \overline{U}_{2k-1} \cup \overline{U}_{2k} \cup \overline{U}_{2k+1};$$

$$I_k = [s_k, t_k] \quad \text{where } s_k = (k-1)/m \text{ and } t_k = k/m.$$

By (1), $\overline{U}_{2k-1} \cap \overline{U}_{2k+1} = \varnothing$ for each $k = 1, \ldots, m$. Hence, by Theorem 1 of [36, p. 210], there is a continuous function f_k from M_k into I_k for each $k = 1, \ldots, m$ such that

$$f_k^{-1}(s_k) = \overline{U}_{2k-1} \quad \text{and} \quad f_k^{-1}(t_k) = \overline{U}_{2k+1}.$$

Hence, letting

$$f(x) = \begin{cases} f_k(x), & \text{if } x \in M_k \text{ for some } k = 1, \ldots, m \\ 1, & \text{if } x \in \overline{U}_k \end{cases}$$

we see using (1) that f is well defined. Also, f is defined on all of X since $\cup \, \mathcal{C} = X$ and, by the definitions of m and the sets M_k,

$$X - \bigcup_{k=1}^{m} M_k \subset U_n.$$

Since each f_k is continuous, f is continuous on X. Thus, since X is a continuum and $0, 1 \in f(X)$, f maps X onto $[0,1]$. If $0 \leq r < 1$, then $f^{-1}(r) \subset \overline{U}_i$ for some i. Hence,

(2) diameter $(f^{-1}(r)) < \epsilon/2$ when $0 \leq r < 1$.

If $2m + 1 = n$, then $f^{-1}(1) = \overline{U}_n$; if $2m + 1 \neq n$, then $2m + 1 = n - 1$ and, thus, $f^{-1}(1) = \overline{U}_{n-1} \cup \overline{U}_n$. Hence, in either case,

(3) diameter $(f^{-1}(1)) < \epsilon$.

Combining (2) and (3) with the other established properties of f, we have that f is an ϵ-map from X onto $[0,1]$. Therefore, we have proved that X is arc-like. Conversely, assume that X is an arc-like continuum. Let $\epsilon > 0$. Let g be an ϵ-map from X onto $[0,1]$. Then, let $\delta > 0$ be as guaranteed by 2.33. Let $\mathcal{D} = \{W_1, \ldots, W_\ell\}$ be a δ-chain in $[0,1]$ covering $[0,1]$. Let

$$V_i = g^{-1}(W_i) \quad \text{for each } i = 1, \ldots, \ell.$$

Then, letting $\mathcal{L} = \{V_1, \ldots, V_\ell\}$, it is easy to see that $\mathcal{L}$ is an ϵ-chain in X covering X. Therefore, we have proved that X is chainable. This completes the proof of 12.11.

You are asked to provide an analogue of 12.11 for a circle-like continua in 12.61. We now give some applications of 12.11.

Recall from 3.21 that knowing the continuous images of a compact metric space is equivalent to knowing the usc decompositions of the space. We have seen several instances where we determined all usc decompositions by first determining all continuous images. For example, we did this for the Cantor set (7.7 and 7.8) and for any Peano continuum (8.19 and 8.20). Now, we shall see the other side of the coin, namely, we shall prove a mapping result (12.14) as a consequence of a decomposition result (12.13). Actually, we used this procedure once before and, in fact, it was done for a special case of 12.13 and 12.14 (see the proof of 8.22 where we used the decomposition result in 6.24). In connection with this discussion, see the paragraph following the proof of 3.21.

We note the following general lemma about chains for use in the proof of 12.13.

12.12 Lemma. Let X be a continuum, and let $\mathcal{C} = \{U_1, \ldots, U_n\}$ be a chain in X covering X. If H is a connected subset of X such that $H \cap U_j \neq \varnothing$ and $H \cap U_\ell \neq \varnothing$ for some $j < \ell$, then

$$H \cap U_k \neq \varnothing \qquad \text{whenever } j < k < \ell.$$

Proof. Fix an integer k such that $j < k < \ell$. Let $V = H \cap (\bigcup_{i=1}^{k-1} U_i)$ and let $W = H \cap (\bigcup_{i=k+1}^{n} U_i)$. Then, V and W are disjoint, nonempty, open subsets of H. Thus, since H is connected,

$$H \neq V \cup W, \quad \text{i.e.,} \quad H \not\subset V \cup W.$$

Therefore, since $\mathcal{C}$ covers X, we must have that $H \cap U_k \neq \varnothing$. This proves 12.12.

12.13 Theorem. Let X be an arc-like continuum, and let $\mathcal{D}$ be a nondegenerate usc decomposition of X (with the decomposition topology). If each member of $\mathcal{D}$ is a subcontinuum of X, then $\mathcal{D}$ is an arc-like continuum.

Proof. By 3.10, $\mathcal{D}$ is a continuum. To show that $\mathcal{D}$ is arc-like, it suffices by 12.11 to show that $\mathcal{D}$ is chainable. Let d and ρ denote metrics for X and $\mathcal{D}$ respectively. Let $\epsilon > 0$. We construct an ϵ-chain Γ in $\mathcal{D}$ covering $\mathcal{D}$ as follows. Let $\pi: X \to \mathcal{D}$ denote the natural map (3.1). Since π is continuous, π is uniformly continuous. Hence, there exists $\delta > 0$ such that

(1) if $d(x,y) < \delta$, then $\rho(\pi(x), \pi(y)) < \epsilon/5$.

By 12.11 X is chainable. Thus, there is a δ-chain $\mathcal{C} = \{U_1, \ldots, U_n\}$ in X covering X. Let $k(1) = 1$. If $k(1) < n$, let

$$k(2) = \max\{j \geq k(1): D \cap U_{k(1)} \neq \varnothing \text{ and } D \cap U_j \neq \varnothing$$
$$\text{for some } D \in \mathcal{D}\}.$$

Note that $k(2) > k(1)$ since $U_1 \cap U_2 \neq \varnothing$. If $k(2) < n$, let

$$k(3) = \max\{j \geq k(2): D \cap U_{k(2)} \neq \varnothing \text{ and } D \cap U_j \neq \varnothing$$
$$\text{for some } D \in \mathcal{D}\}.$$

Note that $k(3) > k(2)$ since $U_{k(2)} \cap U_{k(2)+1} \neq \varnothing$. Hence, continuing in this fashion (if $k(3) < n$) finitely many times, we obtain

$$1 = k(1) < k(2) < \cdots < k(m) = n, \quad k(i) < n \text{ if } i < m.$$

For later use, note (2) and (3) below which follow immediately from the way the integers $k(i)$ are defined:

(2) If $1 \leq i \leq m - 1$, there exists $D \in \mathcal{D}$ such that $D \cap U_{k(i)} \neq \varnothing$ and $D \cap U_{k(i+1)} \neq \varnothing$;

(3) If $1 \leq i \leq m - 2$, there is no $D \in \mathcal{D}$ such that $D \cap U_{k(i)} \neq \varnothing$ and $D \cap U_{k(i+1)+1} \neq \varnothing$.

If s and t are integers such that $1 \leq s \leq t \leq n$, define $\mathcal{V}(s;t)$ by

$$\mathcal{V}(s;t) = \left\{ D \in \mathcal{D}: D \subset \bigcup_{i=s}^{t} U_i \right\}$$

and note by 3.37 that

(4) Each $\mathcal{V}(s;t)$ is open in $\mathcal{D}$.

Now, assume that $m \geq 5$ (the case when $m \leq 4$ will be considered later). Let $\ell \geq 0$ be the unique integer such that

$$m - 4 \leq 3\ell + 1 \leq m - 2.$$

Define Γ by letting $\Gamma = \{\gamma_1, \ldots, \gamma_{\ell+1}\}$ where

$$\gamma_i = \begin{cases} \mathcal{V}(k(3i-2);k(3i+2)), & \text{if } 1 \leq i \leq \ell \\ \mathcal{V}(k(3\ell+1);n), & \text{if } i = \ell + 1. \end{cases}$$

By using (2), (3), and 12.12, we see that

(5) $\gamma_i \neq \varnothing$ for each $i = 1, \ldots, \ell + 1$.

An easy argument using (3) and 12.12 shows that

(6) $\cup \Gamma = \mathcal{D}$

(7) $\gamma_i \cap \gamma_j = \varnothing$ when $|i - j| \geq 2$.

By (4), we have that

(8) γ_i is open in $\mathcal{D}$ for each $i = 1, \ldots, \ell + 1$.

By (5)–(8) and the connectedness of $\mathcal{D}$, we must have that

(9) $\gamma_i \cap \gamma_{i+1} \neq \varnothing$ for each $i = 1, \ldots, \ell$.

Now, by (6)–(9), Γ is a chain in $\mathcal{D}$ covering $\mathcal{D}$. Since $\mathcal{C}$ is a δ-chain in X covering X, it follows easily using (1), (2), and the triangle inequality that mesh(Γ) $< \epsilon$. Therefore, for the case when $m \geq 5$, we have constructed an ϵ-chain Γ in $\mathcal{D}$-covering $\mathcal{D}$. If $m \leq 4$, then, since $\mathcal{C}$ is a δ-chain in X covering X, it follows easily using (1), (2), and the triangle inequality that

diameter ($\mathcal{D}$) $< 4\epsilon/5$

and, hence, $\{\mathcal{D}\}$ is a $(4\epsilon/5)$-chain in $\mathcal{D}$ covering $\mathcal{D}$. This completes the proof of 12.13.

The result in 12.13 (and, hence, the following reformulation of it) is due to Bing (Theorem 3 of [2]).

12.14 Theorem. If X is an arc-like continuum and f is a monotone map from X onto a nondegenerate continuum Y, then Y is arc-like.

Proof. Use the first part of 3.21 and apply 12.13.

In 1971, Lelek asked if a nondegenerate open image of an arc-like continuum must be arc-like (Problem 3 of [41]). The following theorem due to Rosenholtz [57] gives an affirmative answer to this question. We remark that a number of people worked on the question and were somewhat surprised by the simple proof.

12.15 Theorem. If X is an arc-like continuum and f is an open map from X onto a nondegenerate continuum Y, then Y is arc-like.

Proof. By 12.11, it suffices to show that Y is chainable. Let d and ρ denote metrics for X and Y respectively. Let $\epsilon > 0$. We construct an ϵ-chain $\mathcal{L}$ in Y covering Y as follows. Since f is uniformly continuous, there exists $\delta > 0$ such that

(1) if $d(x_1, x_2) < \delta$, then $\rho(f(x_1), f(x_2)) < \epsilon$.

By 12.11, X is chainable. Hence, there is a δ-chain $\mathcal{C} = \{U_1, \ldots, U_n\}$ in X covering X. For convenience, assume that $n \geq 2$. Let

$$
V_j = \begin{cases} f(U_j), & j = 1 \text{ or } 2 \\ f(U_j) - f\left(\bigcup_{i=1}^{j-2} \overline{U}_i\right), & 3 \leq j \leq n. \end{cases}
$$

Some of the sets V_j may be empty. The ones which are nonempty will be the links of our desired chain $\mathcal{L}$, as we shall see following (2)–(4) below. Since f is an open map from X onto Y,

(2) V_j is open in Y for each $j = 1, \ldots, n$.

Clearly, from the definition of the sets V_j,

(3) $V_j \cap V_k = \varnothing$ whenever $|j - k| \geq 2$.

Next, we show that $\bigcup_{j=1}^n V_j = Y$. Let $y \in Y$. Since $f(X) = Y$ and $\bigcup \mathcal{C} = X$, ℓ defined below exists:

$$\ell = \min\{i : f^{-1}(y) \cap U_i \neq \varnothing\}.$$

Note that $f^{-1}(y) \cap U_\ell \neq \varnothing$. Hence, if $\ell = 1$ or 2, $y \in \bigcup_{j=1}^n V_j$. So, for the purpose of proof, assume $\ell \geq 3$. Since $\mathcal{C}$ is a chain in X, $U_\ell \cap (\bigcup_{i=1}^{\ell-2} U_i) = \varnothing$ and U_ℓ is open in X. Thus,

$$U_\ell \cap \left(\bigcup_{i=1}^{\ell-2} \overline{U}_i \right) = \varnothing.$$

Hence, by the definition of ℓ, $f^{-1}(y) \cap (\bigcup_{i=1}^{\ell-2} \overline{U}_i) = \varnothing$. Thus, since $f^{-1}(y) \cap U_\ell \neq \varnothing$, $y \in V_\ell$. Hence, we have proved that

(4) $\displaystyle\bigcup_{j=1}^n V_j = Y.$

Now, let $m = \max\{j \leq n : V_j \neq \varnothing\}$. Then, since $V_1 \neq \varnothing$ and $V_m \neq \varnothing$, it follows easily from (2)–(4) and the connectedness of Y that

(5) $V_j \cap V_{j+1} \neq \varnothing$ for each $j = 1, \ldots, m - 1$.

Let $\mathcal{L} = \{V_1, \ldots, V_m\}$. By (2)–(5), $\mathcal{L}$ is a chain in Y covering Y. Since $\mathcal{C}$ is a δ-chain in X covering X, we see from (1) that mesh $(\mathcal{L}) < \epsilon$. This completes the proof of 12.15.

12.16   Corollary.   If f is an open map from an arc onto a nondegenerate continuum Y, then Y is an arc.

Proof.   Use 12.15 in conjunction with either 8.28 and 12.6 or 8.17 and 8.41.

We remark that 12.16 can be proved in a more direct manner. One such proof is in 9.46, and another is in [47]. We also note that 12.15 and 12.16 have an equivalent formulation in terms of continuous decompositions (by 13.11).

We refer the reader to [57] for some other interesting results related to 12.15 which we do not have time to give a proper treatment to here.

For results about arbitrary continuous images of arc-like continua, see [14], [15], [17], [29], [40], and [46] (also, see 12.68). We remark that the characterization of the continuous images of arc-like continua in [14] and [40] is in terms of natural types of refinements using weak chains (8.11). Thus, it is an analogue of the Hahn-Mazurkiewicz Theorem (8.14).

3. REPRESENTATION BY INVERSE LIMITS

In this section, we prove that a continuum is arc-like if and only if it is an inverse limit of arcs with onto bonding maps (12.19–thus, we provide a proof for a special case of 2.13). We also give two applications of this result (12.20 and 12.22). We remark that there can be few better illustrations of the value of using inverse limits for constructing continua than that given by the proof of 12.22.

The following lemma is a special case of Lemma 2 of [43], and the proof is a straightforward specialization of the ideas in [43].

12.17 Lemma.   Let (X,d) be a nonempty compact metric space, let g_1 be a continuous function from X onto $I = [0,1]$, and let $\eta > 0$. Then there exists $\epsilon = \epsilon(\eta,g_1) > 0$ such that if g_2 is any ϵ-map from X onto I, there is a continuous function $\varphi\colon I \to I$ (φ may not map onto I) such that

$$|g_1(x) - \varphi \circ g_2(x)| < \eta \qquad \text{for all } x \in X.$$

In other words, there is a map φ (depending on g_2) such that the diagram below is η-commutative:

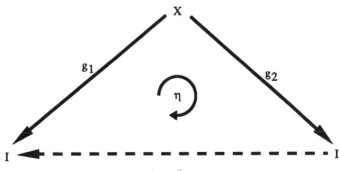

Proof. Fix a positive integer m so that $1/m < \eta/2$. Let $s_i = i/m$ for each $i = 0, 1, \ldots, m$. Since g_1 is uniformly continuous, there exists $\gamma > 0$ such that

(1) if $A \subset X$ and diameter $(A) < \gamma$, then diameter $(g_1(A)) < 1/m$.

Let $\epsilon = \gamma/2$. We show that ϵ is as desired. To prove this, let g_2 be any ϵ-map from X onto I. By 2.33, there exists $\delta > 0$ such that

(2) if $B \subset I$ and diameter $(B) < \delta$, then diameter $(g_2^{-1}(B)) < \gamma$.

Fix a positive integer n such that $1/n < \delta/2$. Let $t_j = j/n$ for each $j = 0, 1, \ldots, n$. Let

$$A_0 = [s_0, s_1), \quad A_i = (s_{i-1}, s_{i+1}) \text{ for } 1 \le i \le m - 1, \quad A_m = (s_{m-1}, s_m],$$
$$B_0 = [t_0, t_1), \quad B_j = (t_{j-1}, t_{j+1}) \text{ for } 1 \le j \le n - 1, \quad B_n = (t_{n-1}, t_n].$$

By (2), diameter $(g_2^{-1}(B_j)) < \gamma$ for each $j = 0, 1, \ldots, n$. Hence, by (1),

(3) diameter $(g_1(g_2^{-1}(B_j))) < 1/m$ for each $j = 0, 1, \ldots, n$.

For each $j = 0, 1, \ldots, n$, there exists, by (3), $A_{i(j)}$ such that

(4) $g_1(g_2^{-1}(B_j)) \subset A_{i(j)}$.

Assume from now on that for each j, $A_{i(j)}$ and, hence, $s_{i(j)}$ is fixed. Define φ on $\{t_0, t_1, \ldots, t_n\}$ by letting $\varphi(t_j) = s_{i(j)}$ for each j, and extend linearly to obtain φ on I by letting

$$\varphi(t) = (1 - \lambda) \cdot s_{i(j)} + \lambda \cdot s_{i(j+1)}, \qquad 0 \le \lambda \le 1,$$
$$\text{when } t = (1 - \lambda) \cdot t_j + \lambda \cdot t_{j+1}.$$

Clearly, $\varphi: I \to I$ is a continuous function. We prove that g_1 is η-close to $\varphi \cdot g_2$. First, since $B_j \cap B_{j+1} \ne \varnothing$ for each $j = 0, 1, \ldots, n - 1$ and since $g_2(X) = I$, clearly

$$g_1(g_2^{-1}(B_j)) \cap g_1(g_2^{-1}(B_{j+1})) \ne \varnothing$$

which, by (4), gives that $A_{i(j)} \cap A_{i(j+1)} \ne \varnothing$; hence,

(5) $|i(j) - i(j + 1)| \le 1$ for each $j = 0, 1, \ldots, n - 1$.

Now, let $x \in X$. Then, $g_2(x) \in B_j$ for some $j = 0, 1, \ldots, n$. Hence,

$$g_1(x) \in g_1(g_2^{-1}(B_j)).$$

Thus, by (4), $g_1(x) \in A_{i(j)}$. Hence,

(6) $|g_1(x) - s_{i(j)}| < 1/m$.

Since $g_2(x) \in B_j$,

$$t_{j-1} \le g_2(x) \le t_j (j \ne 0) \quad \text{or} \quad t_j \le g_2(x) \le t_{j+1} (j \ne n).$$

Hence, from the way φ is defined,

$s_{i(j-1)} \leq \varphi(g_2(x)) \leq s_{i(j)} \ (j \neq 0)$ or

$$s_{i(j)} \leq \varphi(g_2(x)) \leq s_{i(j+1)} \ (j \neq n).$$

Thus, by (5), we have in either case that

(7) $|s_{i(j)} - \varphi(g_2(x))| < 1/m.$

By (6), (7), and the triangle inequality, we see that

$|g_1(x) - \varphi(g_2(x))| < 2/m < \eta.$

This completes the proof of 12.17.

In view of the general nature of the following proposition, it would have perhaps been appropriate to have included it in Chapter II. However, because of the technical conditions in its statement and its immediate application in the proof of 12.19, we have chosen to present it here. It is Lemma 5 of [43].

12.18 Proposition. Let $Y_\infty = \varprojlim\{Y_i, f_i\}_{i=1}^\infty$ where each Y_i is a nonempty compact metric space with metric d_i ($f_i \colon Y_{i+1} \to Y_i$ may not map *onto* Y_i). Let (X, d) be a nonempty compact metric space such that, for each $i = 1, 2, \ldots$, there is an ϵ_i-map g_i from X onto Y_i where $\lim_{i \to \infty} \epsilon_i = 0$. Also, assume that, for each $i = 1, 2, \ldots$, there exists $\delta_i > 0$ such that (1)–(3) below hold:

(1)   If $A \subset Y_j$ and diameter $(A) \leq \delta_j$, then, for all $i < j$, diameter $(f_{i,j}(A)) \leq \delta_i / 2^{j-i}$ [where $f_{i,j} = f_i \circ \cdots \circ f_{j-1}$ if $j > i + 1$ and $f_{i,i+1} = f_i$];

(2)   If $d(x, x') \geq 2\epsilon_i$, then $d_i(g_i(x), g_i(x')) > 2\delta_i$;

(3) $d_i(g_i, f_i \circ g_{i+1}) \leq \delta_i / 2$ [where $d_i(g_i, f_i \circ g_{i+1}) = \mathrm{lub}\{d_i(g_i(x), f_i \circ g_{i+1}(x)) \colon x \in X\}$].

Then, X and Y_∞ are homeomorphic.

Proof. For any $f, g \colon X \to Y_i$, let (as is standard and used in (3) above)

$d_i(f, g) = \mathrm{lub}\{d_i(f(x), g(x)) \colon x \in X\}.$

The idea of the proof is as follows: We define maps $\varphi_i \colon X \to Y_i$ by showing that $\{f_{i,j} \circ g_j(x)\}_{j=i+1}^\infty$ is a Cauchy sequence in Y_i and letting $\varphi_i(x)$ be its limit; we then show $\varphi = (\varphi_1, \varphi_2, \ldots)$ is a homeomorphism from X onto Y_∞. We first prove (a) below:

(a) $d_i(g_i, f_{i,j} \circ g_j) \leq \delta_i$ whenever $i < j$.

We prove (a) by induction on $j - i$. If $j - i = 1$, then (a) follows immediately from (3). Assume inductively that (a) holds whenever $i < j$

and $j - i = k$. Let ℓ and m be such that $\ell < m$ and $m - \ell = k + 1$. Then, since $m - (\ell + 1) = k$, we have by the inductive assumption that

$$d_{\ell+1}(g_{\ell+1}, f_{\ell+1,m} \circ g_m) \leq \delta_{\ell+1}.$$

Hence, by (1) and since $f_\ell \circ f_{\ell+1,m} = f_{\ell,m}$,

$$d_\ell(f_\ell \circ g_{\ell+1}, f_{\ell,m} \circ g_m) \leq \delta_\ell/2.$$

Thus, since, by (3), $d_\ell(g_\ell, f_\ell \circ g_{\ell+1}) \leq \delta_\ell/2$, we have

$$d_\ell(g_\ell, f_{\ell,m} \circ g_m) \leq d_\ell(g_\ell, f_\ell \circ g_{\ell+1}) + d_\ell(f_\ell \circ g_{\ell+1}, f_{\ell,m} \circ g_m)$$
$$\leq \delta_\ell/2 + \delta_\ell/ = \delta_\ell.$$

Therefore, by induction, we have proved (a). Now, for any $i < j < k$, we have by (a) that $d_j(g_j, f_{j,k} \circ g_k) \leq \delta_j$; hence, by (1) and since $f_{i,j} \circ f_{j,k} = f_{i,k}$,

$$d_i(f_{i,j} \circ g_j, f_{i,k} \circ g_k) \leq \delta_i/2^{j-1}.$$

Thus, for any fixed i, we have that for each $x \in X$

$$\{f_{i,j} \circ g_j(x)\}_{j=i+1}^{\infty} \text{ is a Cauchy sequence in } Y_i.$$

Therefore, for each i, we can define a function $\varphi_i: X \to Y_i$ by letting

(b) $\quad \varphi_i(x) = \lim_{j \to \infty} f_{i,j} \circ g_j(x) \quad$ for each $x \in X$.

For later use, note that by (a) and (b) we have:

(c) $\quad d_i(g_i(x), \varphi_i(x)) \leq \delta_i \quad$ for each $x \in X$ and each i.

For any k and ℓ such that $k < \ell$, we have from (b) and the continuity of $f_{k,\ell}$ that, for each $x \in X$,

$$f_{k,\ell}(\varphi_\ell(x)) = f_{k,\ell}(\lim_{j \to \infty} f_{\ell,j} \circ g_j(x)) = \lim_{j \to \infty} f_{k,j} \circ g_j(x) = \varphi_k(x).$$

Hence,

(d) $\quad \varphi_i = f_{i,j} \circ \varphi_j \quad$ whenever $i < j$.

By letting $\varphi(x) = (\varphi_1(x), \varphi_2(x), \ldots)$ for each $x \in X$, we see by (d) that φ is a function from X into Y_∞. We now prove (e) below:

(e) $\quad \varphi$ is continuous.

To prove (e), it suffices to show that each φ_i is continuous. Fix i. Let $p_1 \in X$, and let $\epsilon > 0$. Fix $j > i$ such that $\delta_i/2^{j-1} < \epsilon/3$. By the continuity of g_j, there exists $\alpha > 0$ such that

$$d_j(g_j(p_1), g_j(x)) < \delta_j \quad \text{if } d(p_1, x) < \alpha.$$

Hence, fitting $p_2 \in X$ such that $d(p_1, p_2) < \alpha$, we have by (1) that

(*) $d_i(f_{i,j} \circ g_j(p_1), f_{i,j} \circ g_j(p_2)) \leq \delta_i / 2^{j-1} < \epsilon/3$.

By (c),

$d_j(g_j(p_n), \varphi_j(p_n)) \leq \delta_j$ for each $n = 1$ and 2.

Thus, by (1) and (d),

(#) $d_i(f_{i,j} \circ g_j(p_n), \varphi_i(p_n)) \leq \delta_i / 2^{j-i} < \epsilon/3$ for each $n = 1$ and 2.

By (#), (*), and the triangle inequality, we have that

$d_i(\varphi_i(p_1), \varphi_i(p_2)) < \epsilon$.

Hence, we have proved φ_i is continuous at p_1 and, therefore, we have proved (e). Next, we prove (f) below:

(f) $\varphi(X) = Y_\infty$.

To prove (f), it suffices by (e) and the compactness of X to show that $\varphi(X)$ is dense in Y_∞. Let W be a nonempty open subset of Y_∞. We show that $\varphi(X) \cap W \neq \varnothing$ as follows. Let $\pi_i \colon Y_\infty \to Y_i$ denote the ith projection map for each i. Now, let $y \in W$. By 2.28, there are an i and an $\epsilon > 0$ such that

$y \in \pi_i^{-1}(B_i) \subset W$ where $B_i = \{z \in Y_i \colon d_i(\pi_i(y), z) < \epsilon\}$.

Fix j such that $j > i$ and $\delta_i / 2^{j-i} < \epsilon$. Then, since $g_j(X) = Y_j$, there exists $x \in X$ such that $g_j(x) = \pi_j(y)$. Hence, by (c),

$d_j(\pi_j(y), \varphi_j(x)) = d_j(x), \varphi_j(x)) \leq \delta_j$.

Hence, by (1),

$d_i(f_{i,j} \circ \pi_j(y), f_{i,j} \circ \varphi_j(x)) \leq \delta_i / 2^{j-i} < \epsilon$.

Thus, since $f_{i,j} \circ \pi_j = \pi_i$ (2.2) and, by (d), $f_{i,j} \circ \varphi_j = \varphi_i$,

$d_i(\pi_i(y), \varphi_i(x)) < \epsilon$,   i.e., $\varphi_i(x) \in B_i$.

This means that $\pi_i(\varphi(x)) \in B_i$ and, hence, $\varphi(x) \in \pi_i^{-1}(B_i)$. Thus, $\varphi(x) \in W$. Hence, we have proved that $\varphi(X) \cap W \neq \varnothing$ and, thus, that $\varphi(X)$ is dense in Y_∞. Therefore, we have proved (f). Finally, we prove (g) below:

(g) φ is one-to-one.

To prove (g), let $x_1, x_2 \in X$ such that $x_1 \neq x_2$. Then, since $\lim_{i \to \infty} \epsilon_i = 0$, there exists i such that $d(x_1, x_2) \geq 2\epsilon_i$. By (c),

$d_i(g_i(x_k), \varphi_i(x_k)) \leq \delta_i$ for each $k = 1$ and 2.

Hence: If $\varphi_i(x_1) = \varphi_i(x_2)$, then, by the triangle inequality,

$$d_i(g_i(x_1), g_i(x_2)) \leq 2\delta_i$$

which, since $d(x_1,x_2) \geq 2\epsilon_i$, contradicts (2). Thus, $\varphi_i(x_1) \neq \varphi_i(x_2)$ and, hence, $\varphi(x_1) \neq \varphi(x_2)$. Therefore, we have proved (g). By (e)–(g), φ is a homeomorphism from X onto Y_∞. Therefore, we have proved 12.18.

Now we are ready to prove the following inverse limit representation theorem for arc-like continua.

12.19 Theorem. A continuum X is arc-like if and only if X is an inverse limit of arcs with onto bonding maps.

Proof. Let (X,d) be an arc-like continuum. We use induction to define ϵ_i, δ_i, g_i, and f_i satisfying the hypotheses of 12.18 as follows. Let ϵ_1 be such that $0 < \epsilon_1 \leq 1$. Since X is arc-like, there is an ϵ_1-map g_1 from X onto $I = [0,1]$. Note:

If $d(x_1,x_2) \geq 2\epsilon_1$, then $|g_1(x_1) - g_1(x_2)| > 0$.

Hence, by compactness, there is a $\delta_1 > 0$ such that

   (a) if $d(x_1,x_2) \geq 2\epsilon_1$, then $|g_1(x_1) - g_1(x_2)| > 2\delta_1$.

Let $\eta = \delta_1/2$. Then, there is $\epsilon = \epsilon(\eta, g_1)$ as in 12.17. Let

$$\epsilon_2 = \min\{\epsilon, 1/2\}.$$

Then, since X is arc-like, there is an ϵ_2-map g_2 from X onto I and, hence, by 12.17, there is a continuous function $f_1 \colon I \to I$ such that

   (b) $|g_1(x) - f_1 \circ g_2(x)| < \delta_1/2$ for all $x \in X$.

As in the proof of (a), there is a $\delta_2 > 0$ such that

   (c) if $d(x_1,x_2) \geq 2\epsilon_2$, then $|g_2(x_1) - g_2(x_2)| > 2\delta_2$.

In addition, we may assume by the uniform continuity of f_1 that δ_2 was chosen small enough so that

   (d) if $A \subset I$ and diameter $(A) \leq \delta_2$, then diameter $(f_1(A)) \leq \delta_1/2$.

Note that, by (a)–(d), the appropriate parts of 12.18 are satisfied by ϵ_i, δ_i, and g_i $(i = 1, 2)$ and by f_1. Now, assume inductively that for a given $k \geq 2$ we have defined

$$\epsilon_i, \delta_i, g_i \text{ for } 1 \leq i \leq k \text{ and } f_i \text{ for } 1 \leq i \leq k - 1$$

such that (1)–(3) of 12.18 are satisfied for these values of i and g_i is an ϵ_i-map from X onto I with $\epsilon_i \leq 1/i$ for $1 \leq i \leq k$. Then, let $\eta = \delta_k/2$ and let $\epsilon = \epsilon(\eta, g_k)$ be as in 12.17. Let

$$\epsilon_{k+1} = \min\{\epsilon, 1/(k + 1)\}.$$

Then, since X is arc-like, there is an ϵ_{k+1}-map g_{k+1} from X onto I and hence, by 12.17, there is a continuous function $f_k: I \to I$ such that

(e) $|g_k(x) - f_k \circ g_{k+1}(x)| < \delta_k/2$ for all $x \in X$.

As in the proof of (a), there is a $\delta_{k+1} > 0$ such that

(f) if $d(x_1, x_2) \geq 2\epsilon_{k+1}$, then $|g_{k+1}(x_1) - g_{k+1}(x_2)| > 2\delta_{k+1}$.

Let $f_{i,k+1} = f_i \circ \cdots \circ f_k$ if $i < k$ and let $f_{k,k+1} = f_k$. Since each of the finitely many maps $f_{i,k+1}$, $1 \leq i < k + 1$, is uniformly continuous, we may assume δ_{k+1} was chosen small enough so that

(g) if $A \subset I$ and diameter $(A) \leq \delta_{k+1}$, then diameter $(f_{i,k+1}(A)) \leq \delta_i/2^{k+1-i}$ for each $i < k + 1$.

Now, by induction, we have defined $\epsilon_i, \delta_i, g_i$, and f_i for each $i = 1, 2, \ldots$ so that all the hypotheses of 12.18 are satisfied. Thus, letting

$$Y_\infty = \varprojlim\{Y_i, f_i\}_{i=1}^\infty \quad \text{where } Y_i = I \text{ for each } i,$$

we have by 12.18 that X and Y_∞ are homeomorphic. The bonding maps f_i may not map onto Y_i. However, this is easily remedied: Recall from 2.6 that

$$Y_\infty = \varprojlim\{\pi_i(Y_\infty), f_i | \pi_{i+1}(Y_\infty)\}_{i=1}^\infty \quad \text{and} \quad f_i(\pi_{i+1}(Y_\infty)) = \pi_i(Y_\infty).$$

Note that since Y_∞ is a nondegenerate continuum (because X is), $\pi_i(Y_\infty)$ is a nondegenerate closed subinterval of $Y_i = I$ for all but finitely many i. Therefore, it now follows easily that X is (homeomorphic to) an inverse limit of arcs with onto bonding maps (comp., 2.37). This proves half of 12.19. The other half is a special case of what is proved in the paragraph following 2.13. Therefore, we have proved 12.19.

The theorem just proved is the second part of 2.13 for the case when $\mathcal{P} = \{\text{arc}\}$. With a few generalizations and modifications of what we have done, one can see how to prove the second part of 2.13 in general provided we do not insist the bonding maps be onto. The key generalizations have to do with 12.17 and come from thinking of the proof of 12.17 as follows: The points s_i and t_i are the vertices of small triangulations of $I = [0,1]$, the sets A_i and B_i are the open stars of these vertices, and the map φ is a simplicial map; starting the proof of 12.17 with a small enough triangulation of the range space I, we found a small enough triangulation of the domain space I so that the simplicial map $\varphi: I \to I$ made the diagram in 12.17 η-commutative. By using these ideas in the general setting of triangulations of polyhedra and modifying appropriately the statements and proofs of 12.17 and 12.19, one obtains half of the equivalence in the second part of 2.13 (but without onto bonding maps)—the other half is proved after the statement of 2.13. This more general procedure, together with some additional lemmas which are used to get onto bonding maps, is what is done in [43] to prove the second part of 2.13. We refer the reader to [43] for the details.

We now give two applications of 12.19.

The following theorem was originally proved by Bing (Theorem 4 of [4]) using different methods.

12.20 Theorem. Every arc-like continuum is embeddable in the plane R^2.

Proof. By 12.19, we may apply 2.36.

In relation to 12.20, we remark there are circle-like continua which are not embeddable in the plane. In fact, the dyadic solenoid (2.8) is not even the continuous *image* of any continuum in the plane [18]. Results concerning when a circle-like continuum can be embedded in the plane are in [3]. We note that every circle-like continuum can be embedded in the cartesian product of a simple triod and an arc (Theorem 8 of [3]). In connection with [18], see [29].

We also remark that a proof of 12.20 using 12.19, but not using 2.36, was attempted in [46]. However, the general statement about inverse limits at the beginning of the proof [46, p. 179] is false and cannot be modified so as to apply where it is used later [46, p. 180]. Comp., Theorem 3 of [9].

In previous chapters, we have seen some examples of universal continua (1.11, 10.37, and 10.49). As our next application of 12.19, we show that there is a universal arc-like continuum (12.22). The result is due to Schori [59]. Our proof is a modification of the main idea in [59], but it is more closely related to the general scheme carried out in the proof of the more general Theorem 1 of [45]. However, it differs from what is done in [45] and [59] in that, by virtue of the following lemma, we do not need to use Brown's approximation theorem [7].

12.21 Lemma. There is a countable subset $\mathcal{F}$ of the set of all continuous functions from $I = [0,1]$ into I such that if X is any arc-like continuum, then

$$X = \varprojlim \{X_i, f_i\}_{i=1}^{\infty} \quad \text{where } X_i = I \text{ and } f_i \in \mathcal{F} \text{ for each } i.$$

Proof. The lemma is a consequence of some observations concerning the proofs of 12.17 and 12.19. For any two positive integers n and m, let PL(n,m) denote the set of all continuous functions $f: I \rightarrow I$ such that (a)–(c) below hold for each $i = 0, 1, \ldots, n$:

(a) $f(i/n) = k(i)/m$ for some $k(i) = 0, 1, \ldots, m$;

(b) $|k(i) - k(i + 1)| \leq 1, 0 \leq i \leq n - 1$;

(c) $f|[i/n, (i + 1)/n]$ is linear.

Let $\mathcal{F} = \cup \{$PL(n,m): n and m are positive integers$\}$. It is clear that each PL(n,m) is a finite set. Hence, $\mathcal{F}$ is countable. We see that the map φ produced in the proof of 12.17 is a member of $\mathcal{F}$. Hence, we may assume that the map φ satisfying the conclusion of 12.17 is a member of $\mathcal{F}$. Thus, since the maps f_i in the proof of 12.19 were chosen from the maps guaranteed by 12.17, we can assume that all the maps f_i in the proof of 12.19 are members of $\mathcal{F}$. Furthermore, by the proof of 12.19, the arc-like continuum X is homeomorphic to an inverse limit using these maps f_i as the bonding maps. Therefore, 12.21 now follows.

We note the surprising fact that $\mathcal{F}$ in 12.21 can be taken to consist of only two continuous functions–see [9], [30], and, for some related work, see [61] and [62].

Now, we prove Schori's theorem concerning the existence of a universal arc-like continuum [59]. The existence of universal $\mathcal{P}$-like continua for some other classes $\mathcal{P}$ of polyhedra may be obtained from Theorems 1 and 2 of [45].

12.22 Theorem. There is a universal arc-like continuum A_∞, i.e., A_∞ is an arc-like continuum which contains a topological copy of every arc-like continuum.

Proof. Let $\mathcal{F} = \{f_i: i = 1, 2, \ldots\}$ be as in 12.21. In the diagram below, each vertical line segment denotes a copy of $I = [0,1]$ and id denotes the identity map. There are $i!$ mutually disjoint copies of I in the ith column. For each such copy I', the $i + 1$ maps $f_1, \ldots, f_i$, id each map one of the $i + 1$ copies of I in the $(i + 1)$st column into I'. Let A_1 denote the copy of I in the first column. For each $i \geq 2$, form an arc A_i by attaching at end points a new copy J of I to each two consecutively placed copies of I already in the ith column. It is easy to see how to obtain a continuous function $\alpha_i: A_{i+1} \rightarrow A_i$ which agrees with the maps $f_1, \ldots, f_i$, id wherever they appear. Now, let

$$A_\infty = \lim_{\leftarrow}\{A_i, \alpha_i\}_{i=1}^\infty.$$

By 12.19, A_∞ is an arc-like continuum. It is easy to see using 12.21 that

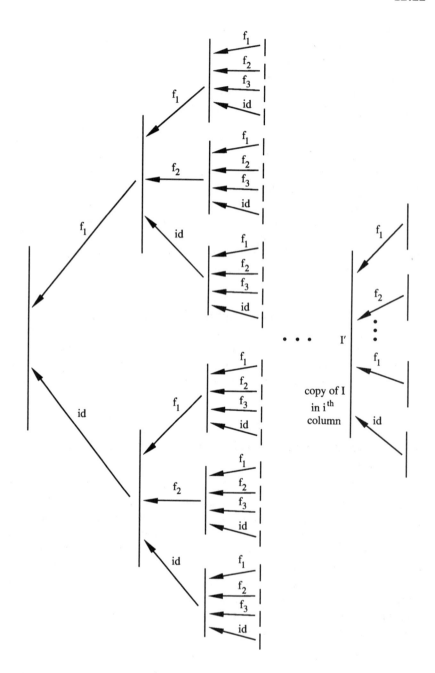

there is a topological copy of any arc-like continuum in A_∞. This completes the proof of 12.22.

4. UNIVERSAL MAPS AND SPAN

Recall (d) of 1.19. This result suggests studying the following types of maps which we shall see have an intimate connection with arc-like continua.

12.23 Definition of Universal Map. Let Y and Z be topological spaces. A continuous function $f: Y \to Z$ is said to be a *universal map* provided that if $g: Y \to Z$ is any continuous function, then $f(p) = g(p)$ for some $p \in Y$.

It is obvious that a topological space Z has the fixed point property if and only if the identity map $i: Z \to Z$ is a universal map. Some other very elementary facts about universal maps are in 12.53.

Our main result about arc-like continua in this section in 12.29 which, by the comment above, implies arc-like continua have the fixed point property (12.30). We remark that it is not known if the property given by 12.29 *characterizes* arc-like continua. This question, phrased differently, and some related questions are mentioned after 12.32.

Universal maps have played an important role in the study of continua. It would thus be appropriate to devote an entire chapter or more to these maps. We shall not do so, but we take this opportunity to prove a few general, fundamental results about them. These results will be applied to obtain 12.29 (which can be proved more directly). They will also be used to prove some other theorems about arc-like continua (e.g., 12.44–12.46 and 12.55).

The following proposition gives a characterization of universal maps to n-cells. In fact, the proposition says that such maps are the same as the Alexander-Hopf maps to n-cells whose existence characterizes those spaces Z of dimension $\geq n$ (see 13.54, 13.55, and Chapter VI of [27]). The "if" part of the proposition is due to Lokuciewski [42], and the other half has been observed by several people (e.g., in 1.1 of [25]).

12.24 Proposition. Let Z be a topological space, and let $f: Z \to B^n$ be continuous. Then, f is universal if and only if

$$f|f^{-1}(S^{n-1}): f^{-1}(S^{n-1}) \to S^{n-1}$$

cannot be extended to a continuous function from Z into S^{n-1}.

Proof. If $f|f^{-1}(S^{n-1}): f^{-1}(S^{n-1}) \to S^{n-1}$ can be extended to a continuous function $F: Z \to S^{n-1}$, then, clearly, $f(z) \neq -F(z)$ for all $z \in Z$; thus, since $-F$ is a continuous function from Z into B^n, f is not universal. Conversely, assume that f is not universal. Then, there is a continuous function $g: Z \to B^n$ such that $f(z) \neq g(z)$. Hence, for each $z \in Z$, the directed line segment beginning at $g(z)$, going through $f(z)$, and ending at a (unique) point $\varphi(z)$ in S^{n-1} exists and is unique. It is easy to see that the function $\varphi: Z \to S^{n-1}$ so defined is a continuous extension of $f|f^{-1}(S^{n-1})$. This completes the proof of 12.24.

Note the following equivalent formulation of 12.24. It will be used in section 6.

12.25 Corollary. The result in 12.24 remains valid with B^n replaced by any n-cell K and S^{n-1} replaced by the manifold boundary M of K (i.e., M is the image of S^{n-1} under any homeomorphism of B^n onto K).

Proof. Left as an exercise (12.54).

The following special case of 12.25 will be used in the proof of 12.29. It is Proposition 8 of [24]. A proof of it without using 12.25 is straightforward and was asked for in (d) of 1.19.

12.26 Corollary. Any continuous function from a connected topological space onto an arc is universal.

Proof. If $f: Z \to A$ is such a function, and p and q are the end points of the arc A, then, by the connectedness of Z and the ontoness of f,

$$f|f^{-1}(\{p,q\}): f^{-1}(\{p,q\}) \to \{p,q\}$$

cannot be extended to a continuous function from Z to $\{p,q\}$. Hence, by 12.25, f is universal. This proves 12.26.

The following general lemma [24, p. 436] will also be used in the proofs of 12.29 and some later results.

12.27 Lemma. Let (X_1,d_1) and (X_2,d_2) be nonempty compact metric spaces, and let $f: X_1 \to X_2$ be a continuous function. If, for each $\epsilon > 0$, there exist a space Z_ϵ and an ϵ-map $f_\epsilon: X_2 \to Z_\epsilon$ such that $f_\epsilon \circ f: X_1 \to Z_\epsilon$ is a universal map, then f is a universal map.

Proof. Let $g: X_1 \to X_2$ be continuous. Then, for each $\epsilon > 0$, since $f_\epsilon \circ f: X_1 \to Z_\epsilon$ is a universal map, there exists $p_\epsilon \in X_1$ such that

$$f_\epsilon(f(p_\epsilon)) = f_\epsilon(g(p_\epsilon)).$$

Thus, since f_ϵ is an ϵ-map,

(1) $d_2(f(p_\epsilon), g(p_\epsilon)) < \epsilon$ for each ϵ.

Let $\epsilon(i) = 1/i$ for each $i = 1, 2, \ldots$. Since $\{p_{\epsilon(i)}\}_{i=1}^{\infty}$ is a sequence in the compact metric space X_1, there is a subsequence of $\{p_{\epsilon(i)}\}_{i=1}^{\infty}$ converging to some point $p \in X_1$. Then, since $\epsilon(i) \to 0$ as $i \to \infty$ and since f and g are continuous, we see using (1) that $f(p) = g(p)$. Therefore, we have proved 12.27.

We note the following corollary which is a simple generalization of the theorem in [42].

12.28 Corollary. If X is a continuum such that, for each $\epsilon > 0$, there is a universal ϵ-map f_ϵ from X to a space Z_ϵ, then X has the fixed point property.

Proof. In 12.27, let $X_1 = X_2 = X$ and let $f: X \to X$ be the identity map. Then, by 12.27, f is a universal map. Therefore, clearly, X has the fixed point property. This proves 12.28.

Now, we prove the following theorem about arc-like continua. The theorem has been known for some time, but it is not clear who first proved it (see [38, p. 210]).

12.29 Theorem. Every continuous function from any continuum onto any arc-like continuum is a universal map.

Proof. Let f be a continuous function from a continuum Y onto an arc-like continuum X. For each $\epsilon > 0$, there is an ϵ-map f_ϵ from X onto $[0,1]$ (2.12). By 12.26, each of the maps $f_\epsilon \circ f$ is universal. Hence, by 12.27, f is a universal map. This proves 12.29.

As a consequence of 12.29, we have the following result due to Hamilton [20] (stronger results are in (c) of 12.55 and in 12.56).

12.30 Corollary. Every arc-like continuum has the fixed point property.

Proof. Let Z be an arc-like continuum. By 12.29, the identity map $i: Z \to Z$ is a universal map. Therefore, clearly, Z has the fixed point property. This proves 12.30.

As we remarked following 10.31, there are tree-like continua without the fixed point property and it is not known if such a tree-like continuum exists in the plane. The result in 12.30 shows that such a tree-like continuum cannot be arc-like.

Intensive investigation has been done on classes of continua defined in terms of various properties of mappings. We shall discuss two aspects of this in the next chapter (13.71 and 13.72). Here, we note that the continua with the property that every mapping of any continuum onto them is universal and some closely related properties have been studied using the following terminology due to Lelek ([38], [39]).

12.31 Definitions of Four Types of Span. Let (X,d) be a continuum. For any two continuous functions f_1 and f_2 from a continuum Z into X, let

$$\text{glb}(f_1,f_2) = \text{glb}\{d(f_1(z),f_2(z)): z \in Z\}.$$

Then, the *surjective span of X*, denoted by $\sigma^*(X)$, is defined as follows:

$$\sigma^*(X) = \text{lub}\{\text{glb}(f_1,f_2): f_1 \text{ and } f_2 \text{ map some continuum } Z \text{ onto } X\}.$$

The *surjective semispan of X*, denoted by $\sigma_0^*(X)$, is defined in the same way but by requiring only one of the maps f_1, $f_2: Z \to X$ to map Z onto X. Now, the *span of X* and the *semispan of X*, denoted by $\sigma(X)$ and $\sigma_0(X)$ respectively, are defined so as to take into account the subcontinua of X, namely:

$$\sigma(X) = \text{lub}\{\sigma^*(Y): Y \text{ is a subcontinuum of } X\};$$
$$\sigma_0(X) = \text{lub}\{\sigma_0^*(Y): Y \text{ is a subcontinuum of } X\}.$$

Note that these notions depend on the specific metric d on X; however, we have suppressed this dependence since there will be no confusion in what we do. It is clear that the condition $\tau(X) = 0$ where $\tau = \sigma^*, \sigma_0^*, \sigma$, or σ_0 is a topological invariant (where X is a continuum).

By considering the maps $f_1, f_2: S^1 \to S^1$ given by $f_1(z) = z$ and $f_2(z) = -z$, we see that each of $\sigma^*(S^1)$, $\sigma_0^*(S^1)$, $\sigma(S^1)$, and $\sigma_0(S^1)$ is equal to the diameter of S^1. In general, note that the diameter of a continuum (X,d) may be thought of as being

$$\text{lub}\{\text{glb}(f_1,f_2): f_1 \text{ and } f_2 \text{ map some space } Z \text{ into } X\}.$$

Thus, the notions in 12.31 may be considered to be analogues of diameter using continuous functions on continua with equal or related ranges. Clearly, for any continuum (X,d),

$$0 \leq \sigma^*(X) \leq \sigma(X) \leq \sigma_0(X) \leq \text{diameter } (X).$$
$$0 \leq \sigma^*(X) \leq \sigma_0^*(X) \leq \sigma_0(X) \leq \text{diameter } (X).$$

Some interesting geometric results are in [64]–[66].

We note the following result which relates 12.29 to the notions in

12.31. Some other basic facts about the notions in 12.31 are in 12.57–12.60.

12.32 Corollary. If X is an arc-like continuum, then each of $\sigma^*(X)$, $\sigma_0^*(X)$, $\sigma(X)$, and $\sigma_0(X)$ is equal to zero.

Proof. By 12.29 and 12.1, $\sigma_0(X) = 0$. Therefore, all of 12.32 now follows.

The result in 12.29 specifically says that if X is an arc-like continuum, then $\sigma_0^*(X) = 0$. At the present time, it is not known if $\sigma_0^*(X) = 0$ or $\sigma^*(X) = 0$ implies X is arc-like. It is also not known whether or not $\sigma(X) = 0$ implies X is arc-like (Problem 1 of [41])—see 12.59 and 12.60. Regarding these problems, we note that $\sigma(X) = 0$ if and only if $\sigma_0(X) = 0$ (Theorem 6 of [10]). In a forthcoming paper of Kato, Oversteegen, and Tymchatyn (Houston Journal) it is shown that if $\sigma^*(X) = 0$, then X is tree-like and a-triodic. Other results about these problems are in, e.g., [10] and [53].

Our next result about arc-like continua is in 12.44. It is a hyperspace theorem which is related in spirit to 12.29 and which is general enough to yield two different types of results about arc-like continua as easy corollaries (12.45 and 12.46). For the purpose of proving 12.44, we need a few results from homotopy theory. In the interest of keeping our development as self-contained as possible, we prove these results in the next section. They are not all given in full generality, but rather they are presented in a generality appropriate to the context in which they will be used.

5. A BRIEF EXCURSION TO HOMOTOPY THEORY

12.33 Definition of Homotopy. Let X and Y be topological spaces. A *homotopy* is a continuous function from $X \times [0,1]$ into Y. If $f,g: X \to Y$, then we say f *is homotopic to* g provided that there is a homotopy $h: X \times [0,1] \to Y$ such that $h(x,0) = f(x)$ and $h(x,1) = g(x)$ for all $x \in X$.

Intuitively, saying that two maps are homotopic means that either one of the maps can be continuously deformed to the other (however, one must be careful about interpreting this statement too literally—e.g., see 12.62). Clearly, the relation of being homotopic is an equivalence relation.

12.34 Definition of Essential, Inessential Map. Let X and Y be topological spaces. A map $f: X \to Y$ is said to be *essential* provided that f

is not homotopic to any constant map of X into Y. A map $f: X \to Y$ is said to be *inessential* provided that f is not essential.

Note that a map $f: X \to Y$ may be inessential and yet there may be a constant map $g: X \to Y$ such that f is not homotopic to g. In fact, assuming Y is a Hausdorff space, this happens if and only if Y is not arcwise connected (since all constant maps from X into Y are homotopic to each other if and only if Y is arcwise connected, as follows easily using 8.28).

We note the following elementary proposition.

12.35 Proposition. If $f: X \to Y$ is inessential and $A \subset X$, then $f|A: A \to Y$ is inessential.

Proof. If $h: X \times [0,1] \to Y$ is a homotopy such that for all $x \in X$ and some $q \in Y$.

$$h(x,0) = f(x) \quad \text{and} \quad h(x,1) = q,$$

then consideration of $h|A \times [0,1]$ shows that $f|A: A \to Y$ is inessential. This proves 12.35.

There is a subtlety in 12.35—see (a) of 12.63.

Note the following important definition, which is a special case of the notion of a lift in the theory of covering spaces.

12.36 Definition of Continuous Log. We first define the map exp: $R^1 \to S^1$ by

$$exp(t) = (\cos(t), \sin(t)) \quad \text{for all } t \in R^1.$$

Now, let X be a topological space and let $f: X \to S^1$ be continuous. Then, we say that f has a *continuous log* provided that there is a continuous function $\varphi: X \to R^1$ such that $f = exp \circ \varphi$, i.e., such that the diagram below commutes.

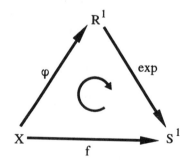

The reason for the terminology in 12.36 is clear since φ is a continuous log of f if and only if for each $x \in X$, $\varphi(x)$ is a value of $(1/i) \log(f(x))$ where i is the complex number $(0,1)$ and log is the multi-valued complex log function.

We determine which maps from compact metric spaces to S^1 have continuous logs in 12.38.

12.37 Lemma. Let X be a topological space and let $f, g: X \to S^1$ be continuous. If f has a continuous log and

$$\|f(x) - g(x)\| < 2 \qquad \text{for all } x \in X,$$

then g has a continuous log.

Proof. The condition that $\|f(x) - g(x)\| < 2$ for all $x \in X$ means that the map $g/f: X \to S^1$ never takes on the value $(-1,0)$. Let

$$e = \exp|(-\pi, \pi): (-\pi, \pi) \to S^1 - \{(-1,0)\}$$

and note that e is a homeomorphism of $(-\pi, \pi)$ onto $S^1 - \{(-1,0)\}$. Now, letting $\varphi: X \to R^1$ be a continuous log of f, define $\alpha: X \to R^1$ by

$$\alpha(x) = e^{-1}(g(x)/f(x)) + \varphi(x) \qquad \text{for all } x \in X.$$

Then, since $\exp(\alpha(x)) = [g(x)/f(x)]f(x) = g(x)$ for all $x \in X$, we see that α is a continuous log of g. Therefore, we have proved 12.37.

The geometry behind the proof above should be clear—since f and g are never diametrically opposite one another, there is, for each x, a unique smallest angle such that one obtains $g(x)$ upon rotating $f(x)$ through that angle. The quantity

$$e^{-1}(g(x)/f(x))$$

measures that angle in radian measure.

The following result was originally due to Eilenberg in greater generality (namely, with no restrictions on X—see [13, p. 68] or [37, pp. 426–427]).

12.38 Proposition. Let X be a compact metric space and let $f: X \to S^1$ be continuous. Then, f is inessential if and only if f has a continuous log.

Proof. First, assume that f has a continuous log φ. Then, defining $h: X \times [0,1] \to S^1$ by

$$h(x,t) = \exp(t\varphi(x)) \qquad \text{for all } (x,t) \in X \times [0,1],$$

we see that f is inessential. Next, to prove the converse, assume that f is inessential. Let $k: X \times [0,1] \to S^1$ be a homotopy such that, for each $x \in X$,

$$k(x,0) = f(x) \quad \text{and} \quad k(x,1) = q, \text{ some fixed } q \in S^1.$$

For each $t \in [0,1]$, let $k_t: X \to S^1$ be defined by

$$k_t(x) = k(x,t) \quad \text{for all } x \in X.$$

Since k is uniformly continuous, there are finitely many points $t(i) \in [0,1]$,

$$0 = t(1) < t(2) < \cdots < t(n) = 1,$$

such that, for each $i = 1, 2, \ldots, n - 1$,

(1) $\quad \|k_{t(i)}(x) - k_{t(i+1)}(x)\| < 2 \quad$ for all $x \in X$.

Since $k_{t(n)} = k_1$ is a constant map, clearly $k_{t(n)}$ has a continuous log. Hence, by (1) and 12.37, $k_{t(n-1)}$ has a continuous log. Thus, by (1) and 12.37, $k_{t(n-2)}$ has a continuous log (provided, of course, that $n \geq 3$). Continuing in this fashion, using (1) in order to apply 12.37 each time, we see after finitely many successive steps that $k_{t(1)} = f$ has a continuous log. This completes the proof of 12.38.

It is immediate from 12.38 that if two maps from X (compact metric) to S^1 are homotopic and one of them has a continuous log, then so does the other. Some simple exercises concerning what we have covered in this section so far are in 12.63 and 12.64.

Let us also note the following consequence of 12.38.

12.39 Brouwer Theorem for B^2. A 2-cell has the fixed point property.

Proof. It suffices to prove the theorem for B^2 (by (a) of 1.19). Let $j: S^1 \to S^1$ be the identity map. Suppose j has a continuous log φ. Then, since $j = \exp \circ \varphi$ (12.36), φ must be one-to-one and, hence, φ is an embedding of S^1 in R^1 which is impossible. Thus, j does not have a continuous log. Hence, by 12.38,

(1) $\quad j: S^1 \to S^1$ is essential.

Let $i: B^2 \to B^2$ be the identity map. We show that i is universal by using 12.24. Note that B^2 is contractible (defined in 8.49; simply let $h(z,t) = (1 - t) \cdot z$ for all $(z,t) \in B^2 \times [0,1]$). Thus, any continuous function $f: B^2 \to S^1$ is inessential (use $f \circ h$ where h is any contraction of B^2). Hence, by (1) and 12.35, we see that $j: S^1 \to S^1$ cannot be extended to a continuous function $f: B^2 \to S^1$. Thus, since $j = i|i^{-1}(S^1)$, we have by 12.24 that i is universal. Therefore, clearly, B^2 has the fixed point property (see the comment immediately following 12.23). This proves 12.39.

Note the following lemma for use in the next section.

12.40 Lemma. Let Y_1 and Y_2 be continua such that $Y_1 \cap Y_2 = \{p_1, p_2\}$ where $p_1 \neq p_2$. Let $S = A_1 \cup A_2$ be a simple closed curve where A_1 and A_2 are arcs such that $A_1 \cap A_2 = \{q_1, q_2\}$. Let $g: Y_1 \cup Y_2 \to S$ be a continuous function such that

$$g(Y_i) \subset A_i \quad \text{and} \quad g(p_i) = q_i \quad \text{for each } i = 1 \text{ and } 2.$$

Then, g is an essential map.

Proof. For convenience, we prove the lemma under the assumption that $S = S^1$ (we leave it to the reader to show that the lemma as stated then follows easily). Suppose that

$$g: Y_1 \cup Y_2 \to S^1 \text{ is inessential.}$$

Then, by 12.38, g has a continuous log φ. Let $K = \varphi(Y_1) \cap \varphi(Y_2)$. Since K is the intersection of two intervals in R^1, K is connected. Hence,

(1) $\exp(K)$ is connected.

For each $i = 1$ and 2, $p_i \in Y_1 \cap Y_2$ and, hence, $q_i \in g(Y_1 \cap Y_2)$. Thus, since

$$g(Y_1 \cap Y_2) = \exp(\varphi(Y_1 \cap Y_2)) \subset \exp(K),$$

we have that

(2) $q_1, q_2 \in \exp(K)$.

Clearly, any connected subset of S^1 containing two points contains one of the two arcs with these points as end points. Hence, by (1) and (2),

$A_1 \subset \exp(K)$ or $A_2 \subset \exp(K)$, say $A_1 \subset \exp(K)$.

Thus, since (by assumption) $g(Y_1) \subset A_1$, we have that $g(Y_1) \subset \exp(K)$. Therefore, since $g(Y_1) = \exp(\varphi(Y_1))$,

$\exp(\varphi(Y_1)) \subset \exp(K)$.

Hence, since $\exp(K) = \exp(\varphi(Y_1) \cap \varphi(Y_2)) \subset \exp(\varphi(Y_2))$, we have that

$\exp(\varphi(Y_1)) \subset \exp(\varphi(Y_2))$.

Thus, since $g(Y_i) = \exp(\varphi(Y_i))$ for each $i = 1$ and 2, $g(Y_1) \subset g(Y_2)$.

However, this is impossible (according to the hypotheses of the lemma). Therefore, we must have that

$g: Y_1 \cup Y_2 \to S^1$ is essential.

This completes the proof of 12.40.

It is convenient to use the language in the following definition which is a generalization of the notion of being contractible (defined in 8.49).

12.41 Definition of Contractibility with Respect to a Space. For topological spaces X and Y, we say that X is *contractible with respect to Y* provided that every continuous function $f: X \to Y$ is inessential.

The reader should verify that a space is contractible (8.49) if and only if it is contractible with respect to every space.

The following hyperspace result is of independent interest and will be used in the next section. It is due to Krasinkiewicz (1.9 of [33], stated only for $C(X)$). The proof we give uses less machinery than the proof in [33].

12.42 Proposition. If X is any continuum, then $C(X)$ and 2^X are contractible with respect to any metric absolute neighborhood retract; hence, $C(X)$ and 2^X are contractible with respect to any simple closed curve.

Proof. We do the proof for $C(X)$ and, as we shall see, the proof for 2^X is analogous. By (b) of 8.39, $X = \bigcap_{i=1}^\infty X_i$ where each X_i is a Peano continuum and $X_i \supset X_{i+1}$ for each i. Hence,

(1) $C(X) = \bigcap_{i=1}^{\infty} C(X_i)$ and $C(X_i) \supset C(X_{i+1})$ for each i.

Now, let Y be a metric absolute neighborhood retract and let $f: C(X) \to Y$ be continuous. Then, there is an open subset U of $C(X_1)$ with $C(X) \subset U$ and such that f can be extended to a continuous function $F: U \to Y$ ([37, pp. 332–333], or see Theorem 3.2 of [26, p. 84]). By (1) and 4.17, we may use 1.7 to obtain an n such that $C(X_n) \subset U$. Let

$g = F|C(X_n): C(X_n) \to Y$.

By the theorem in 8.49, $C(X_n)$ is contractible. Thus, g is inessential (see the comment following 12.41). Hence, by 12.35,

$g|C(X): C(X) \to Y$ is inessential.

Thus, since $g|C(X) = f, f: C(X) \to Y$ is inessential. Therefore, we have proved 12.42 for $C(X)$. The proof for 2^X is easily done the same way by replacing $C(X)$ by 2^X and $C(X_i)$ by 2^{X_i}. This completes the proof of 12.42.

For continua, the property of being contractible with respect to all compact metric absolute neighborhood retracts is equivalent to having trivial shape in the sense of Borsuk [6] (some other properties equivalent to having trivial shape are conveniently listed in 2.1 of [34]). Thus, 12.42 shows that $C(X)$ and 2^X have trivial shape when X is any continuum.

Applications of the material we have covered in this section are in the next section, the exercises at the end of this chapter, and the next chapter. For a much more thorough treatment of the specialized aspects of homotopy we have discussed, see, e.g., [13], [67, pp. 219–238], and appropriate parts of Chapters 8–10 of [37]. Also, see 13.76.

6. RETURN TO ARC-LIKE CONTINUA

We prove 12.44 and then derive two consequences of it. It is convenient to have the following lemma which is a special case of 12.44.

12.43 Lemma. If g is any continuous function from a continuum X onto $I = [0,1]$, then $\hat{g}: C(X) \to C(I)$, defined as in 4.27, is universal.

Proof. Recall from Section 4 of Chapter IV that $C(I)$ is a 2-cell and note that its manifold boundary S is given by $S = A_1 \cup A_2$ where

$$A_1 = \{A \in C(I): 0 \in A \text{ or } 1 \in A\}, \quad A_2 = \{\{t\}: t \in I\}.$$

Let $p_1, p_2 \in X$ such that $g(p_1) = 0$ and $g(p_2) = 1$. By the Order Arc Theorem in the discussion near the end of 5.25, there are order arcs $\alpha_i \subset C(X)$ from $\{p_i\}$ to X for each $i = 1$ and 2. Let

$$Y_1 = \alpha_1 \cup \alpha_2, \quad Y_2 = \{\{x\}: x \in X\}.$$

We see that $Y_1 \cap Y_2 = \{\{p_1\}, \{p_2\}\}$, $A_1 \cap A_2 = \{\{0\}, \{1\}\}$, and that $\hat{g}|Y_1 \cup Y_2$ maps $Y_1 \cup Y_2$ to S (continuously, by 4.27) and satisfies the hypotheses of 12.40. Hence, by 12.40,

$$\hat{g}|Y_1 \cup Y_2: Y_1 \cup Y_2 \to S \text{ is essential.}$$

Thus, by 12.35 (since $Y_1 \cup Y_2 \subset (\hat{g})^{-1}(S)$),

$$\hat{g}|(\hat{g})^{-1}(S): (\hat{g})^{-1}(S) \to S \text{ is essential.}$$

Hence, by 12.42 and 12.35, $\hat{g}|(\hat{g})^{-1}(S): (\hat{g})^{-1}(S) \to S$ cannot be extended to a continuous function which maps $C(X)$ to S. Therefore, by 12.25, $\hat{g}$: $C(X) \to C(I)$ is universal. This proves 12.43.

12.44 Theorem. If f is any continuous function from a continuum X onto any arc-like continuum Y, then $\hat{f}$: $C(X) \to C(Y)$ is universal.

Proof. The key results to be used are 12.27 and 12.43. For the purpose of using 12.27, let $\epsilon > 0$. Then, since Y is arc-like, there is an ϵ-map f_ϵ from Y onto $[0,1] = I$ (2.12). Note that by 4.39,

$\hat{f}_\epsilon$: $C(Y) \to C(I)$ is an ϵ-map.

Now, let $g = f_\epsilon \circ f$ which maps X onto I. Hence, by 12.43,

$\hat{g}$: $C(X) \to C(I)$ is universal.

Clearly, from the definition in 4.27, $\hat{g} = \hat{f}_\epsilon \circ \hat{f}$. Thus, we have shown that for each $\epsilon > 0$ there is an ϵ-map $\hat{f}_\epsilon$: $C(Y) \to C(I)$ such that $\hat{f}_\epsilon \circ \hat{f}$: $C(X) \to C(I)$ is universal. Therefore, by 12.27, $\hat{f}$ is universal. This proves 12.44.

The theorem just proved is in [52, p. 226], where it was observed that the following two well-known and diverse results follow easily from it.

12.45 Theorem (Segal [60]). If Y is an arc-like continuum, then $C(Y)$ has the fixed point property.

Proof. If i: $Y \to Y$ is the identity map, then $\hat{i}$: $C(Y) \to C(Y)$ is the identity map and, by 12.44, $\hat{i}$ is universal. Therefore, by the comment immediately following 12.23, we have proved 12.45.

We remark that at this time it is not known if 2^Y has the fixed point property for all arc-like continua Y. If Y can be represented as an inverse limit of arcs with open onto bonding maps, then 2^Y has the fixed point property by 3.4 of [51] (such is the case, e.g., when Y is the Buckethandle Continuum (2.9)). We note that if 12.43 could be proved with $\hat{g}$ replaced by g^*: $2^X \to 2^I$ (4.27), then 12.44 would be true with f replaced by f^*: $2^X \to 2^Y$ and, hence, 12.45 would be true with $C(Y)$ replaced by 2^Y.

Some results similar to 12.45 are in 12.55. In particular, we note that 12.45 remains valid when Y is a circle-like continuum ((e) of 12.55).

Our second immediate consequence of 12.44 is the following theorem.

12.46 Theorem (Read [54]). If f is any continuous function from a continuum X onto any arc-like continuum Y, then each subcontinuum of Y is the image of a subcontinuum of X under f.

Proof. By 12.44, $\hat{f}$: $C(X) \rightarrow C(Y)$ is universal. Therefore, by (a) of 12.53, $\hat{f}$ maps $C(X)$ onto $C(Y)$. This proves 12.46.

A few general comments related to 12.46 are appropriate. Whenever a natural induced map arises, it is important and often productive to inquire about what properties of the original map are carried over to be properties of the induced map (indeed, this kind of inquiry is the cornerstone of much algebraic topology and homological algebra). From this point of view, 12.46 is of fundamental importance since it gives a sufficient condition on Y in order that surjectivity be preserved when going from f: $X \rightarrow Y$ to $\hat{f}$: $C(X) \rightarrow C(Y)$. We remark that surjectivity is not always so preserved—for example, let f be the map exp in 12.36 restricted to the interval $[0,2\pi]$ (such an example was asked for in 4.27; also, compare 12.46 with the question at the end of 4.27). A further discussion of the significance of 12.46 is best postponed until Chapter XIII. There it will be seen that investigating the mapping property in 12.46 comes about naturally as a consequence of weakening a property of open maps (the resulting types of maps are called weakly confluent—see the discussion at the beginning of Section 6 of Chapter XIII, and see 13.71).

EXERCISES

12.47 Exercise. Prove that every arc-like continuum is contractible with respect to any metric absolute neighborhood retract (i.e., arc-like continua have trivial shape). [Hint: Use 12.19, 2.3, and 1.7.] One consequence is that no arc-like continuum in R^2 (comp., 12.20) separates R^2, but Theorem 10 of [37, p. 470] (or 2.1 of [11, p. 357]) is also used for the proof. A partial generalization of the result in this exercise is in (e) of 12.69.

12.48 Exercise. Some continua are both arc-like and circle-like. In fact, in general, if $\mathcal{P}_1$ and $\mathcal{P}_2$ are two collections of compact metric spaces and if there exist $P_1 \in \mathcal{P}_1$ and $P_2 \in \mathcal{P}_2$ such that each of P_1 and P_2 is a continuous image of the other, then there is a compact metric space which is both $\mathcal{P}_1$-like and $\mathcal{P}_2$-like (why?). Prove that the Buckethandle Continuum (2.9) is both arc-like and circle-like. [Hint: Consider the maps f: $[0,1] \rightarrow S^1$ and g: $S^1 \rightarrow [0,1]$ given by

$$f(t) = \exp(2\pi t), \ 12.36, \quad \text{and} \quad g(x,y) = (1/\pi)\cos^{-1}(x).]$$

The next two exercises also concern continua which are simultaneously arc-like and circle-like. These types of continua are studied in [8].

12.49 Exercise. Prove that a continuum X is arc-like if and only if for each $\epsilon > 0$, there is an inessential ϵ-map $f_\epsilon \colon X \to S^1$. [Hint: Use 12.47 for half and 12.38 for the other half.] Thus, circle-like continua which are *not* arc-like are distinguished from those which *are* arc-like (12.48) by a homotopy condition on ϵ-maps to S^1 for small ϵ.

12.50 Exercise. Prove that if X is a continuum which is both arc-like and circle-like, then X is either indecomposable or the union of two indecomposable continua. [Hint: By 12.5, $X = \mathrm{irr}(p,q)$. Assume X is decomposable. Then, there are Y and Z satisfying 11.53. Assume, as we may, $p \in Y$ and $q \in Z$. Now, suppose $Y = A \oplus B$ (11.23) with $p \in A$. Then, $A \cap Z = \varnothing$ (why?). By 11.8, $X - (A \cup Z)$ is connected. This last fact can be contradicted by using an ϵ-map f_ϵ from X onto S^1 for an appropriately chosen ϵ.]

The result in this exercise is half of the characterization in Theorem 7 of [8]. The hint is based on what is done in the proof of 3.4 of [35].

12.51 Exercise. Prove that each nondegenerate proper subcontinuum of any circle-like continuum is arc-like. [Hint: Easy using only 2.12.]

12.52 Exercise. Regarding the discussion following the proof of 12.4, prove that the p-adic solenoid Σ_p (2.8), $p \geq 2$, is a-triodic, hereditarily unicoherent, and not arc-like. [Hint: Use 12.51, 12.2, 12.4, the result in 2.8, and, to show Σ_p is not arc-like, use 12.30 and the first result in 2.16.]

12.53 Exercise. Prove the general results about universal maps (12.23) in (a)–(d) below where X, Y, and Z are topological spaces.
  (a) If $f \colon X \to Y$ is a universal map, then $f(X) = Y$.
  (b) If $f \colon X \to Y$ is a universal map, then Y has the fixed point property. Thus, Y has the fixed point property if and only if Y is a universal image of some space.
  (c) If $f \colon X \to Y$ and $g \colon Y \to Z$ are continuous and $g \circ f \colon X \to Z$ is a universal map, then g is a universal map.
  (d) If $f \colon X \to Y$ is continuous and $h \colon Y \to Z$ is a homeomorphism of Y onto Z, then f is universal if and only if $h \circ f$ is universal. The same equivalence holds for f and $f \circ h$ ($h \colon X \to Y, f \colon Y \to Z$).
In relation to (c) and (d), we remark that the composition of two universal maps need not be universal (e.g., see 2.1 of [22]).

12.54 Exercise. Show how 12.25 follows from 12.24.

12.55 Exercise. We have illustrated how to use universal maps to obtain two fixed point theorems (12.30 and 12.45). Using this technique, prove (a)–(e) below. In each case, first prove an analogue of 12.43 for natural induced maps and then use 12.28.

    (a)   The cone (see 3.28) over any arc-like continuum has the fixed point property. [Hint: First prove the cone over any continuum is contractible.]

    (b)   The suspension (see 3.29) over any arc-like continuum has the fixed point property. [Hint: Use the result in the hint for (a) and 12.38 to prove the suspension over any continuum is contractible with respect to S^1 (do (c) of 12.63 first).]

    (c)   The cartesian product of any two arc-like continua has the fixed point property. [Hint: First prove that if continua X and Y are each contractible with respect to S^1, so is $X \times Y$ (for a map $f: X \times Y \to S^1$, use 12.38 on $f|\{x\} \times Y$, each $x \in X$, and on $f|X \times \{y_0\}$, some fixed $y_0 \in Y$, to obtain a continuous log for f). Then, recall 12.47.]

    (d)   The cone over any circle-like continuum has the fixed point property. [Hint: By virtue of (a), one can make use of 12.49.] Clearly, (b) and (c) do not hold when arc-like is replaced by circle-like.

    (e)   The hyperspace $C(X)$ of a circle-like continuum X has the fixed point property. [Hint: By virtue of 12.45, one can make use of 12.49.]

The results in (a)–(e) each contains 12.39 as a special case. We remark that (c) is true for *any* cartesian product of arc-like continua [12]. The proof suggested for (e) is based on the proof in [35, p. 160].

12.56 Exercise. Prove the following result which is clearly stronger than 12.30: If X is any arc-like continuum and $F: X \to 2^X$ is continuous, then there is a point $p \in X$ such that $p \in F(p)$. [Hint: Suppose that $x \notin F(x)$ for any $x \in X$. Then, using 12.11, obtain an ϵ-chain $\mathcal{C} = \{U_1, \ldots, U_n\}$ covering X such that if $x \in U_i$ and $F(x) \cap U_j \neq \varnothing$, then $|i - j| \geq 2$. Consider

$$\left\{ x \in X \colon \text{if } x \in U_i, \text{ then } F(x) \subset \bigcup_{j=i+2}^{n} U_j \right\}.\,]$$

The theorem is due to Ward [63]. A similar result for dendrites is in 10.54. Also, see [56] and [58].

12.57 Exercise. This exercise shows how the four types of span defined in 12.31 can be computed using only the two projection maps π_i: $X \times X \to X$ where $\pi_i(x_1,x_2) = x_i$ for each $i = 1$ and 2. Prove that if (X,d) is a continuum, then, letting $\mathrm{glb}(\cdot,\cdot)$ be as in 12.31,

$\sigma^*(X) = \mathrm{lub}\{\mathrm{glb}(\pi_1|Z,\pi_2|Z)\colon Z$ is a subcontinuum of $X \times X$ and
$\pi_1(Z) = \pi_2(Z) = X\}$

$\sigma_0^*(X) = \mathrm{lub}\{\mathrm{glb}(\pi_1|Z,\pi_2|Z)\colon Z$ is a subcontinuum of $X \times X$ and
$\pi_1(Z) = X\}$.

Thus, $\sigma(X)$ and $\sigma_0(X)$ can also be computed using only π_1 and π_2. The result in this exercise has been observed in [38] and [39].

12.58 Exercise. For any continuum (X,d) and any subcontinuum A of X (with the metric inherited from d), it is immediate from 12.31 that $\sigma(A) \leq \sigma(X)$ and $\sigma_0(A) \leq \sigma_0(X)$. However, the analogous inequalities using σ^* and σ_0^* do not always hold. For example, let $X = A \cup B$ where

$A = \{(x,y) \in R^2\colon x^2 + y^2 = 1/4\}, \quad B = \{(x,0) \in R^2\colon -1/2 \leq x \leq 0\}$

with the usual Euclidean metric. Prove that

$\sigma^*(A) = 1 = \sigma_0^*(A)$ and $\sigma^*(X) = 1/2 = \sigma_0^*(X)$.

[Hint: Show that $\sigma^*(A) = 1$, $\sigma^*(X) \geq 1/2$, and $\sigma_0^*(X) \leq 1/2$.] The example is 1.4 of [39].

12.59 Exercise. Prove that if X is a continuum such that $\sigma_0^*(X) = 0$, then, for any continuous function $F\colon X \to C(X)$, there exists $p \in X$ such that $p \in F(p)$. [Hint: Find a way to use 12.57.] Thus, if $\sigma_0^*(X) = 0$, X has the fixed point property. These results remain true with σ_0^* replaced by σ since $\sigma_0^*(X) \leq \sigma_0(X)$ and $\sigma_0(X) = 0$ if $\sigma(X) = 0$ (Theorem 6 of [10]). Therefore, recalling 12.56, we see that the results in this exercise give some evidence supporting affirmative answers to the open problems discussed following 12.32. The next exercise gives more such evidence.

12.60 Exercise. Prove that if $\sigma^*(X) = 0$, then X is not a weak triod (11.24) and X is unicoherent. [Hint: For $X = X_1 \oplus X_2 \oplus X_3$ in 11.24, let $p_i \in X_i - (X_j \cup X_k)$ and

$Z_i = [\{p_i\} \times (X_j \cup X_k)] \cup [(X_j \cup X_k) \times \{p_i\}], \quad \{i,j,k\} = \{1,2,3\}$

and let $Z = \cup_{i=1}^3 Z_i$; use 12.57. For X not unicoherent, i.e., $X = A \cup B$ with A and B continua and $A \cap B = P|Q$ (6.2), let $p \in P$ and $q \in Q$. Use 5.6 to obtain continua C and D such that $p \in C \subset A$, $C - B \neq \varnothing$, $C \cap$

$Q = \varnothing, p \in D \subset B, D - A \neq \varnothing$, and $D \cap Q = \varnothing$. Let $a \in C - B$ and let $b \in D - A$. Let

$$Y = [\{q\} \times (C \cup D)] \cup [\{a\} \times B] \cup [\{b\} \times A]$$

and let $Z = Y \cup \{(x,x') \in X \times X: (x',x) \in Y\}$; use 12.57.]

By the result in this exercise we see that if $\sigma(X) = 0$, then X is a-triodic (use 11.25) and hereditarily unicoherent. Hence, by Theorem 11 of [4] (stated after the proof of 12.4), we see that any hereditarily decomposable continuum X with $\sigma(X) = 0$ must be arc-like. This gives an affirmative answer for the class of hereditarily decomposable continua to Problem 1 of [41] (stated following 12.32). This fact is well known. Our hint above is the proof of Theorem 3 of [10].

12.61 Exercise. Prove the straightforward analogue of 12.11 for circle-like continua using your own appropriate definition of what should be meant by a circular chain (in analogy with 12.8).

12.62 Exercise. After giving the definition in 12.33, we said that two maps being homotopic meant *intuitively* that either one of the maps can be continuously deformed to the other. Whatever the descriptive phrase "continuously deformed" means, it should have something to do with the topology on the function space (if you did not think so at the time, examine the homotopy $h(x,t) = (1 - t) \cdot x$ for $(x,t) \in R^1 \times [0,1]$). For simplicity, assume that X and Y are metric spaces. Let Y^X denote the set of all continuous functions from X into Y. If Y is compact with metric d, then a "natural" topology for Y^X comes from the supremum metric ρ on Y^X,

$$\rho(f,g) = \text{lub}\{d(f(x),g(x)): x \in X\}, \quad \text{all } f, g \in Y^X.$$

However, this is not a good topology with respect to our intuitive comments about homotopy–verify the facts in (a):

(a) If $X = R^1$ and $Y = S^1$, then (Y^X,ρ) is not connected even though every $f \in Y^X$ is homotopic to a constant map. [Hint: Consider all $f \in Y^X$ which have a bounded continuous log (12.36)–use 12.37.]

Another "natural" topology for Y^X is the *compact-open topology* T_{co}, i.e.: T_{co} is the topology generated by all the sets $[C,U] = \{f \in Y^X: f(C) \subset U\}$, where C is a compact subset of X and U is open in Y. Prove (b) remembering that X and Y are metric spaces.

(b) For $f, g \in Y^X$ and $f \neq g$, f is homotopic to g if and only if there is an arc in (Y^X,T_{co}) having end points f and g. [Hint: The proof is natural; you will eventually use 8.28.]

We remark that the compact-open topology agrees with the topology obtained from ρ when X is compact [37, p. 89].

12.63 Exercise. This exercise concerns the material in Section 5 preceding 12.39 and has nothing directly to do with arc-like continua. Prove (a)–(f) below.

(a) A map $f: X \to Y$ can be inessential and yet, for some $A \subset X$ (even for $A = X$), $f|A: A \to f(A)$ can be essential (give an example). Thus, writing $f|A: A \to Y$ in 12.35 was not frivolous.

(b) If $f: Z \to S^1$ (Z any space) has continuous logs φ_1 and φ_2, then $\{z \in Z: \varphi_1(z) = \varphi_2(z)\}$ is both open and closed in Z.

(c) If X is a connected topological space and $f: X \to S^1$ has continuous logs φ_1 and φ_2, then there is an integer k such that $\varphi_1(x) - \varphi_2(x) = 2k\pi$ for all $x \in X$. Thus, a continuous log for f is unique once a particular value at some point is specified.

(d) Two maps $f, g: X \to S^1$ are homotopic if and only if $f/g: X \to S^1$ is inessential.

(e) If A is a closed subset of a compact metric space X and $f: A \to S^1$ is inessential, then f can be extended to a continuous function $f^*: X \to S^1$. [Hint: Use 12.38 and the Tietze Extension Theorem [36, p. 127].] See (f).

(f) The ability to extend maps to S^1 is an invariant of homotopy class, i.e.: If A is a closed subset of a compact metric space X and if $f, g: A \to S^1$ are homotopic and one of them can be extended to a continuous function mapping X to S^1, then the other one can be so extended. [Hint: Assume g can be extended, and then use (d) and (e) to extend f/g.]

Part (f) is a very special case of the Homotopy Extension Theorem in 8.1 of [5, p. 94] which has now been proved in full generality [49].

12.64 Exercise. We have seen that arc-like continua are contractible with respect to S^1 (12.47) and hereditarily unicoherent (12.2). Prove that if X is a continuum which is contractible with respect to S^1, then X is unicoherent. [Hint: Suppose that $X = A \cup B$ where A and B are continua and $A \cap B = P|Q$ (6.2). Let $g: X \to [0, \pi]$ be continuous such that $g(P) = 0$ and $g(Q) = \pi$. Define $f: X \to S^1$ by

$$f(x) = \begin{cases} \exp(g(x)), & x \in A \\ \exp(-g(x)), & x \in B \end{cases}$$

and use 12.38 and (c) of 12.63.] We remark that the hint shows that the result is true for connected, normal spaces since 12.38 is valid for all spaces X [37, pp. 426–427].

Give an example of an hereditarily decomposable continuum which is unicoherent but not contractible with respect to S^1. Compare with (e) of 12.69.

Regarding this exercise, we note that being unicoherent is equivalent to being contractible with respect to S^1 for Peano continua [37, p. 438].

12.65 Exercise. For any continuum X, $C(X)$ and 2^X are continua (5.25) and size functions (Whitney maps) always exist (4.33). In this exercise, we show that all subcontinua of X of the same size, so-called Whitney levels, form a continuum in $C(X)$, and we obtain a result for the case when X is arc-like. Prove (a)–(d) below.

   (a)   For any continuum X, $C(X)$ and 2^X are unicoherent. [Hint: Use 12.42 and 12.64.]

   (b)   If X is a continuum and $f: C(X) \to R^1$ is continuous such that $f(A) \leq f(B)$ whenever $A \subset B$, then f is monotone (8.21). [Hint: Use (a) and the result about order arcs following (e) of 5.25.]

   (c)   If X is a continuum and $\mu: C(X) \to R^1$ is a size function, then $\mu^{-1}(t)$ is a continuum whenever $0 \leq t \leq \mu(X)$.

   (d)   If X is an arc-like continuum and $\mu: C(X) \to R^1$ is a size function, then $\mu^{-1}(t)$ is an arc-like continuum for $0 \leq t < \mu(X)$. [Hint: Fix such a t, and fix $\epsilon > 0$. Find a $\delta > 0$ such that for a δ-map f_δ of X onto $[0,1]$, the map given by $g(A) =$ midpoint of A, each $A \in \mu^{-1}(t)$, is an ϵ-map of $\mu^{-1}(t)$ onto a subarc of $[0,1]$.]

Part (c), hence parts (b) and (d), is false for 2^X [50, p. 466]. Part (d) remains true with arc-like replaced by *proper circle-like* (circle-like and not arc-like (see 12.48)) [35, 6.2 (b)]. However, part (d) is not true with arc-like replaced by circle-like [55, p. 581]. Much work has been done on Whitney levels—see Chapter XIV of [50]. On the other hand, T. West and I have recently written the first paper concerned with maps satisfying the type of inequality in (b). Our paper is entitled "Size levels for arcs."

12.66 Exercise. Arc-like continua are contractible with respect to S^1 (12.47) and their nondegenerate monotone or open images are again arc-like (12.14 and 12.15). Prove that any monotone or open map from any continuum onto S^1 is essential. [Hint: Find a way to use 12.38, 8.22, and 12.16.]

12.67 Exercise. For specific use in 12.68, prove that any continuous function from $[0,\infty)$ to S^1 has a continuous log. [Hint: $[0,\infty)$ is contractible, but 12.38 as stated does not apply. So, instead, use 12.38 by considering the closed intervals $[0,n]$, $n = 1, 2, \ldots$, and find a way to use (c) of 12.63 for a given continuous function $f:[0,\infty) \to S^1$.]

The hint is easily modified to obtain the same result with $[0,\infty)$ replaced by, e.g., R^n for any $n = 1, 2, \ldots$.

12.68 Exercise. Arc-like continua are embeddable in R^2 (12.20) and, thus, cannot be mapped continuously onto the dyadic solenoid [18]. We give an example of a simple continuum Y in R^2 which is not the continuous image of any continuum contractible with respect to S^1 (thus, Y is a planar continuum which is not the continuous image of any arc-like continuum (by 12.47)). The continuum Y is drawn in Figure 12.68 and is defined analytically by

$$Y = S^1 \cup \mathbb{S} \quad \text{where } \mathbb{S} = \{[1 - (1/t)] \exp(t): t \geq 2\}.$$

Define $r: Y \to S^1$ by $r(y) = y/\|y\|$ for all $y \in Y$. Prove that if f is a continuous function from a continuum X onto Y, then

$$r \circ f: X \to S^1 \text{ is essential.}$$

[Hint: Let f be as above, but suppose $r \circ f: X \to S^1$ is inessential. Let $Z = f^{-1}(\mathbb{S})$, and let $r' = r|\mathbb{S}: \mathbb{S} \to S^1$. There are continuous logs φ for $r \circ f$ (12.38) and α for r' (12.67). Clearly, $\varphi|Z$ and $\alpha \circ f|Z$ are continuous logs for $r \circ f|Z$. Hence, letting

$$E = \{z \in Z: \varphi(z) = \alpha \circ f(z)\},$$

we may assume $E \neq \varnothing$ (why?). By (b) of 12.63, E is open in Z. Hence, E is open in X. Thus, there is an $x \in \bar{E} - E$ (why?). Let $z_i \in E$ such that $z_i \to x$ as $i \to \infty$. Show $\{\alpha \circ f(z_i)\}_{i=1}^{\infty}$ is both bounded and unbounded in R^1.]

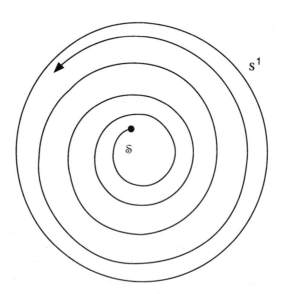

Figure 12.68

We remark that there are simple tree-like continua in R^2 which are not the continuous image of any arc-like continuum [14, p. 393]. The result in this exercise is due to Fort [18, p. 542] who used a slightly different continuum than Y. The hint is from a proof in [16].

12.69 Exercise. In (e) of this exercise, we obtain a generalization of 12.47 for hereditarily decomposable continua. Let X be a compact metric space, and let $f: X \to S^1$ be continuous. Let

$$\mathcal{E}_f = \{A \in 2^X : f | A : A \to S^1 \text{ is essential}\}.$$

Prove (a)–(f) below, (e) and (f) being of particular interest.

   (a) $\mathcal{E}_f$ is a closed subset of 2^X. [Hint: Prove first that if $f|K: K \to S^1$ has a continuous log φ for some $K \in 2^X$, then there exists U open in X such that $U \supset K$ and $f|U: U \to S^1$ has a continuous log (use (c) of 12.63). Then, (a) follows using 12.38.]

   (b) If $\mathcal{E}_f \neq \varnothing$, then $\mathcal{E}_f$ has a minimal member M. [Hint: Use (a) to apply 4.34.]

   (c) If M is as in (b), then M is a continuum and, if A is any subcontinuum of M, then $M - A$ is connected. [Hint: Use 6.3.]

   (d) If M is as in (b), then M is either indecomposable or nonunicoherent. [Hint: Use (c).]

   (e) If X is an hereditarily decomposable and hereditarily unicoherent continuum, then X is contractible with respect to S^1. [Hint: Use (b) and (d).] See comments below.

   (f) Every dendroid is contractible with respect to S^1. [Hint: Follows from (e) on recalling 11.54.] Give an example of a noncontractible dendroid (comp., 10.51).

By 12.2, (e) is a generalization of 12.47 for the class of hereditarily decomposable continua. The result would be false if hereditary unicoherence were replaced by unicoherence (an example was asked for in 12.64). Continua satisfying the hypotheses in (e) are called λ-dendroids. An application of (e) is in the proof of 13.40.

12.70 Exercise. Let M denote an n-cell, $n \geq 2$, or the Hilbert cube, and let d denote a metric for M. Let

$$\mathcal{Q} = \{A \in C(M): A \text{ is an arc-like continuum}\}.$$

Prove that $\mathcal{Q}$ is a dense G_δ in $C(M)$. [Hint: Only arcs need to be used to show $\mathcal{Q}$ is dense in $C(M)$. To show $\mathcal{Q}$ is a G_δ in $C(M)$, consider, for each

$i = 1, 2, \ldots,$ $\mathscr{F}_i = \{K \in C(M): K$ is not contained in the union of any 2^{-i}-chain in M (12.8)$\}$.]

The result shows that "most" continua in M are arc-like. Now, let $\mathscr{K}$ be as in (d) of 1.23. Then, by (d) of 1.23 and the Baire theorem [37, p. 9], $\mathscr{K} \cap \mathcal{Q}$ is a dense G_δ in $C(M)$. Therefore, since Theorem 1 of [2] says

$$\mathscr{K} \cap \mathcal{Q} = \{Y \in C(M): Y \text{ is a pseudo-arc}\},$$

we see that "most" continua in M are pseudo-arcs (Theorem 2 of [2]).

REFERENCES

1. R. H. Bing, A homogeneous indecomposable plane continuum, Duke Math. J., 15(1948), 729–742.

2. R. H. Bing, Concerning hereditarily indecomposable continua, Pacific J. Math., 1(1951), 43–51.

3. R. H. Bing, Embedding circle-like continua in the plane, Can. J. Math., 4(1962), 113–128.

4. R. H. Bing, Snake-like continua, Duke Math. J., 18(1951), 653–663.

5. Karol Borsuk, *Theory of Retracts*, Monografie Mat., Vol. 44, Polish Scientific Publishers, Warszawa, Poland, 1967.

6. Karol Borsuk, *Theory of Shape*, Monografie Mat., Vol. 59, Polish Scientific Publishers, Warszawa, Poland, 1975.

7. Morton Brown, Some applications of an approximation theorem for inverse limits, Proc. Amer. Math. Soc., 11(1960), 478–483.

8. C. E. Burgess, Chainable continua and indecomposability, Pacific J. Math., 9(1959), 653–659.

9. H. Cook and W. T. Ingram, Obtaining AR-like continua as inverse limits with only two bonding maps, Glasnik Math., 4(1969), 309–311.

10. James Francis Davis, The equivalence of zero span and zero semi-span, Proc. Amer. Math. Soc., 90(1984), 133–138.

11. James Dugundji, *Topology*, Allyn and Bacon, Inc., Boston, Mass., 1967.

12. Eldon Dyer, A fixed point theorem, Proc. Amer. Math. Soc., 7(1956), 662–672.

13. Samuel Eilenberg, Transformations continues en circonférence et la topologie du plan, Fund. Math., 26(1936), 61–112.

14. Lawrence Fearnley, Characterizations of the continuous images of the pseudo-arc, Trans. Amer. Math. Soc., 111(1964), 380–399.

15. Lawrence Fearnley, Topological operations on the class of continuous images of all snake-like continua, Proc. London Math. Soc., 15(1965), 289–300.

16. Steve Ferry, Shape equivalence does not imply CE equivalence, Proc. Amer. Math. Soc., 80(1980), 154–156.

17. Gary A. Feuerbacher, Weakly chainable circle-like continua, Fund. Math., 106(1980), 1–12.

18. M. K. Fort, Images of plane continua, Amer. J. Math., 81(1959), 541–546.

19. J. B. Fugate, Decomposable chainable continua, Trans. Amer. Math. Soc., 123(1966), 460–468.

20. O. H. Hamilton, A fixed point theorem for pseudo-arcs and certain other metric continua, Proc. Amer. Math. Soc., 2(1951), 173–174.

21. George W. Henderson, Proof that every compact decomposable continuum which is topologically equivalent to each of its nondegenerate subcontinua is an arc, Annals of Math., 72(1960), 421–428.

22. W. Holsztyński, On the composition and products of universal mappings, Fund. Math., 64(1969), 181–188.

23. W. Holsztyński, On the product and composition of universal mappings of manifolds into cubes, Proc. Amer. Math. Soc., 58(1976), 311–314.

24. W. Holsztyński, Universal mappings and fixed point theorems, Bull. Pol. Acad. Sci., 15(1967), 433–438.

25. W. Holsztyński, Universality of the product mappings onto products of I^n and snake-like spaces, Fund. Math., 64(1969), 147–155.

26. Sze-Tsen Hu, *Theory of Retracts*, Wayne State Univ. Press, Detroit, Mich., 1965.

27. Witold Hurewicz and Henry Wallman, *Dimension Theory*, Princeton Univ. Press, Princeton, N.J., 1948.

28. W. T. Ingram, An atriodic tree-like continuum with positive span, Fund. Math., 77(1972), 99–107.

29. W. T. Ingram, Concerning non-planar circle-like continua, Can. J. Math., 19(1967), 242–250.

30. V. Jarník and V. Knichal, Sur l'approximation des fonctions continues par les superpositions de deux fonctions, Fund. Math., 24(1935), 206–208.

31. B. Knaster, Un continu dont tout sous-continu est indécomposable, Fund. Math., 3(1922), 247–286.

32. B. Knaster and C. Kuratowski, Problem 2, Fund. Math., 1(1920), 223.

33. J. Krasinkiewicz, Certain properties of hyperspaces, Bull. Pol. Acad. Sci., 21(1973), 705–710.

34. J. Krasinkiewicz, Curves which are continuous images of tree-like continua are movable, Fund. Math., 89(1975), 233–260.

35. J. Krasinkiewicz, On the hyperspaces of snake-like and circle-like continua, Fund. Math., 83(1974) 155–164.

36. K. Kuratowski, *Topology*, Vol. I, Acad. Press, New York, NY, 1966.

37. K. Kuratowski, *Topology*, Vol. II, Acad. Press, New York, NY, 1968.

38. A. Lelek, Disjoint mappings and the span of spaces, Fund. Math., 55(1964), 199–214.

39. A. Lelek, On the surjective span and semispan of connected metric spaces, Colloq. Math., 37(1977), 35–45.

40. A. Lelek, On weakly chainable continua, Fund. Math., 51(1962), 271–282.

41. A. Lelek, Some problems concerning curves, Colloq. Math., 23(1971), 93–98.

42. O. W. Lokuciewski, On a theorem on fixed points, Ycπ. Mat. Hayκ, 12 3(75), (1957), 171–172 (Russian).

43. Sibe Mardešić and Jack Segal, ε-Mappings onto polyhedra, Trans. Amer. Math. Soc., 109(1963), 146–164.

44. M. Mazurkiewicz, Problem 14, Fund. Math., 2(1921), p. 286.

45. Michael C. McCord, Universal $\mathcal{P}$-like compacta, Mich. Math. J., 13(1966), 71–85.

46. J. Mioduszewski, A functional conception of snake-like continua, Fund. Math., 51(1962), 179–189.

47. S. S. Mitra, Continuous open maps on the unit interval, Amer. Math. Monthly, 76(1969), 817–818.

48. E. E. Moise, An indecomposable plane continuum which is homeomorphic to each of its nondegenerate subcontinua, Trans. Amer. Math. Soc., 63(1948), 581–594.

49. Kiiti Morita, On generalizations of Borsuk's homotopy extension theorem, Fund. Math., 88(1975), 1–6.

50. Sam B. Nadler, Jr., *Hyperspaces of Sets*, Monographs and Textbooks in Pure and Applied Math., vol. 49, Marcel Dekker, Inc., New York, N.Y., 1978.

51. Sam B. Nadler, Jr., Induced universal maps and some hyperspaces with the fixed point property, Proc. Amer. Math. Soc., 100(1987), 749–754.

52. Sam B. Nadler, Jr., Universal mappings and weakly confluent mappings, Fund. Math., 110(1980), 221–235.

53. Lex G. Oversteegen and E. D. Tymchatyn, On span and weakly chainable continua, Fund. Math., 122(1984), 159–174.

54. David R. Read, Confluent and related mappings, Colloq. Math., 29(1974), 233–239.

55. James J. Rogers, Jr., Whitney continua in the hyperspace $C(X)$, Pacific J. Math., 58(1975), 569–584.

56. Ronald H. Rosen, Fixed points for multi-valued functions on snake-like continua, Proc. Amer. Math. Soc., 10(1959), 167–173.

57. Ira Rosenholtz, Open maps of chainable continua, Proc. Amer. Math. Soc., 42(1974), 258–264.

58. Helga Schirmer, Coincidences and fixed points of multifunctions into trees, Pacific J. Math., 34(1970), 759–767.

59. Richard M. Schori, A universal snake-like continuum, Proc. Amer. Math. Soc., 16(1965), 1313–1316.

60. J. Segal, A fixed point theorem for the hyperspace of a snake-like continuum, Fund. Math., 50(1962), 237–248.

61. S. Subbiah, A dense subsemigroup of $S(R)$ generated by two elements, Fund. Math., 117(1983), 85–90.

62. Saraswathi Subbiah, Some finitely generated subsemigroups of $S(X)$, Fund. Math., 86(1975), 221–231.

63. L. E. Ward, A fixed point theorem, Amer. Math. Monthly, 65(1958), 271–272.

64. Thelma West, On the spans and width of simple triods, Proc. Amer. Math. Soc., 105(1989), 776–786.

65. Thelma West, Spans of an odd triod, Topology Proc., 8(1983), 347–353.

66. Thelma West, Spans of simple triods. Proc. Amer. Math. Soc., 102(1988), 407–415.

67. Gordon Thomas Whyburn, *Analytic Topology*, Amer. Math. Soc. Colloq. Publ., Vol. 28, Amer. Math. Soc., Providence, R.I., 1942.

XIII
Special Types of Maps

One cannot study continua without considering special types of maps. Thus, it has been both appropriate and unavoidable that in previous chapters we have discussed some particular kinds of maps and obtained results about them. In this chapter, we shall introduce some new types of maps, extend some previous results and ideas, and, as we have done before, show how special types of maps can be useful in examining properties of continua.

Since such a large number of special types of maps have been studied, we cannot even *attempt* to be comprehensive in our treatment. Instead, we shall concentrate on only a few types of maps. We present some of their basic properties, including their relationships to one another, and give some applications. For a more systematic and thorough treatment, see [32]. Also, an especially readable treatment of some results of others about monotone and open maps, and about so-called "common models," is in [33, pp. 46–52]. In this connection, also see [36].

1. MONOTONE-LIGHT FACTORIZATION

Recall that a monotone map is a continuous function whose point inverses are connected (8.21). At the other end of the spectrum are the light maps:

13.1   Definition of Light Map.   Let S_1 and S_2 be topological spaces. A continuous function f from S_1 onto S_2 is called a *light map* provided that $f^{-1}(y)$ is totally disconnected (7.10) for each $y \in S_2$.

In spite of or, perhaps, because of the antipathy between the notions of a monotone map and a light map, there is an interesting and useful connection between them which we shall prove in 13.3.

13.2   Lemma.   Let (X,T) be a compact metric space, and let $\mathcal{D}$ be a usc decomposition of X. Let

$$\mathcal{C} = \{C \subset X : C \text{ is a component of some } D \in \mathcal{D}\}.$$

Then, $\mathcal{C}$ is a usc decomposition of X.

Proof.   We show $\mathcal{C}$ satisfies 3.5. Let $C \in \mathcal{C}$ and let $U \in T$ such that $C \subset U$. Let $D \in \mathcal{D}$ such that $C \subset D$. By 3.8, D is compact. Hence, by 5.2, $D = F_1 \cup F_2$ where F_1 and F_2 are disjoint compact sets with $C \subset F_1$ and $D - U \subset F_2$ (F_2 may be empty). Clearly, $F_1 \subset U$. Thus, by the normality of X, there exists $G \in T$ such that $F_1 \subset G$, $\overline{G} \subset U$, and $\overline{G} \cap F_2 = \varnothing$. Note that (1)–(3) hold:

(1) $Bd(G) \cap D = \varnothing$;

(2) $C \subset G$;

(3) $G \subset U$.

Let $H = G \cup (X - \overline{G})$. Clearly, $H \in T$ and, by (1), $D \subset H$. Thus, since $\mathcal{D}$ is usc, there exists (by 3.5) $W \in T$ with $D \subset W$ such that (4) below holds:

(4)   If $A \in \mathcal{D}$ and $A \cap W \neq \varnothing$, then $A \subset H$.

Now, let $V = W \cap G$. Clearly, $V \in T$ and, since $C \subset D \subset W$ and $C \subset G$ (by (2)), $C \subset V$. Hence, the proof of 13.2 will be finished once we prove (*) below:

(*)   If $C' \in \mathcal{C}$ and $C' \cap V \neq \varnothing$, then $C' \subset U$.

To prove (*), let $C' \in \mathcal{C}$ such that $C' \cap V \neq \varnothing$. Let $D' \in \mathcal{D}$ such that $C' \subset D'$. Then, $D' \cap V \neq \varnothing$ and, hence, $D' \cap W \neq \varnothing$. Thus, by (4), $D' \subset H$. Hence, since $C' \subset D'$ and $H = G \cup (X - \overline{G})$,

(5) $C' \subset G \cup (X - \overline{G})$.

Since $C' \cap V \neq \varnothing$, $C' \cap G \neq \varnothing$. Thus, since C' is connected, we see from (5) that $C' \subset G$. Hence, by (3), $C' \subset U$. This proves (*) and, therefore, completes the proof of 13.2.

The following theorem is called a factorization theorem because a given map f is "factored into" two maps (meaning, f is written as a composition of two maps—a more formal definition of a factorization is in 13.58). The theorem is due to Eilenberg [12].

13.3 Monotone-Light Factorization Theorem. Let X and Y be compact metric spaces, and let f be a continuous function from X onto Y. Then, there exist a compact metric space M, a monotone map m of X onto M, and a light map ℓ of M onto Y such that $f = \ell \circ m$.

Proof. Let $\mathcal{D}_f = \{f^{-1}(y): y \in Y\}$, and let $\mathcal{C} = \{C: C \text{ is a component}$ of some $f^{-1}(y) \in \mathcal{D}_f\}$. By 3.21, $\mathcal{D}_f$ is usc. Hence, by 13.2, $\mathcal{C}$ is usc. Therefore, letting M denote $\mathcal{C}$ with the decomposition topology, we have by the second part of 3.21 that M is a compact metric space. Now, let

$m: X \to M$ be the natural map (i.e., $m = \pi$ (3.1))

and define $\ell: M \to Y$ by letting, for each $z \in M$,

$\ell(z) =$ the unique point in $f(m^{-1}(z))$.

For each $x \in X$: $\ell(m(x)) =$ the unique point in $f(m^{-1}(m(x)))$; thus, since $x \in m^{-1}(m(x))$, $\ell(m(x)) = f(x)$. Hence, $f = \ell \circ m$. Also, m is continuous (3.1), ℓ is continuous (by 3.22 since m is a quotient map), and, clearly, $m(X) = M$ and $\ell(M) = Y$. Thus, it remains only to prove that ℓ is light. Let $y \in Y$, and let K be a component of $\ell^{-1}(y)$. By 8.46, $m^{-1}(K)$ is connected. Thus, since

$m^{-1}(K) \subset m^{-1}(\ell^{-1}(y)) = f^{-1}(y),$

$m^{-1}(K) \subset C$ for some $C \in \mathcal{C}$. Hence, noting that $m^{-1}(K)$ is a union of members of $\mathcal{C}$ (see 3.6), we have that $m^{-1}(K) = C$. Thus, K is a one-point set. Therefore, we have proved that ℓ is light. This completes the proof of 13.3.

We note that the factorization in 13.3 is topologically unique in the sense of 13.58.

The value of the Monotone-Light Factorization Theorem comes in part from the following considerations. First, the "middle space" M in 13.3 is sometimes simpler than X or is sometimes a member of a well-studied class of continua. If such is the case, facts about M or the light maps on M can often be used to obtain information about X or the maps on X. Second, when a class of continua is invariant under monotone maps, investigating images of members of that class is, by 13.3, reduced to investigating their light images. Third, special types of maps can sometimes be characterized, by using 13.3, as being compositions of other, between known types of maps; then, facts about the latter types of maps can be used to obtain facts about the former types of maps. We shall illustrate the usefulness of some of these ideas later—other illustrations may be found in, e.g., the proof of II.6 of [10, p. 413] and the proof of 2.3 of [46, p. 188].

We finish this section with a very simple application of 13.3. Recall that we have shown every Peano continuum is a continuous image of $[0,1]$ (8.14).

13.4 Corollary. Every nondegenerate Peano continuum is a light image of $[0,1]$.

Proof. Let Y be a nondegenerate Peano continuum. By 8.14, there is a continuous function f from $[0,1]$ onto Y. Then, by 13.3, $f = \ell \circ m$ where m is a monotone map of $[0,1]$ onto a continuum M and ℓ is a light map of M onto Y. Clearly, M is nondegenerate (since ℓ maps M onto Y). Hence, by 8.22, M is an arc. Thus, there is a homeomorphism h from $[0,1]$ onto M, and we see that $\ell \circ h$ is a light map of $[0,1]$ onto Y. This proves 13.4.

2. OPEN MAPS AND CONTINUOUS DECOMPOSITIONS

Recall that a function f from X onto Y is called an open map provided that f is continuous and f takes open sets in X open open sets in Y (2.18). An important characterization of open maps $f: X \to Y$ between compact metric spaces is that if p and q are close in Y, then $f^{-1}(p)$ and $f^{-1}(q)$ are close in 2^X with the Hausdorff metric (due to Eilenberg [13, p. 174]):

13.5 Theorem. Let X and Y be compact metric spaces. A continuous function f from X onto Y is an open map if and only if

$$f^{-1}(y_i) \to f^{-1}(y) \text{ in } 2^X \text{ whenever } y_i \to y \text{ in } Y.$$

Proof. Assume first that f is an open map. Let $y_i \to y$ in Y. Let $A_i = f^{-1}(y_i)$ for each i, and let $A = f^{-1}(y)$. By a simple sequence argument using the continuity of f, or directly by (b) of 7.15 and 7.16, we have that

(a) $\lim \sup A_i \subset A$.

We show that $A \subset \lim \inf A_i$. Let $z \in A$, and let U be open in X such that $z \in U$. Then, $y = f(z) \in f(U)$ and, since f is an open map, $f(U)$ is open in Y. Thus, since $y_i \to y$, there is an N such that $y_i \in f(U)$ for all $i \geq N$. Hence, $A_i \cap U \neq \varnothing$ for all $i \geq N$. Therefore, we have proved that $z \in \lim \inf A_i$. Hence,

(b) $A \subset \lim \inf A_i$.

Now, by (a) and (b) and by 4.10, $\lim A_i = A$. Thus, by 4.11, $A_i \to A$ in 2^X (with respect to the Hausdorff metric). This proves half of 13.5. To prove the other half, assume that the continuous function f from X onto Y is not an open map. Then, there is an open subset W of X such that $f(W)$ is not open in Y. Since $f(W)$ is not open in Y, there exist $y \in f(W)$ and $y_i \in Y - f(W)$, $i = 1, 2, \ldots$, such that $y_i \to y$ in Y. Let $x \in W$ such that $f(x) = y$. Since $y_i \notin f(W)$ for any i, clearly

$$f^{-1}(y_i) \cap W = \varnothing \qquad \text{for each } i.$$

Thus, since $x \in W$ and W is open in X, $x \notin \lim \sup f^{-1}(y_i)$. Thus, since $x \in f^{-1}(y)$,

$$f^{-1}(y) \not\subset \lim \sup f^{-1}(y_i).$$

Hence, by 4.11, $f^{-1}(y_i) \not\to f^{-1}(y)$ in 2^X. Therefore, since $y_i \to y$ in Y, we have proved the other half of 13.5.

Now, let X be a compact metric space. Let $\mathcal{D}$ be a usc decomposition of X (3.5), and let $\pi: X \to \mathcal{D}$ be the natural map (3.1). We consider two topologies on $\mathcal{D}$–the decomposition topology $T(\mathcal{D})$ in 3.1, and the subspace topology $T_V(\mathcal{D})$ which $\mathcal{D}$ inherits as a subset of 2^X (by 3.8, each $D \in \mathcal{D}$ is a point of 2^X). Recall that $(\mathcal{D}, T(\mathcal{D}))$ is a compact metric space (by the second part of 3.21). Therefore, we may apply 13.5 to see that $T(\mathcal{D}) = T_V(\mathcal{D})$ if and only if $\pi: X \to \mathcal{D}$ is an open map with respect to $T(\mathcal{D})$ (for the easy proof, recall from 4.25 that $T(\mathcal{D}) \subset T_V(\mathcal{D})$). In general, as remarked in 4.25, decompositions for which the decomposition topology is identical to the relativized Vietoris topology are called continuous decompositions. However, this terminology can be motivated from a more conceptual point of view, and we shall do so below. Our discussion leads to the definition in 13.9. As we shall see, the discussion is analogous to, and may be viewed as a continuation of, previous discussions relating usc decompositions to usc functions.

Recall from (a) of 7.15 that a function $F: X \to 2^Y$ is usc if and only if $\{x \in X: F(x) \subset U\}$ is open in X for all U open in Y. Thus, the following definition gives a dual notion to usc.

13.6 Definition of LSC Function. Let X and Y be topological spaces, and let (as in (1) of 4.1).

$2^Y = \{A: A \text{ is a nonempty closed subset of } Y\}$.

A function $F: X \to 2^Y$ is said to be *lower semi-continuous*, written *lsc*, provided that $\{x \in X: F(x) \subset C\}$ is closed in X for all C closed in Y.

The terminology in 13.6 comes about naturally by considering real functions of a real variable (just as does usc–see 3.25): Recall that a function $f: R^1 \to R^1$ is called an *lsc function* provided that

$$f(p) \le \lim_{t \to p} \inf f(t), \quad \text{each } p \in R^1.$$

For a function $f: R^1 \to [0,\infty)$, define $F: R^1 \to 2^{[0,\infty)}$ by

$$F(x) = \{(x,y) \in R^2: 0 \le y \le f(x)\}$$

for each $x \in R^1$. Then, it follows that f is an lsc function if and only if F is lsc in the sense of 13.6. We note that just as a real function $f: R^1 \to R^1$ is continuous if and only if it is both usc (3.25) and lsc, so the following result is true.

13.7 Proposition. Let X and Y be topological spaces, and let $F: X \to 2^Y$. Then, F is continuous with respect to the Vietoris topology T_V (4.24) if and only if F is both usc (7.1) and lsc (13.6).

Proof. The base for T_V in 4.24 is, by 4.4, generated by the sets $\Gamma(U)$ and $\Lambda(U)$ in 4.3. By using these sets, 13.7 follows easily.

Now, recall from the paragraph following 7.1 that a closed partition $\mathcal{D}$ of a space X is a usc decomposition of X if and only if the natural map π (3.1), thought of as a function from X into 2^X, is usc in the sense of 7.1. Thus, the following definition may be considered to be analogous to the notion of a usc decomposition.

13.8 Definition of LSC Decomposition. A closed partition $\mathcal{D}$ of a topological space X is said to be a *lower semi-continuous (lsc) decomposition of X* provided that the natural map π (3.1), thought of as a function from X into 2^X, is lsc in the sense of 13.6.

Now, keeping the discussion above in mind and, in particular, recalling 13.7, we arrive at the following definition naturally.

13.9   Definition of Continuous Decomposition.   A closed partition $\mathcal{D}$ of a topological space X is said to be a *continuous decomposition of X* provided that the natural map π (3.1), thought of as a function from X into 2^X, is continuous with respect to the Vietoris topology T_V (4.24), i.e., provided that $\mathcal{D}$ is both a usc and an lsc decomposition.

Furthermore, we have the following theorem.

13.10   Theorem.   Let X be a compact metric space, and let $\mathcal{D}$ be a usc decomposition of X. Let $T(\mathcal{D})$ denote the decomposition topology on $\mathcal{D}$, and let $T_V(\mathcal{D})$ denote the Vietoris topology on $\mathcal{D}$ which $\mathcal{D}$ inherits as a subspace of 2^X. Then (1)–(3) below are equivalent:

(1) $\mathcal{D}$ is a continuous decomposition;
(2) $\pi: X \to \mathcal{D}$ (3.1) is an open map with respect to $T(\mathcal{D})$;
(3) $T(\mathcal{D}) = T_V(\mathcal{D})$.

Proof.   Since $T(\mathcal{D})$ is the largest topology on $\mathcal{D}$ such that π is continuous (3.1) and since, by 4.25, $T(\mathcal{D}) \subset T_V(\mathcal{D})$, we see immediately from 13.9 that (1) and (3) are equivalent. Therefore, since the ingredients for the proof of the equivalence of (2) and (3) were given in the paragraph following the proof of 13.5, this completes the proof of 13.10.

13.11   Corollary.   Let X and Y be compact metric spaces, and let f be a continuous function from X onto Y. Then, f is an open map if and only if

$$\mathcal{D}_f = \{f^{-1}(y): y \in Y\}$$

is a continuous decomposition of X.

Proof.   Since f is a closed map onto Y, (b) of 3.23 allows us to apply (a) of 3.23 to obtain a (unique) homeomorphism h from Y onto $\mathcal{D}_f$, with the decomposition topology $T(\mathcal{D}_f)$, such that $h \circ f = \pi$. Then, clearly, f is an open map if and only if π is an open map with respect to $T(\mathcal{D}_f)$. Therefore, since (by 3.21) $\mathcal{D}_f$ is a usc decomposition of X, 13.11 now follows from the equivalence of (1) and (2) in 13.10.

We conclude this section with the following observations. Let X be a compact metric space. By 13.11 and 3.21, any open metric image of X is a continuous decomposition of X. Conversely, by 13.10 and 3.9, any continuous decomposition of X is an open metric image of X. In other words: The study of the open metric images of X is equivalent to the study of the continuous decompositions of X. Thus, for example, (c) of 9.46 says that a continuous decomposition of a graph is a graph, and 12.15 says that a nondegenerate continuous decomposition of an arc-like continuum is arc-like. Other such determinations of continuous decompositions come from, e.g., (d) and (e) of 9.46 and from the result we shall obtain in 13.41.

3. INVERSION OF CONNECTED SETS

In general, continuous functions from one continuum onto another do not have the property that each subcontinuum of the range is the image of a subcontinuum of the domain (e.g., consider the map exp in 12.36 restricted to $[0,2\pi]$, or any map of any hereditarily decomposable continuum onto B^2 and use 1.10 and 4.37). However, as we shall see, all open maps between compact metric spaces *do* have this property—in fact, they have the stronger property in the following definition due to Charatonik [4].

13.12   Definition of Confluent Map.   Let S_1 and S_2 be topological spaces. A continuous function $f: S_1 \to S_2$ is said to be *confluent* provided that for any subcontinuum B of S_2 and any component A of $f^{-1}(B)$, $f(A) = B$.

As we shall see in 13.14 and 13.15, confluent maps between compact metric spaces are simultaneously a generalization of open maps and monotone maps. We note the following simple lemma for use in the proof of 13.14.

13.13   Lemma.   Let S_1 and S_2 be topological spaces, and let f be an open map of S_1 onto S_2. If $Z \subset S_2$, then

$$f|f^{-1}(Z): f^{-1}(Z) \to Z$$

is an open map.

Proof.   Let $g = f|f^{-1}(Z)$, and let W be open in $f^{-1}(Z)$. Then, there is U open in S_1 such that $U \cap f^{-1}(Z) = W$. It follows easily that $g(W) = f(U) \cap Z$. Therefore, since f is an open map, $g(W)$ is open in Z. This proves 13.13.

The following theorem is due to Whyburn (1.5 of [48]).

13.14   Theorem.   Every open map of one compact metric space onto another is confluent.

Proof. Let f be an open map of a compact metric space X onto a compact metric space Y. Suppose that f is not confluent. Then, there is a subcontinuum B of Y and a component A of $f^{-1}(B)$ such that $f(A) \neq B$ (13.12). Thus, since $f(A) \subset B$, there exists $p \in B$ such that $A \cap f^{-1}(p) = \emptyset$. Therefore, since A is a component of $f^{-1}(B)$, we have by 5.2 that

$$f^{-1}(B) = G|H \text{ (6.2)} \quad \text{with } A \subset G \text{ and } f^{-1}(p) \subset H.$$

We see that $f(G)$ is closed in B [since G, being closed in $f^{-1}(B)$, is compact], $f(G)$ is open in B [by 13.13 since G is open in $f^{-1}(B)$], $f(G) \neq \emptyset$ [since $G \neq \emptyset$], and $f(G) \neq B$ [since $p \notin f(G)$]. These facts contradict the connectedness of B. Therefore, we have proved 13.14.

For the case of open maps between Peano continua, we shall obtain more information in 13.24. At the moment, let us note the following result and the two definitions which follow it.

13.15 Theorem. Every monotone map of one compact metric space onto another is confluent.

Proof. The theorem follows immediately from 8.46.

13.16 Definition of Weakly Monotone Map. Let X and Y be continua. A continuous function f from X onto Y is said to be *weakly monotone* provided that for any subcontinuum B of Y having nonempty interior in Y and for any component A of $f^{-1}(B)$, $f(A) = B$.

13.17 Definition of Quasi-Monotone Map. Let X and Y be continua. A continuous function f from X onto Y is said to be *quasi-monotone* provided that for any subcontinuum B of Y having nonempty interior in Y, $f^{-1}(B)$ has only finitely many components and each of these components maps onto B under f.

13.18 Proposition. Monotone maps between continua are quasi-monotone, and quasi-monotone maps are weakly monotone. Also, confluent maps are weakly monotone.

Proof. Obvious from definitions (recall 8.46 from the first part).

A notion similar to the ones in 13.16 and 13.17 was first studied by Wallace in [45]. There, he considered only maps between Peano continua and called maps satisfying his notion by the name quasi-monotone. His definition is actually formally closer to 13.16 than to 13.17—for maps between Peano continua, 13.16 and 13.17 are equivalent (13.25) and are equivalent to confluence (13.23) and to Wallace's notion (see 13.63). Nowadays, quasi-monotone has come always to have the meaning given in 13.17.

We now show that weakly monotone maps, hence quasi-monotone maps, of continua onto Peano continua are confluent (13.20 and 13.21).

13.19 Lemma. Let Y be a Peano continuum, and let B be a subcontinuum of Y. Then, there is a sequence $\{B_i\}_{i=1}^{\infty}$ of Peano subcontinua of Y such that $\cap_{i=1}^{\infty} B_i = B$ and, for each i, the interior B_i° of B_i in Y contains B.

Proof. Fix a positive integer i. We define B_i as follows. By the first part of 8.4, Y has property S. Hence, by 8.9, there are finitely many connected open subsets $U_i, \ldots, U_n$ of Y such that for each $k = 1, \ldots, n$

U_k has property S, diameter $(U_k) < 1/i$,

$$U_k \cap B \neq \varnothing, \quad B \subset \bigcup_{i=1}^{n} U_k.$$

Let $B_i = \cup_{k=1}^{n} \overline{U}_k$. Having thus defined B_i for each i, we see that the sequence $\{B_i\}_{i=1}^{\infty}$ has the desired properties (the fact that B_i is a Peano continuum follows easily by first recalling 8.5 to know that each $\overline{U}_k$ is a Peano continuum). This completes the proof of 13.19.

13.20 Theorem. Every weakly monotone map of any continuum onto a Peano continuum is confluent.

Proof. Let f be a weakly monotone map of a continuum X onto a Peano continuum Y. Let B be a subcontinuum of Y, and let A be a component of $f^{-1}(B)$. Let $\{B_i\}_{i=1}^{\infty}$ be as in 13.19. Noting that (since $B_i \supset B$) $f^{-1}(B_i) \supset A$, let A_i denote the component of $f^{-1}(B_i)$ containing A. By 4.18, there is a subsequence $\{A_{i(j)}\}_{j=1}^{\infty}$ of $\{A_i\}_{i=1}^{\infty}$ such that

(1) $A_{i(j)} \to A^*$ where A^* is a subcontinuum of X.

Since each $A_{i(j)} \supset A$, (1) gives us that

(2) $A^* \supset A$

Since each B_i is a continuum and $B_i^{\circ} \neq \varnothing$ (13.19) and since f is weakly monotone, we have that

(3) $f(A_{i(j)}) = B_{i(j)}$ for each j.

Also, note that since $\cap_{j=1}^{\infty} B_{i(j)} = B$,

(4) $B_{i(j)} \rightarrow B$.

By (1) and 4.27, $f(A_{i(j)}) \rightarrow f(A^*)$. Hence, by (3) and (4),

(5) $f(A^*) = B$.

By (2) and (5), $A \subset A^* \subset f^{-1}(B)$. Thus, since A is a component of $f^{-1}(B)$ and, by (1), A^* is a continuum, we must have that $A^* = A$. Hence, by (5), $f(A) = B$. Therefore, we have proved 13.20.

13.21 Corollary. Every quasi-monotone map of any continuum onto a Peano continuum is confluent.

Proof. Follows immediately from 13.18 and 13.20.

Let us note that a confluent map, even an open map, of a continuum onto a Peano continuum need not be quasi-monotone, as the example in 13.37 can be used to show.

We complete this section with the following useful inversion theorem for confluent maps onto Peano continua. Applications of it are, e.g., in the proofs of 13.23 and 13.26.

13.22 Theorem. Let f be a confluent map of a continuum X onto a Peano continuum Y. If U is any connected open subset of Y and H is any component of $f^{-1}(U)$, then $f(H) = U$.

Proof. Let $p \in H$ (we may assume U, hence H, is nonempty), and let $q = f(p)$. There are subcontinua B_i of U, $i = 1, 2, \ldots$, such that $q \in B_i$ for each i and $\bigcup_{i=1}^{\infty} B_i = U$ [since the one-point compactification $U^* = U \cup \{\infty\}$ is a Peano continuum (see the proof of 8.27), we may apply 8.45 to obtain open subsets V_i^* of U^*, $i = 1, 2, \ldots$, such that $\{\infty\} = \bigcap_{i=1}^{\infty} V_i^*$, $U^* - V_i^*$ is connected for each i, and $q \in U^* - V_i^*$ for each i; then, take $B_i = U^* - V_i^*$ for each i]. Noting that $p \in f^{-1}(B_i)$ for each i, let A_i denote the component of $f^{-1}(B_i)$ containing p for each i. Then, since f is confluent,

(1) $f(A_i) = B_i$ for each i.

For each i, since $f^{-1}(B_i) \subset f^{-1}(U)$, clearly $A_i \subset f^{-1}(U)$ and, since $p \in A_i \cap H$, $A_i \cap H \neq \varnothing$. Thus, since each A_i is connected and H is a component of $f^{-1}(U)$, we must have that $A_i \subset H$ for each i. Hence, using (1),

$$f(H) \supset f\left(\bigcup_{i=1}^{\infty} A_i\right) = \bigcup_{i=1}^{\infty} f(A_i) = \bigcup_{i=1}^{\infty} B_i = U$$

and, therefore, $f(H) = U$. This proves 13.22.

We remark that the necessary condition for f to be confluent in 13.22 is also sufficient–see 13.63.

4. CHARACTERIZATIONS OF SPECIAL MAPS ON PEANO CONTINUA

The following result shows that the converse of 13.21 is true provided that we assume the domain of the confluent map is a Peano continuum (comp., comment about 13.37 following the proof of 13.21). This converse result is due to Charatonik [4, p. 215]. The proof we give is somewhat different than the proof in [4] in that we make use of 13.22.

13.23 Theorem. If f is a mapping of a Peano continuum X onto a (Peano) continuum Y, then f is confluent if and only if f is quasi-monotone.

Proof. Assume that f is confluent. Let B be a subcontinuum of Y such that the interior $B°$ of B in Y is nonempty. Then, to prove f is quasi-monotone, it suffices (by recalling the definitions in 13.12 and 13.17) to prove (*) below:

(*) $f^{-1}(B)$ has only finitely many components.

Let $p \in B°$. Since Y is a Peano continuum (8.17), there is a connected open subset U of Y such that $p \in U \subset B°$ (8.1). Then, by 13.22,

(1) $f^{-1}(p) \cap H \neq \emptyset$ for each component H of $f^{-1}(U)$.

Since X is a Peano continuum, each component of $f^{-1}(U)$ is open in X [(ii) of 8.1]. Thus, since $f^{-1}(p)$ is a compact subset of $f^{-1}(U)$,

(2) $f^{-1}(p) \subset \bigcup_{i=1}^{n} H_i$

where each H_i is a component of $f^{-1}(U)$ and $n < \infty$.

Note that by 13.12, and since $p \in B$,

(3) $f^{-1}(p) \cap A \neq \emptyset$ for each component A of $f^{-1}(B)$.

By (1) and (2), we see that $H_1, \ldots, H_n$ must be all the components of $f^{-1}(U)$. Since $H_i \subset f^{-1}(B)$ for each i, we see from (2) and (3) that each component of $f^{-1}(B)$ must intersect, hence contain, at least one H_i. It now follows easily that (*) holds. Thus, we have proved f is quasi-monotone. This proves half of 13.23. Since the other half is in 13.21 (recall 8.17), we have therefore proved 13.23.

We now give two immediate consequences of the result above and some results in the previous section.

13.24 Corollary. Every open map of one Peano continuum onto another is quasi-monotone.

Proof. Use 13.14 and 13.23.

Let us note that for maps between Peano continua, 13.24 is a stronger result than 13.14 by 13.21. Also, it is important in 13.24 that the domain be a Peano continuum (recall the comment following 13.21).

13.25 Corollary. A map between Peano continua is weakly monotone if and only if it is quasi-monotone.

Proof. Half follows from 13.20 and 13.23, and the other half is in 13.18.

Next, we show that the converse of 13.24 is true for light maps between Peano continua. This result, together with the Monotone-Light Factorization Theorem (13.3), will then be used to obtain a characterization of quasi-monotone maps between Peano continua (13.29).

13.26 Theorem. If f is a light map of a Peano continuum X onto a (Peano) continuum Y, then f is a quasi-monotone map if and only if f is an open map.

Proof. To prove the half not in 13.24, assume that f is quasi-monotone (and light). To show that f is an open map, let W be a nonempty open subset of X and let $q \in f(W)$. Let

(1) $p \in W \cap f^{-1}(q)$.

Note that since f is light, $f^{-1}(q)$ is a compact, totally disconnected set (13.1). Hence, using (1), 7.11, and the normality of X, we see that there is an open subset V of X such that (2) and (3) below hold.

(2) $p \in V \subset W$

(3) $Bd(V) \cap f^{-1}(q) = \varnothing$.

Noting that, by (3), $q \in Y - f[Bd(V)]$, let U denote the component of $Y - f[Bd(V)]$ containing q. Noting that, by (1), $p \in f^{-1}(U)$, let H denote the component of $f^{-1}(U)$ containing p. Then, recalling from (ii) of 8.1 that U is open in Y and, from 13.21, that f is confluent, we have by 13.22 that

(4) $f(H) = U$.

Since $U \subset Y - f[Bd(V)]$, we have by (4) that $H \cap Bd(V) = \varnothing$. Also, since $p \in H$ and, by (2), $p \in V$, we have that $H \cap V \neq \varnothing$. Thus, since H is connected, we must have that $H \subset V$. Hence, using (4) and (2),

$$U = f(H) \subset f(V) \subset f(W).$$

Thus, since $q \in U$ and U is open in Y, it follows that we have proved $f(W)$ is open in Y. Therefore, we have proved f is an open map. This completes the proof of 13.26.

The following general proposition and the corollary which follows it will be used several times.

13.27 Proposition. Let X and Y be continua, let f_1 be a continuous function from X onto a continuum Z, let f_2 be a continuous function from Z onto Y, and let $f = f_2 \circ f_1$.

$$\begin{array}{ccc} & f & \\ X & \longrightarrow & Y \\ f_1 \searrow & \circlearrowleft & \nearrow f_2 \\ & Z & \end{array}$$

Then, (1) and (2) below hold:

(1)   If f_1 and f_2 are confluent, then f is confluent.
(2)   If f is confluent, then f_2 is confluent.

Proof. To prove (1), let B be a subcontinuum of Y and let A be a component of $f^{-1}(B) = f_1^{-1}[f_2^{-1}(B)]$. Then, $f_1(A)$ is contained in a component K of $f_2^{-1}(B)$. Note that

$$A \subset f_1^{-1}(K) \quad \text{and} \quad f_1^{-1}(K) \subset f_1^{-1}[f_2^{-1}(B)] = f^{-1}(B).$$

Thus, since A is a component of $f^{-1}(B)$, A must be a component of $f_1^{-1}(K)$. Now, by the confluence of f_1 and f_2,

$$f(A) = f_2[f_1(A)] = f_2[K] = B.$$

This proves (1). To prove (2), let B be a subcontinuum of Y and let L be a component of $f_2^{-1}(B)$. Let M be a component of $f_1^{-1}(L)$. Then, since

$$M \subset f_1^{-1}(L) \subset f_1^{-1}[f_2^{-1}(B)] = f^{-1}(B),$$

M is contained in a component Q of $f^{-1}(B)$. Note that $f_1(Q)$ is a connected subset of $f_2^{-1}(B)$ and that, since $Q \supset M$, $f_1(Q) \cap L \neq \emptyset$. Thus, since L is a component of $f_2^{-1}(B)$, $f_1(Q) \subset L$. Also, since f is confluent, $f(Q) = B$. Hence:

$$B = f(Q) = f_2[f_1(Q)] \subset f_2[L]$$

and, thus, $f_2(L) = B$. This proves (2). Therefore, we have proved 13.27.

13.28 Corollary. If X and Y, hence Z (8.17), in 13.27 are Peano continua, then (1) and (2) of 13.27 remain true when confluent is replaced throughout by quasi-monotone or by weakly monotone.

Proof. Immediate from 13.23, 13.25, and 13.27.

Regarding 13.27 and 13.28, the reader should see 13.61 and 13.62.

The following characterization of a certain type of map by factorization into other special types of maps is but one of a number of such results in the literature.

13.29 Theorem. Let X and Y be Peano continua, and let f be a continuous function from X onto Y. Then, f is quasi-monotone (equivalently, by 13.23 and 13.25, confluent or weakly monotone) if and only if $f = f_2 \circ f_1$ where f_1 is a monotone map of X onto a continuum M and f_2 is a light, open map of M onto Y.

Proof. Assume $f = f_2 \circ f_1$ as above. Then, f_1 is quasi-monotone by 13.18 and, since M is a Peano continuum (8.17), f_2 is quasi-monotone by 13.24. Hence, by (1) of 13.28, f is quasi-monotone. To prove the converse, assume f is a quasi-monotone map. Let $f = \ell \circ m$ be a monotone-light factorization of f with "middle space" M (13.3). Then, since f is quasi-monotone, we have by (2) of 13.28 that ℓ is quasi-monotone. Thus, since ℓ is also light and M is a Peano continuum (8.17), we have by 13.26 that ℓ is an open map. Therefore, $f = f_2 \circ f_1$ where $f_1 = m$ and $f_2 = \ell$ have the desired properties. This completes the proof of 13.29.

The theorem below indicates a way 13.29 can be used. We shall give a specific application of it in 13.31.

13.30 Theorem. A topological property is invariant under quasi-monotone, confluent, or weakly monotone maps on Peano continua if and only if it is invariant under both monotone and light open maps on Peano continua.

Proof. Immediate from 13.29 (and 8.17).

Suppose it is known that a certain type of continuum is invariant under images by all members of certain classes of maps. Then, clearly, it is invariant under images by all mixed finite compositions of the maps in the union of these classes. Thus, we have the following consequence of some results from previous chapters and 13.30.

13.31 Theorem. If X is an arc (or a graph, respectively) and if f is a quasi-monotone or, equivalently (13.23, 13.25), a confluent or weakly monotone map of X onto a nondegenerate continuum Y, then Y is an arc (or a graph, respectively). Also, if X is a simple closed curve, Y is a simple closed curve or an arc.

Proof. By 8.22, 9.45, and (c)–(e) of 9.46, we may apply 13.30 to prove 13.31.

We have shown in 12.14 and 12.15 that the property of being arc-like is invariant under monotone maps and open maps. However, the arc being the only Peano continuum which is arc-like (8.41), we cannot apply 13.29 to conclude, e.g., that nondegenerate confluent images of arc-like continua must be arc-like. In fact, at the present time, it is not known if this is true (Problem 4 of [30]). However, it *is* known that the confluent images of tree-like (hence, of arc-like) continua must be tree-like [37]. We shall prove a special case of this result in 13.40.

We conclude this section by turning our attention briefly to the following general notion due to Eilenberg [11]. Of particular interest is the corollary in 13.36.

13.32 Definition of Multicoherence Degree. Let X be a continuum. Let $\mathcal{C} = \{(A,B): A$ and B are subcontinua of X and $A \cup B = X\}$. For each $(A,B) \in \mathcal{C}$, let $r(A,B)$ denote one less than the number of components of $A \cap B$ (if $A \cap B$ has infinitely many components, let $r(A,B) = \infty$). Then, the *multicoherence degree of X*, denoted by $r(X)$, is defined by letting

$$r(X) = \mathrm{lub}\{r(A,B): (A,B) \in \mathcal{C}\} \quad \text{or} \quad r(X) = \infty$$

depending on whether $\{r(A,B): (A,B) \in \mathcal{C}\}$ is bounded above or not.

It is clear that a continuum X is unicoherent (10.34) if and only if $r(X) = 0$. Also, note, e.g., that $r(S^1) = 1$ and, if X is a theta curve (9.32), $r(X) = 2$.

We now prove that the multicoherence degree of a continuum cannot be raised by a quasi-monotone map. The theorem is due to Eilenberg [11, p. 163]; the proof we give is due to Whyburn (see §8 of [46, p. 164]).

13.33 Theorem. If X is any continuum and f is a quasi-monotone map of X onto a continuum Y, then $r(Y) \leq r(X)$.

Proof. Let $Y = A \cup B$ where A and B are subcontinua of Y. We prove (a) below (notation is as in 13.32):

(a) $r(A,B) \leq r(X)$.

If A or B has empty interior in Y, say A does, then $B = Y$; hence, $A \cap B = A$ is connected, so $r(A,B) = 0 \leq r(X)$ and, thus, (a) holds. Therefore, to prove (a), we may assume that both A and B have nonempty interior in Y. Then, by 13.17, the set $\mathscr{F} = \{K: K$ is a component of $f^{-1}(A)$ or of $f^{-1}(B)\}$ may be enumerated in one-to-one fashion as follows:

$$\mathscr{F} = \{K_1, \ldots, K_m, K_{m+1}, \ldots, K_n\}, \quad n < \infty,$$

where K_i is a component of $f^{-1}(A)$ for $1 \leq i \leq m$ and K_i is a component of $f^{-1}(B)$ for $m + 1 \leq i \leq n$, so $f(K_i) = A$ for $1 \leq i \leq m$ and $f(K_i) = B$ for $m + 1 \leq i \leq n$. Noting that $\cup \mathscr{F} = X$, we have by 1.18 that there exists $\ell, 1 \leq \ell \leq n$, such that $\cup_{i \neq \ell} K_i$ is a subcontinuum of X. Hence, letting $C = \cup_{i \neq \ell} K_i$, we see that K_ℓ and C are subcontinua of X and $K_\ell \cup C = X$. Thus, to prove (a), it suffices to prove

(b) $r(A,B) \leq r(K_\ell, C)$.

Note that (b) says $K_\ell \cap C$ has at least as many components as $A \cap B$ has— hence, to prove (b), it suffices by the continuity of f to prove

(c) $f(K_\ell \cap C) = A \cap B$.

To prove (c), assume that $1 \leq \ell \leq m$, i.e., K_ℓ is a component of $f^{-1}(A)$ (the proof when $m + 1 \leq \ell \leq n$ is similar). Then, $K_\ell \cap K_i = \emptyset$ whenever $1 \leq i \leq m$ and $i \neq \ell$. Thus,

$$K_\ell \cap C = K_\ell \cap \left(\bigcup_{i \neq \ell} K_i \right) = \bigcup_{i \neq \ell} (K_\ell \cap K_i)$$

$$= \bigcup_{i = m+1}^{n} (K_\ell \cap K_i) \subset f^{-1}(A \cap B)$$

and, hence, $f(K_\ell \cap C) \subset A \cap B$. To prove the reverse containment, let $y \in A \cap B$. Then, since $f(K_\ell) = A$, $y = f(x)$ for some $x \in K_\ell$. Since $f(x) = y \in B$, $x \in f^{-1}(B)$ and, hence, $x \in K_j$ for some j such that $m + 1 \le j \le n$. Thus, since $j \ne \ell$, $x \in C$. We now have that $x \in K_\ell \cap C$ and, hence, $y \in f(K_\ell \cap C)$. This proves the reverse containment and, thus, completes the proof of (c). Now: (c) gives (b), (b) gives (a), and, since A and B were arbitrary, (a) proves 13.33.

13.34 Corollary. If X is any unicoherent continuum and f is a quasi-monotone map of X onto a continuum Y, then Y is unicoherent.

Proof. The result is immediate from 13.33 and a comment following 13.32.

13.35 Corollary. If X is any continuum and f is a monotone map of X onto a continuum Y, then $r(Y) \le r(X)$. In particular, if X is unicoherent, Y is unicoherent.

Proof. Use 13.18 to apply 13.33 and 13.34.

We remark that 13.35 can be easily proved in a more direct manner (see 13.65).

13.36 Corollary. If X is a Peano continuum and f is an open map or a confluent map of X onto a (Peano) continuum Y, then $r(Y) \le r(X)$. In particular, if X is unicoherent, Y is unicoherent.

Proof. Recall 8.17, and use 13.23 and 13.24 to apply 13.33 and 13.34.

As an application of 13.35, let us prove that S^n is unicoherent whenever $n \ge 2$. Fix an $n \ge 2$. Then, there is a monotone map of B^n onto S^n (e.g., use the natural map π (3.1) of B^n onto B^n/S^{n-1} (3.14)). Thus, by 13.35, $r(S^n) \le r(B^n)$. Since B^n is contractible, B^n is contractible with respect to S^1; hence, by 12.64, B^n is unicoherent, i.e., $r(B^n) = 0$. Therefore, since $r(S^n) \le r(B^n)$, $r(S^n) = 0$, i.e., S^n is unicoherent. We note that this proves S^n, $n \ge 2$, is contractible with respect to S^1 by using [28, p. 438]. This also follows easily from (d) or (j) of 13.76.

It is important to note the following example in connection with 13.36. The example is from [4, p. 216].

13.37 Example. Even though the multicoherence degree of a continuum cannot be raised by a quasi-monotone map (13.33), we show here that it may be raised by an open, hence confluent map (comp., 13.36). Let X be the continuum pictured in Figure 13.37; X is all points in R^2 having polar coordinates (r,θ) where

$$r = 1, \quad r = 2, \quad \text{or} \quad r = (2 + e^\theta)/(1 + e^\theta), \quad -\infty < \theta < +\infty.$$

Define $f: X \rightarrow S^1$ by letting $f(x) = x/\|x\|$ for all $x \in X$. Then, it is easy to check that f is an open map, X is unicoherent so $r(X) = 0$, and $r(S^1) = 1$.

5. MORE ON CONFLUENT MAPS

Confluent maps are defined in 13.12. We have proved a number of results about them in the last two sections. For the most part, these results concerned relationships between confluent maps and other special types of maps (which were, perhaps, more familiar to the reader because confluent maps have been studied only since 1964). In this section, we present a few results about confluent maps apart from their connection with other types of maps. Of course, any result about confluent maps between compact metric spaces automatically yields results about open maps and monotone maps (by 13.14 and 13.15).

13.38 Lemma. Let X and Y be continua, and let f be a confluent map of X onto Y. If Z is a subcontinuum of Y and K is a component of $f^{-1}(Z)$, then

$$f|K: K \rightarrow Z$$

is confluent.

Proof. Let $g = f|K: K \rightarrow Z$. To prove g is confluent, let B be a subcontinuum of Z and let A be a component of $g^{-1}(B)$. Then, noting that

$$g^{-1}(B) = f^{-1}(B) \cap K$$

and recalling that K is a component of $f^{-1}(Z)$, an easy argument shows that A must be a component of $f^{-1}(B)$. Thus, since f is confluent, $f(A) = B$. Hence, since $g(A) = f(A)$, $g(A) = B$. Therefore, we have proved g is confluent. This proves 13.38.

The following theorem will be used in the proof of 13.40. We note that, by 13.14 and 13.15, it is a stronger result than the one you were asked to prove in 12.66.

13.39 Theorem. Any confluent map of a continuum onto a simple closed curve is essential (in the sense of 12.34).

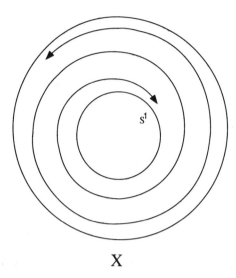

Figure 13.37

Proof. Let *f* be a confluent map of continuum *X* onto S^1 (without loss of generality). Suppose that *f* is inessential. Then, by 12.38, *f* has a continuous log φ, i.e. (12.36), $f = \exp \circ \varphi$. Since *X* is a continuum, φ maps *X* onto a closed and bounded interval *J* in R^1. Let

$$\alpha = \exp | J : J \to S^1.$$

Clearly, $f = \alpha \circ \varphi$ and, since $\varphi(X) = J$ and $f(X) = S^1$, $\alpha(J) = S^1$. Hence, by (2) of 13.27, α is a confluent mapping of the arc *J* onto S^1. This contradicts 13.31. Therefore, *f* must be essential and we have proved 13.39.

By using previous results about monotone and open images, we determined the confluent images of a few special types of continua in 13.31. We now use 13.39 to determine some more such images. We note the following standard terminology: A λ-*dendroid* is an hereditarily decomposable, hereditarily unicoherent continuum. Note that by (a) of 10.58 and 11.54, every dendroid is a λ-dendroid. In particular, then, every dendrite is a λ-dendroid (every dendrite is a dendroid by 8.23 and 10.35).

13.40 Theorem. If *X* is a λ-dendroid and *f* is a confluent map of *X* onto a continuum *Y*, then *Y* cannot contain a simple closed curve. Hence, if *X* is a dendroid (or a dendrite), *Y* is a dendroid (or a dendrite, respectively.).

Proof. Let X and $f: X \to Y$ be as in the first part, and suppose that Y contains a simple closed curve S. Let K be a component of $f^{-1}(S)$, and let

$$g = f|K: K \to S.$$

Then, by 13.38, we may apply 13.39 to see that

(*) $g: K \to S$ is essential.

Since any subcontinuum of a λ-dendroid is obviously a λ-dendroid, K is a λ-dendroid. Hence (*) contradicts (e) of 12.69. Therefore, Y does not contain a simple closed curve. This proves the first part of the theorem. Now, let X be a dendroid. Then, by the confluence of f, (a) of 10.58, and 8.28 we see that every subcontinuum of Y is arcwise connected. Thus, since Y does not contain a simple closed curve (as is proved above), it follows easily that Y is hereditarily unicoherent. Hence, Y is a dendroid (by the definition in 10.58). Finally, let X be a dendrite. Then, since X is a dendroid (by 8.23 and 10.35), Y is a dendroid (as was just proved). Also, by 8.17, Y is a Peano continuum. Hence, by 10.35, Y is a dendrite. This completes the proof of 13.40.

In Theorem XIV of [4, p. 217], Charatonik proves that the confluent images of λ-dendroids must be λ-dendroids. His proof, in effect, uses the first part of 13.40 together with a decomposition theorem of Miller [38, p. 187]. He obtains, as a corollary [4, p. 219], the result about dendroids in 13.40. We remark that these results follow easily from the more recent result in [37] since λ-dendroids are tree-like [8].

We note the following three corollaries.

13.41 Corollary. If X is a dendroid (or a dendrite) and f is a monotone or an open map onto a continuum Y, then Y is a dendroid (or a dendrite, respectively).

Proof. By 13.15 and 13.14, f is confluent. Therefore, 13.41 follows from 13.40.

13.42 Corollary. Any continuous decomposition of a dendroid (or a dendrite) is a dendroid (or a dendrite, respectively).

Proof. Let $\mathcal{D}$ be any such continuous decomposition. Then, since $\mathcal{D}$ is usc (13.9), $\mathcal{D}$ is a continuum by 3.10. Also, by 13.10, $\pi: X \to \mathcal{D}$ is an open map (onto $\mathcal{D}$). Therefore, 13.42 now follows from 13.41.

13.43 Corollary. If X is a tree and f is a confluent map of X onto a continuum Y, then Y is a tree.

Proof. By 13.31, Y is a graph and, by 13.40, Y is a dendrite. Hence, by the definitions in 10.1 and 9.25, Y is a tree. This proves 13.43.

We now know, e.g., that confluent images of graphs, dendrites, and trees are again graphs (13.31), dendrites (13.40), and trees (13.43), respectively. These results do not give a precise delineation of the confluent images of any particular graph, dendrite, or tree. The next theorem is useful for this purpose. Applications of it are discussed after its proof.

Recall the definition of branch point in 10.21. Also, note that a continuum is said to be *hereditarily arcwise connected* provided that each of its subcontinua is arcwise connected. The following theorem is from [10, 1.1].

13.44 Branch Point Covering Theorem. Let X be an hereditarily arcwise connected continuum, and let f be a confluent map of X onto a continuum Y. Then: Y is hereditarily arcwise connected, and the branch point of any simple triod T in Y is the image under f of the branch point of some simple triod T' in X; also, T' may be chosen so that $f(T') = T$.

Proof. The fact that Y is hereditarily arcwise connected follows immediately from the assumption that X is, the confluence of f, and 8.28. To prove the second part of the theorem, let p be the branch point of a simple triod T in Y. Let a_1, a_2, and a_3 be the end points of T (6.25). Let A be the arc in T from a_1 to a_2, and let B be the arc in T from p to a_3. Let K be a component of $f^{-1}(A)$. Since f is confluent, $f(K) = A$. Let $a_1', a_2' \in K$ such that $f(a_1') = a_1$ and $f(a_2') = a_2$. Since X is hereditarily arcwise connected, there is an arc A' in K from a_1' to a_2'. It follows easily that

(1) $f(A') = A$.

Hence, there exists $q' \in A'$ such that $f(q') = p$. Now, noting that $q' \in f^{-1}(B)$, the component L of $f^{-1}(B)$ containing q' exists. Since f is confluent, $f(L) = B$. Let $a_3' \in L$ such that $f(a_3') = a_3$. Since X is hereditarily arcwise connected, there is an arc Q' in L from a_3' to q'. Now, considering Q' to be ordered by $<$ so that $a_3' < q'$, let p' denote the first point of Q' in A'. Then, let B' denote the subarc of Q' from a_3' to p'. Let

$T' = A' \cup B'$.

Clearly, T' is a simple triod with branch point p'. We show $f(p') = p$ and $f(T') = T$. Since $f(L) = B$ and $B' \subset Q' \subset L$, $f(B') \subset B$. Thus, since $f(a_3') = a_3$ and, by (1), $f(p') \in A$, we see that

(2) $f(B') = B$.

Since $p' \in A' \cap B'$, we have using (1) and (2) that

$$f(p') \in f(A' \cap B') \subset f(A') \cap f(B') = A \cap B = \{p\}.$$

Hence, $f(p') = p$. Finally, noting that $T = A \cup B$, we have by (1) and (2) that $f(T') = T$. This completes the proof of 13.44.

   In 13.31, we determined that any nondegenerate confluent image of an arc must be an arc and that any nondegenerate confluent image of a simple closed curve must be a simple closed curve or an arc. Notice how much more transparent these results are using 13.44 and (b) of 8.40. In 13.31, we also showed that any confluent image of a graph is a graph. Notice that, by 13.44, the image graph cannot have more points of order ≥ 3 than the domain graph has (though order of points can be increased—see (b) of 13.64). In particular, it can be seen that any given graph has only finitely many topologically distinct confluent images. We ask the reader to determine these for some special graphs in (a) and (b) of 13.64.
   The reader should notice that we have not discussed a complete determination of the confluent images of any given non-Peano continuum. Not much is known about this (the only such result seems to be the one in [40]); also, see 13.69).
   We finish this section by stating (without proof) and briefly discussing the following theorem of Lelek [29].

   13.45   Theorem.   Let X and Y be continua, and let f be a confluent map of X onto Y. If $g: Y \to S^1$ is continuous such that $g \circ f: X \to S^1$ is inessential, then $g: Y \to S^1$ is inessential.

   Note that 13.45 gives immediately that if X is contractible with respect to S^1 (12.41), then Y is contractible with respect to S^1. More generally, 13.45 says that f induces an isomorphism f^* from $H^1(Y)$ into $H^1(X)$ as defined in 13.76. Thus, Lelek's theorem extends to confluent maps some earlier results in the literature for monotone maps and open maps—see (d) and (e) of 13.76.

6. WEAKLY CONFLUENT MAPS

Let X and Y be compact metric spaces and let f be a continuous function from X onto Y. We shall discuss the three conditions listed below in the light of some previous results and notions.

   (m)   For each subcontinuum B of Y, $f^{-1}(B)$ is a continuum.

   (c)   For each subcontinuum B of Y, each component of $f^{-1}(B)$ maps onto B under f.

   (w)   For each subcontinuum B of Y, some component of $f^{-1}(B)$ maps onto B under f.

Clearly, (m) implies (c) and (c) implies (w). Also, note that (m) means f is monotone by 8.46, (c) means f is confluent by 13.12, and open maps satisfy (c) by 13.14. Thus, (w) is a simultaneous generalization of monotone maps, open maps, and confluent maps. This generalization is a natural continuation of what we have studied since it is the next logically weakest property of maps to consider without imposing conditions on B (as, e.g., in 13.16 and 13.17). We now give a name to maps satisfying (w).

   13.46   **Definition of Weakly Confluent Map.**   Let X and Y be compact metric spaces. A continuous function $f: X \rightarrow Y$ is said to be *weakly confluent* provided that each subcontinuum of Y is the image of a subcontinuum of X under f.

   Let us note that the condition in 13.46 is equivalent to saying that the induced map $\hat{f}: C(X) \rightarrow C(Y)$, defined in 4.27, maps $C(X)$ onto $C(Y)$. In this connection, we refer the reader to the comments following 12.46. Also, note that 12.46 says that every continuous function from a continuum onto an arc-like continuum is weakly confluent (see 13.71).

   The notion of being weakly confluent is general enough so that weakly confluent maps can be used to prove theorems about continua in a number of situations. Part of the value of knowing when there is a weakly confluent map between two continua is best expressed in the following theorem.

   13.47   **Theorem.**   Let X and Y be compact metric spaces, and let $f: X \rightarrow Y$ be weakly confluent. Then any property which is shared by all the nondegenerate subcontinua of X, and which is an invariant of continuous functions, is also shared by all the nondegenerate subcontinua of Y.

   Proof.   Obvious from 13.46.

   13.48   **Corollary.**   Let X and Y be continua, and let $f: X \rightarrow Y$ be weakly confluent. If X is hlc or hereditarily arcwise connected, then Y is hlc or hereditarily arcwise connected, respectively.

   Proof.   Apply 8.17 and 8.28 to 13.47.

One of the most important techniques utilizing weakly confluent maps may be described in general terms as follows: To obtain information about a given compact metric space X or its subcontinua, first show there is a weakly confluent map of X onto a certain compact metric space Y; then, use properties of Y or its subcontinua to obtain information about X or its subcontinua. One of the most beautiful and best illustrations of this is the proof of 13.57. We now obtain results for use in its proof, several of which are of independent interest. The proof may be summarized in relation to the technique just mentioned as follows. Two general results, 13.52 and 13.55, are applied to obtain the existence theorem for weakly confluent maps onto n-cells in 13.56. Then, using that B^2 contains a nondegenerate indecomposable continuum (1.10), 13.56 and 4.37 are used to prove 13.57.

13.49 Lemma. If Y is a Peano continuum, then the set $\mathfrak{I}(Y)$ of all trees in Y is dense in $C(Y)$.

Proof. Let H denote the Hausdorff metric for $C(Y)$ defined specifically as in 4.1. Let $B \in C(Y)$, and let $\epsilon > 0$. Then, by 13.19, there is a Peano subcontinuum P of Y such that $H(B,P) < \epsilon/2$. Let

$$F = \{p_1, \ldots, p_n\}, \qquad n < \infty \text{ and } p_i \neq p_j \text{ for } i \neq j,$$

be an $\epsilon/2$-dense subset of P (i.e., $F \subset P$ and each point of P is within $\epsilon/2$ of a point of F). If $n = 1$, let $T = \{p_1\}$ and observe that $H(B,T) < \epsilon/2$. If $n > 1$, we construct $T \in \mathfrak{I}(Y)$ as follows. By 8.23, there is an arc A_1 in P from p_1 to p_2. If $F \not\subset A_1$, then, by 8.23, there is an arc A_2 is P irreducible from some point of $F - A_1$ to A_1 (in the sense of 11.29). If $F \not\subset A_1 \cup A_2$, then, by 8.23, there is an arc A_3 in P irreducible from some point of $F - (A_1 \cup A_2)$ to $A_1 \cup A_2$. If $F \not\subset A_1 \cup A_2 \cup A_3$, then continue this process until, after finitely many steps, $F \subset \bigcup_{i=1}^k A_i$ ($k \leq n - 1$). Then, let

$$T = \bigcup_{i=1}^k A_i.$$

Clearly, T is a tree (9.25). Since $F \subset T \subset P$ and F is $\epsilon/2$-dense in P, $H(P,T) < \epsilon/2$. Thus, since $H(B,P) < \epsilon/2$, we have that $H(B,T) < \epsilon$. Therefore, it follows that we have proved 13.49.

An *absolute retract* [2] is a metric space M such that whenever M is embedded as a closed subset M' of a metric space Y, M' is a retract of Y (defined in 1.21). An *absolute extensor* is a metric space M such that any continuous function from a closed subset of a metric space Z into M can be extended to a continuous function of Z into M. We note that being an absolute retract is equivalent to being an absolute extensor, even for normal spaces (by, e.g., Theorem 3.2 of [25, p. 84]). We shall follow the prevalent practice of using the phrase "absolute retract" in the statements of results even when the property of absolute retracts used in their proofs is that they are absolute extensors.

Recall the definition of a universal map in 12.23.

13.50 Lemma. Let Q be a compact metric space, let M be an absolute retract, and let $f: Q \to M$ be a universal map. Then, there is a component K of Q such that

$$f|K: K \to M$$

is a universal map.

Proof. Let $\{K_\lambda: \lambda \in \Lambda\}$ denote the collection of all the components K_λ of Q. For each $\lambda \in \Lambda$, let

$$f_\lambda = f|K_\lambda: K_\lambda \to M.$$

Suppose that $f_\lambda: K_\lambda \to M$ is not universal for any $\lambda \in \Lambda$. Then, for each $\lambda \in \Lambda$, there is a continuous function $g_\lambda: K_\lambda \to M$ such that $f_\lambda(x) \neq g_\lambda(x)$ for all $x \in K_\lambda$ (12.23). Since M is an absolute extensor, each map g_λ can be extended to a continuous function $\bar{g}_\lambda: Q \to M$. For each $\lambda \in \Lambda$, $f(x) \neq \bar{g}_\lambda(x)$ for all $x \in K_\lambda$ and, hence, for all x in some open subset W_λ of Q such that $W_\lambda \supset K_\lambda$. Thus, for each $\lambda \in \Lambda$, we see using 5.2 that there is a simultaneously closed and open subset U_λ of Q such that

(*) $K_\lambda \subset U_\lambda$ and $f(x) \neq \bar{g}_\lambda(x)$ for all $x \in U_\lambda$.

Since Q is compact, there are finitely many of these sets U_λ, denoted by $U_{\lambda(1)}, \ldots, U_{\lambda(n)}$, such that $Q = \bigcup_{i=1}^{n} U_{\lambda(i)}$. Let

$$V_1 = U_{\lambda(1)} \quad \text{and} \quad V_j = U_{\lambda(j)} - \bigcup_{i=1}^{j-1} U_{\lambda(i)} \text{ for each } j = 2, \ldots, n.$$

Now, noting that $V_1, \ldots, V_n$ are mutually disjoint, open subsets of Q and that $Q = \bigcup_{j=1}^{n} V_j$, we obtain a well-defined, continuous function $g: Q \to M$ by letting

$$g(x) = \bar{g}_{\lambda(j)}(x) \quad \text{if } x \in V_j.$$

Furthermore, by (*), $f(x) \neq g(x)$ for all $x \in Q$. Hence, $f: Q \to M$ is not universal. This contradicts an assumption in our lemma. Therefore, $f_\lambda: K_\lambda \to M$ must be universal for some $\lambda \in \Lambda$. This proves 13.50.

For use in the proof of 13.52, we extend the lemma in 13.50 as follows.

13.51 Lemma. Let X and Y be compact metric spaces, let $f: X \to Y$ be a universal map, and let $M \subset Y$ be an absolute retract. Then, there is a component K of $f^{-1}(M)$ such that

$$f|K: K \to M$$

is a universal map and, hence, $f(K) = M$.

Proof. Let $Q = f^{-1}(M)$ and let $f_Q = f|Q: Q \to M$. We show that f_Q: $Q \to M$ is universal. To do this, let $g: Q \to M$ be a continuous function. Since M is an absolute extensor, g can be extended to a continuous function $\bar{g}: X \to M$. Then, since $f: X \to Y$ is universal, $f(p) = \bar{g}(p)$ for some $p \in X$ (12.23). Since $\bar{g}(p) \in M$ and $f(p) = \bar{g}(p)$, we have that $f(p) \in M$ and, thus, $p \in Q$. Hence,

$$f(p) = f_Q(p) \quad \text{and} \quad \bar{g}(p) = g(p).$$

Thus, since $f(p) = \bar{g}(p)$, $f_Q(p) = g(p)$. Therefore, we have shown that $f_Q: Q \to M$ is universal. Hence, by 13.50, there is a component K of Q such that $f_Q|K: K \to M$ is universal. Thus, since $K \subset Q$, $f|K: K \to M$ is universal. Hence, by (a) of 12.53, $f(K) = M$. This completes the proof of 13.51.

We now prove the following theorem relating universal maps and weakly confluent maps [42, 2.6].

13.52 Theorem. Every universal map of a compact metric space onto a Peano continuum is weakly confluent.

Proof. Let X be a compact metric space, let Y be a Peano continuum, and let $f: X \to Y$ be a universal map (onto Y by (a) of 12.53). Let $\mathfrak{I}(Y)$ be as in 13.49. We note that each T in $\mathfrak{I}(Y)$ is an absolute retract by, e.g., (iv) of [28, p. 339]. Hence, by 13.51, there is, for each $T \in \mathfrak{I}(Y)$, a subcontinuum K_T of X such that $f(K_T) = T$. In other words,

$$\hat{f}[C(X)] \supset \mathfrak{I}(Y), \quad \hat{f}: C(X) \to C(Y) \text{ as in 4.27.}$$

Thus, since $C(X)$ is compact (4.17) and $\hat{f}$ is continuous (4.27), 13.49 gives us that $\hat{f}[C(X)] = C(Y)$. Therefore, f is weakly confluent (by the comment following 13.46). This proves 13.52.

In regard to the hypothesis in 13.52 that the range be a Peano continuum, see 13.79.

The next several theorems involve the notion of dimension. The only fact from dimension theory which we shall use is the characterization in 13.54. However, in order that the reader not familiar with dimension be able to gain some geometric feeling for the notion, we include the following inductive definition and some simple exercises (13.78).

13.53 Definition of Dimension. All spaces here are separable metric spaces, and n denotes an integer ≥ 0. If X is a space, then the dimension of X is written $\dim(X)$ and, if $p \in X$, the dimension of X at p is written $\dim_p(X)$. First, define $\dim(X) = -1$ to mean $X = \varnothing$. Now, assume inductively that we have defined $\dim_p(X) \leq n - 1$ and $\dim(X) \leq n - 1$ for some $n \geq 0$. Then define $\dim_p(X) \leq n$ to mean p has arbitrarily small open neighborhoods in X whose boundaries have $\dim \leq n - 1$, and define $\dim(X) \leq n$ to mean $\dim_p(X) \leq n$ for all $p \in X$. Now define $\dim_p(X) = n$ to mean $\dim_p(X) \leq n$ and $\dim_p(X) \nleq n - 1$, and define $\dim(X) = n$ to mean $\dim(X) \leq n$ and $\dim(X) \nleq n - 1$. Finally, define $\dim(X) = \infty$ to mean $\dim(X) \nleq n$ for any n.

It is beyond the scope of this book to cover enough dimension theory for a self-contained proof of the theorem stated in 13.54. This is, in fact, the only theorem stated in the main body of any chapter of this book which is not proved and which is used later. Dimension theory is a beautiful subject, and the classical book by Hurewicz and Wallman [26] is such an elegant treatment of the fundamental results that every topologist, if not every mathematician, should read it.

13.54 Proposition. Let X be a separable metric space. Then, $\dim(X) \leq n$ if and only if for each closed subset C of X and each continuous function $f: C \to S^n$, f can be extended to a continuous function $g: X \to S^n$.

Proof. See [26, p. 83].

Recall 12.24. Using this result, we shall see that the following proposition is simply a reformulation of 13.54.

13.55 Proposition. Let X be a separable metric space. Then, $\dim(X) \geq n$ if and only if there is a universal map of X onto B^n.

Proof. If there is a universal map of X onto B^n, then, by 12.24, we see that we can use 13.54 to obtain that $\dim(X) > n - 1$. Conversely, assume that $\dim(X) > n - 1$. Then, by 13.54, there is a closed subset C of X and a continuous function $f: C \to S^{n-1}$ such that

(*) f cannot be extended to a continuous function $g: X \to S^{n-1}$.

Since f maps C into B^n and B^n is an absolute extensor (by (i) of [26, p. 339]), f can be extended to a continuous function $F: X \to B^n$. Note that

$$C \subset F^{-1}(S^{n-1})$$

and, thus, that $F|F^{-1}(S^{n-1})$ is an extension of f mapping $F^{-1}(S^{n-1})$ to S^{n-1}. Hence, by (*),

$$F|F^{-1}(S^{n-1}): F^{-1}(S^{n-1}) \to S^{n-1}$$

cannot be extended to a continuous function from X into S^{n-1}. Therefore, by 12.24, $F: X \to B^n$ is universal. This completes the proof of 13.55.

We now reap the reward of the work which began with 13.49 by obtaining the next two theorems [35].

13.56 Mazurkiewicz's Theorem I.   If X is any compact metric space and dim$(X) \geq n$, then there is a weakly confluent map of X onto B^n.

Proof.   Follows immediately from 13.55 and 13.52.

13.57 Mazurkiewicz's Theorem II.   Any compact metric space of dimension ≥ 2 contains a nondegenerate indecomposable continuum.

Proof.   Let X be a compact metric space such that dim$(X) \geq 2$. By 13.56, there is a weakly confluent map f of X onto B^2. By 1.10, B^2 contains a nondegenerate indecomposable continuum C. There is a subcontinuum K of X such that $f(K) = C$ (13.46). Therefore, by 4.37, K contains a nondegenerate indecomposable continuum. This proves 13.57.

Three comments about 13.57 are important. First, the theorem itself is not about mappings and, thus, its proof is a splendid illustration of the use of mappings to uncover information about the structure of spaces. Second, the theorem attests to the abundance of indecomposable continua, as did 1.17 from a different point of view, and, in doing so, it indicates the important role indecomposable continua must play in continuum theory. Third, a more powerful theorem is available, namely: Every $(n + 1)$-dimensional compact metric space contains an n-dimensional hereditarily indecomposable continuum. This result is Theorem 5 of [1] (together with Theorem VI 8 of [26, p. 94]).

Some exercises which include more information about weakly confluent maps are in 13.66–13.75.

EXERCISES

13.58 Exercise. All spaces considered here are compact metric spaces. A *factorization* of a continuous function f from X onto Y is an ordered triple (M, f_1, f_2) where M is a space (called the *middle space*), f_1 maps X continuously onto M, f_2 maps M continuously onto Y, and $f = f_2 \circ f_1$. If $\mathcal{C}_1$ and $\mathcal{C}_2$ are two classes of maps, then a factorization (M, f_1, f_2) of f is called a $\mathcal{C}_1$-$\mathcal{C}_2$ *factorization* provided that $f_1 \in \mathcal{C}_1$ and $f_2 \in \mathcal{C}_2$. When appropriate, we simply refer to pertinent properties of maps in $\mathcal{C}_1$ and $\mathcal{C}_2$ rather than to $\mathcal{C}_1$ and $\mathcal{C}_2$ (e.g., we speak of a monotone-light factorization (13.3)). A $\mathcal{C}_1$-$\mathcal{C}_2$ factorization (M, f_1, f_2) of $f: X \to Y$ is said to be *topologically unique* provided that if (M', f_1', f_2') is any $\mathcal{C}_1$-$\mathcal{C}_2$ factorization of f, then there is a homeomorphism h of M' onto M such that $f_1 = h \circ f_1'$ and $f_2 = f_2' \circ h^{-1}$. Prove (a) and (b) below.

    (a)   The monotone-light factorization guaranteed by 13.3 is topologically unique.

    (b)   If f is an open map of X onto Y (compact metric spaces), then f has a topologically unique monotone-light open factorization.

13.59 Exercise. All maps here are understood to be continuous functions between continua. One way to generate a new class of maps is to consider two classes $\mathcal{C}_1$ and $\mathcal{C}_2$ of maps, and then let

$$\mathcal{C}_2 \circ \mathcal{C}_1 = \{f_2 \circ f_1 : f_1 \in \mathcal{C}_1 \text{ and } f_2 \in \mathcal{C}_2\}.$$

If $\mathcal{C}_1$ is the class of all monotone maps and $\mathcal{C}_2$ is the class of all open maps, then a member of $\mathcal{C}_2 \circ \mathcal{C}_1$ is called an *OM-map* and a member of $\mathcal{C}_1 \circ \mathcal{C}_2$ is called an *MO-map*. Prove that f is an OM-map if and only if f can be factored as $f = h \circ g$ where g is a monotone map and h is a light, open map (comp., 13.29 where only maps between Peano continua were considered). See 13.60.

13.60 Exercise. We give some examples concerning the notions in 13.59. Define $\alpha: [-2, 2] \to [0, 1]$ by

$$\alpha(t) = \begin{cases} |t| - 1, & 1 \le |t| \le 2 \\ 0, & |t| \le 1 \end{cases}$$

and prove that α is both an MO-map and an OM-map (note that α is neither monotone nor open). Define $\beta: [0, 1] \to [0, 1]$ by

$$\beta(t) = \begin{cases} 3t, & 0 \le t \le 1/3 \\ 2 - 3t, & 1/3 \le t \le 2/3 \\ 0, & 2/3 \le t \le 1 \end{cases}$$

and prove that β is an OM-map but is not an MO-map. [Hint: β is an OM-map by 13.59 and 13.29. Next suppose $\beta = g \circ f$ where f is an open map of $[0,1]$ onto a continuum Z and g is a monotone map of Z onto $[0,1]$. Let $p = f(0)$ and let $q = f(1)$. By (d) of 9.46, Z is an arc and, by (b) of 9.46, p and q are of order $= 1$ in Z. Note that $g(p) = g(q)$, and show that $p = q$ using that g is monotone. Then, letting

$$V = f^{-1}(f(U)) \quad \text{where } U = (2/3,1],$$

show that V is a neighborhood of 0 and $f(V) = \{0\}$.]
    The first example is from [32, 4.10], and the second is from [31, 3.4]. We note that every MO-map is an OM-map [31, 3.2].

13.61 Exercise. Prove 13.27 for quasi-monotone maps, thus generalizing part of 13.28. See 13.62.

13.62 Exercise. Prove part (2) of 13.27 for weakly monotone maps, thus generalizing part of 13.28 (see 13.61). However, part (1) of 13.27 is not valid for weakly monotone maps: Consider the continua X, Y, and Z in the figure drawn below,

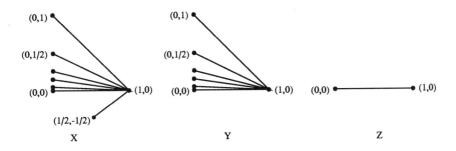

and let f from X onto Y and g from Y onto Z be given by

$$f(x,y) = \begin{cases} (x,y), & y \geq 0 \\ (x,0), & y \leq 0 \end{cases} \quad \text{and} \quad g(x,y) = (x,0).$$

Then, f is quasi-monotone and g is open, thus f and g are weakly monotone (13.14 and 13.18), but $g \circ f$ is not weakly monotone.

The example is from [32, 5.9] (where there is a slight error in the definition of f).

In regard to 13.27, the following more general ideas are of importance. A class $\mathcal{C}$ of maps is said to have the *composition property* provided that if f, $g \in \mathcal{C}$, then $g \circ f \in \mathcal{C}$. A class $\mathcal{C}$ of maps is said to have the *composition factor property* provided that if $f \in \mathcal{C}$ and $f = g \circ h$ (g and h continuous), then $g \in \mathcal{C}$. We have seen many illustrations of both these properties. More are in [32]—see the table in [32, pp. 48–49], and see (c) of 12.52.

13.63   Exercise.   Let f be a continuous function from a continuum X onto a Peano continuum Y. In 13.22, we proved that if f is confluent, then (#) below holds:

(#)   If U is any connected open subset of Y and H is any component of $f^{-1}(U)$, then $f(H) = U$.

Prove, conversely, that if (#) holds, then f is confluent (X and Y as above). [Hint: Observe from the proof of 13.19 that we may assume each B_i° in 13.19 is connected and dense in B_i. Then, use (#) and simple modifications of the proof of 13.20 to show that f is confluent.]

Hence, for maps between Peano continua, (#) is equivalent to being quasi-monotone by 13.23. In [45], Wallace defines quasi-monotone for maps between Peano continua by (#) but for U being any connected subset of Y having nonempty interior in Y. This ends up being equivalent to (#) [45, 2.3].

13.64   Exercise.   Do (a)–(c) below by making use of 13.31, 13.36, 13.43, and 13.44 when appropriate.

(a)   Prove that if X is a simple n-od (9.8) and f is a confluent map of X onto a nondegenerate continuum Y, then Y is a simple k-od with $k \leq n$ and vertex $f(v)$ or Y is an arc.

(b)   Determine all the continua which are confluent images of each of the continua defined in 9.32.

(c)   Determine all the continua which are confluent images of a star (defined in Step 1 of 10.37).

13.65   Exercise.   Prove 13.35 in a simple way only using 8.46 and the definition in 13.32 (and not using 13.33 and 13.34).

13.66   Exercise.   Prove that if X is an hereditarily decomposable continuum and f is a weakly confluent map of X onto a continuum Y, then Y is hereditarily decomposable; hence (assuming Y is nondegenerate), $\dim(Y) = 1$. [Hint: Use 4.37. Recall 13.57.]

13.67 Exercise. Prove that if X is a continuum and $f: X \rightarrow S^1$ is an essential map (12.34), then f is weakly confluent. [Hint: Recall 4.29. Show $\hat{f}: C(X) \rightarrow C(S^1)$ in 4.27 is universal using 12.24, 12.42, and 12.35. Then, apply (a) of 12.53 to $\hat{f}$.]
The result is due to Feuerbacher [16, p. 6] with a different proof. See 13.68 and 13.69.

13.68 Exercise. In relation to 13.67, give the examples asked for in (a) and (b) below and prove (c).
- (a) Show that the converse of 13.67 is false by giving an example of an inessential map of S^1 onto S^1 which is weakly confluent.
- (b) Show that 13.67 cannot be directly extended to non-acyclic graphs by giving an example of an essential map of a figure eight (9.32) onto itself which is not weakly confluent.
- (c) Prove that any continuum can be weakly confluently mapped onto S^1. See 13.69.

13.69 Exercise. We characterize weakly confluent maps of continua onto S^1 as follows. If f is an inessential map of a continuum X onto S^1, then the *angular span of* f, written $AS(f)$, is defined by letting $\varphi: X \rightarrow R^1$ be a continuous log of f (φ exists by 12.38) and letting

$$AS(f) = \text{diameter } (\varphi(X));$$

note that $AS(f)$ is independent of φ by (c) of 12.63. Prove the following theorem:
THEOREM: Let X be a continuum, and let f be a continuous function from X onto X^1. Then, f is weakly confluent if and only if either f is essential or f is inessential with $AS(f) \geq 4\pi$.
[Hint: Use 13.67, 12.46 for the case when Y is a closed and bounded interval, and prove that 13.27 remains true when confluent is replaced throughout by weakly confluent.]
The result is part of J. A. Brooks' master's thesis [3], in which he characterized all weakly confluent maps defined on the $\sin(1/x)$-continuum (1.5) as follows. Let f be a continuous function from the $\sin(1/x)$-continuum X onto a continuum Y; then, f is weakly confluent if and only if one of (1)–(4) holds: (1) Y is an arc, (2) Y is a simple closed curve and $AS(f) \geq 4\pi$, (3) Y is a compactification of $[0, +\infty)$ with an arc as remainder, or (4) Y is a compactification of $[0, +\infty)$ with S^1 as remainder and $AS(f|J) \geq 4\pi$ where $J = \overline{W} - W$ (W as in 1.5). This result generalizes [40]. In connection with this, see [17].

13.70 Exercise. Prove that if X is an arc or a simple closed curve, then a nondegenerate continuum Y is a weakly confluent image of X if and only if Y is an arc or a simple closed curve (in either case). [Hint: Show Y contains no simple triod and apply (b) of 8.40.]

The result is obtained in [10, II.3] as a simple corollary to a branch point covering theorem [10, II.2] for weakly confluent maps on hlc continua similar to 13.44. We note that the part of [10, II.3] which says that the (nondegenerate) weakly confluent image of a fan is a fan or an arc is false since S^1 is such an image by (c) of 13.68. However, even if one adds S^1 to the list, the result is still false as we now show: In [10, III.1], an example is given of a weakly confluent map G of X onto T, where X is the cone over the Cantor Middle-Third set and T is the simple triod given by

$$T = \{(x,y) \in R^2 : y = 0 \text{ and } |x| \le 1, \text{ or } x = 0 \text{ and } 0 \le y \le 1\},$$

such that $G^{-1}((-1,0)) = $ the vertex v of X; now, by attaching a copy Y of X to X at v and another copy Z of X to T at $(-1,0)$, a weakly confluent map is easily obtained (using G on X) from $X \cup Y$ onto $T \cup Z$, and clearly $X \cup Y$ is a fan and, in fact, $X \cup Y$ is homeomorphic to X. In view of this example, it would be of interest to know what continua are weakly confluent images of fans and, in particular, of the cone over the Cantor set. With respect to some other types of images of the cone over the Cantor set, see [5].

We note that it is shown in [10, II.6] that weakly confluent images of graphs must be graphs (comp., 13.31).

13.71 Exercise. From the definition of weakly confluent (13.46), we see that 12.46 says any continuous function from any continuum onto an arc-like continuum is weakly confluent. A continuum Y is said to be in *Class* (W) provided that any continuous function from any continuum onto Y is weakly confluent. Thus, 12.46 says all arc-like continua are in Class (W). Work (a)–(d) below.

(a) If Y is in Class (W), then Y is not a triod (in the sense of 11.22). [Hint: Let Y be a triod, $Y - Z = \bigcup_{i=1}^3 U_i$ (see 11.22). Let L and M be copies of $Z \cup U_1 \cup U_2$ and $Z \cup U_2 \cup U_3$, respectively, such that $L \cap M$ consists of just one point p and p corresponds to a point of U_2. There is a natural map from the continuum $L \cup M$ onto Y which is not weakly confluent.] Comp., (c).

(b) If Y is in Class (W), then Y is unicoherent. [Hint: Similar in spirit to the hint for (a).] Comp., (d).

(c) Find a compactification Y of $[0, +\infty)$ with a simple triod as the remainder so that Y is in Class (W). [Hint: Compactify $[0, +\infty)$ so that every subcontinuum of the remainder is a limit of subcontinua of $[0, +\infty)$.] Hence, continua in Class (W) need not be a-triodic (comp., (a)).

(d) Prove that the continuum Y pictured in 12.68 is in Class (W). Thus, continua in Class (W) need not be hereditarily unicoherent (comp., (b)).

Note that by (a), (b) and 11.34, continua in Class (W) must be irreducible. The idea of studying Class (W) originated in Lelek's topology seminar at the University of Houston on November 8, 1972. Two interesting general characterizations of those continua in Class (W) have now been obtained. One is that Whitney levels in $C(X)$ irreducibly cover X, and the other, which is more geometric, is in terms of C^*-smooth continua. These characterizations are due to Grispolakis and Tymchatyn [19]. The Whitney level property was first studied in [27], but it was Bruce Hughes who first discovered it was connected with Class (W)–see 14.73.21 of [41]. C^*-smooth continua were first studied in Chapter XV of [41]. There are many other, more specialized results about Class (W) which are of interest. A survey of some of them is in [18].

Finally, we remark that notions defined analogously to Class (W) using other special types of maps have been studied. We shall consider one such notion in 13.72.

13.72 Exercise. A continuum Y is said to be in *Class* (C) provided that every continuous function from any continuum onto Y is confluent (comp., 13.71). Prove the following simple characterization for Class (C): A continuum Y is in Class (C) if and only if Y is hereditarily indecomposable. [Hint: First, assume Y is an hereditarily indecomposable continuum. Let f be a continuous function from a continuum X onto Y, let Q be a subcontinuum of Y, and let K be a component of $f^{-1}(Q)$. Assuming, as we may, that $K \neq X$, use 5.5 to obtain subcontinua K_i of X such that $K_i \supset K$, $K_i \neq K$, and $\cap_{i=1}^{\infty} K_i = K$. Show that $f(K_i) \supset Q$ for each i, and use this to prove $f(K) = Q$. Conversely, assume Y is a continuum containing a decomposable subcontinuum A, $A = B \oplus C$ (11.23). Let $p \in B - C$. Let B' be a copy of B such that $B' \cap Y = \varnothing$, and let h be a homeomorphism from B onto B'. Let

$$X = Y \cup_f B' \quad \text{where } f = h|\{p\}: \{p\} \to \{h(p)\} \ (3.18).$$

By 3.20, X is a continuum. Define a continuous function from X onto Y which is not confluent.]

The characterization is due to Lelek and Read [31, 5.7], the "if" part having been done by Cook [7, p. 243].

13.73 Exercise. For use in 13.74 and 13.75, prove that if X is a compact metric space and f is a weakly confluent map of X onto B^2, then $f|f^{-1}(S^1): f^{-1}(S^1) \to S^1$ is essential (12.34).
[Hint: Let $Y = S^1 \cup \mathfrak{S}$ and $r: Y \to S^1$ be as in 12.68. Let $\{Y_i\}_{i=1}^{\infty}$ be a sequence of subcontinua of Y such that $Y_i \supset S^1$ for each i and $\cap_{i=1}^{\infty} Y_i = S^1$. There are subcontinua K_i of X such that $f(K_i) = Y_i$ for each i (13.46). Assume that $\{K_i\}_{i=1}^{\infty}$ converges to a subcontinuum K of X (4.18). By 12.68,

$$r \circ f|K_i: K_i \to S^1 \text{ is essential.}$$

Use this and (a) of 12.69 to show that $r \circ f|K: K \to S^1$ is essential. Note that $r \circ f|K = f|K$ and, hence, the result now follows easily (use 12.35).]

13.74 Exercise. Prove (a) below which includes a partial converse of 13.52 and which shows that, for compact metric spaces which are contractible with respect to S^1, the weakly confluent maps onto B^2 are the same as the better known Alexander-Hopf maps onto B^2 (see the paragraph preceding 12.24). Then, do (b) and, as an application of (a), prove (c).

(a) If X is a compact metric space which is contractible with respect to S^1 and f is a continuous function from X onto B^2, then f is universal if and only if f is weakly confluent. [Hint: Half is by 13.52. For the other half, use 13.73 and 12.24 recalling 12.35.]

(b) Show that the equivalence in (a) is false for maps onto B^3 by giving an example of a monotone map of B^3 onto B^3 which is not universal. [Hint: B^3 is the cone over both B^2 and S^2.]

(c) If X is a compact subset of B^2 which is contractible with respect to S^1 and if f is a weakly confluent map of X onto B^2, then f has a fixed point. [Hint: Trivial using (a).]

The special case of (c) for open maps on unicoherent Peano continua in B^2 was proved in [20]. The material here (and in 13.73) is from [42].

13.75 Exercise. Prove that if X is a compact metric space and $\dim(X) \geq 2$, then X contains a subcontinuum which is not contractible with respect to S^1. [Hint: Use 13.56, 13.73, and 12.69.]

13.76 Exercise. This exercise is concerned mostly with the discussion following 13.45. After giving appropriate definitions and notation, you shall be asked to prove three special cases of 13.45 ((d)–(f) below).

For any topological space Z, let $C(Z,S^1)$ be the set of all continuous functions from Z to S^1, and let $I(Z,S^1) = \{f \in C(Z,S^1): f \text{ is inessential}\}$

(12.34). We consider $C(Z,S^1)$ as being a commutative group under pointwise complex multiplication.

(a) Prove that $I(Z,S^1)$ is a subgroup of $C(Z,S^1)$.

Now, using (a), let $H^1(Z) = C(Z,S^1)/I(Z,S^1)$, the quotient group. If $C(Z,S^1) = I(Z,S^1)$, then we write $H^1(Z) \approx 0$. Note that $H^1(Z) \approx 0$ means that Z is contractible with respect to S^1 (by 12.41). We also remark that the elements of $H^1(Z)$ are in 1-1 correspondence with the elements of the first Čech cohomology group of Z with integer coefficients when Z is a paracompact normal space [8, p. 226].

Let X and Y be topological spaces, and let $f: X \to Y$ be continuous. Define $\bar{f}: C(Y,S^1) \to C(X,S^1)$ by letting $\bar{f}(g) = g \circ f$ for all $g \in C(Y,S^1)$.

(b) Prove that $\bar{f}$ is a homomorphism of $C(Y,S^1)$ into $C(X,S^1)$ and that $\bar{f}(I(Y,S^1)) \subset I(X,S^1)$.

By (b), there is a natural induced homomorphism $(\bar{f})^*$, denoted simply by f^* (not to be confused with f^* in 4.27), of $H^1(Y)$ into $H^1(X)$; f^* is defined by letting $f^*([g]) = [\bar{f}(g)]$ for each $[g] \in H^1(Y)$.

(c) Prove that f^* is an isomorphism of $H^1(Y)$ into $H^1(X)$ if and only if whenever $g \circ f \in I(X,S^1)$ for some $g \in C(Y,S^1)$, then $g \in I(Y,S^1)$.

Note that, by (c), 13.45 says that if f is any confluent map of a continuum X onto a continuum Y, then f^* is an isomorphism of $H^1(Y)$ into $H^1(X)$. We shall not prove this theorem, but you are asked to do (d)-(f) below.

(d) Prove that if f is a monotone map of a continuum X onto a continuum Y, then f^* is an isomorphism of $H^1(Y)$ into $H^1(X)$. [Hint: To use (c), let $g \in C(Y,S^1)$ such that $g \circ f \in I(X,S^1)$. By 12.38, $g \circ f$ has a continuous log φ. Show that 3.22 can be applied to $\varphi \circ f^{-1}$, use 12.38 again, and then apply (c).]

(e) Same as (d) but for f an open map of X onto Y. [Hint: Start with g and φ as in the hint in (d). For each $y \in Y$, let

$$\alpha(y) = \text{glb}\{\varphi(x): x \in f^{-1}(y)\}.$$

Show $g(y) = \exp(\alpha(y))$ for each $y \in Y$, and use 13.5 to prove α is continuous. Then, use 12.38 and apply (c).]

(f) Prove 13.45 for the case when X and Y are Peano continua. [Hint: Use (d), (e), and 13.29.] Note that, on using the hint, we have proved 13.45 for arbitrary continua X and Y when f is an OM-map or an MO-map (defined in 13.59).

As simple applications of (d)-(f), prove (g)-(i) below.

(g) $H^1(S^n) \approx 0$ for any $n \geq 2$. [Hint: This was proved in the paragraph following 13.36—however, a result in [28, p. 438] was used which we have not proved. Recall other facts from the same paragraph which allow you to apply (d) to obtain the result.]

(h) $H^1(P^n) \approx 0$ for any $n \geq 2$, where P^n is projective n-space (3.11). [Hint: Use (e) and (g).]

(i) There is no confluent map of S^n for any $n = 1, 2, \ldots$ onto the torus $S^1 \times S^1$ or the solid torus $S^1 \times B^2$. [Hint: For $n = 1$, use 13.31. For $n \geq 2$, recall (1) of the proof of 12.39 to know $H^1(S^1) \neq 0$. Use this and (d) to see that $H^1(S^1 \times S^1) \neq 0$ and $H^1(S^1 \times B^2) \neq 0$. Then, use (c), (f), and (g).]

Prove (j) below.

(j) If X is a continuum such that $X = Y \cup Z$ where Y and Z are closed in X, $H^1(Y) \approx 0$, $H^1(Z) \approx 0$, and $Y \cap Z$ is connected, then $H^1(X) \approx 0$. [Hint: Use 12.38 and (c) of 12.63.] Note that this result can be used to prove (g).

We remark that (d) and (e) for the case when $H^1(X) \approx 0$ go back to Eilenberg [13, p. 165 and p. 174]. We also remark that (j) is a special case of the exactness of the Mayer–Vietoris cohomology sequence [15, p. 43].

13.77 Exercise. An *n-to-1 map* is a continuous function f from X onto Y such that for each $y \in Y$, $f^{-1}(y)$ consists of exactly n points (n being a fixed, positive integer). Work (a)–(d) below.

(a) Give an example of a 3-to-1 map of $[0,1]$ onto S^1. (We remark there is no 2-to-1 map of $[0,1]$ onto any continuum [22].)

(b) Prove that if X is a continuum with the fixed point property, then there is no 2-to-1 open map of X onto any continuum Y. [Hint: Use 13.5.]

(c) Prove that if Y is a continuum which is not hereditarily unicoherent, then, for each n, there is an n-to-1 map of some continuum X_n onto Y. [Hint: Make use of 3.18 and 3.20 to construct X_n so the map is natural.] This is used to characterize dendrites in [43].

(d) Prove that if there is a continuous function f from a continuum X onto a continuum Y such that $f^{-1}(y)$ has exactly n components for each $y \in Y$, then there is an n-to-1 map of some continuum onto Y. [Hint: Use 13.3.]

Some of many papers on n-to-1 maps are [6], [21]–[24], [39], [43] where (c) and the idea for (d) are, and [44]. In a related direction, see [47]. Also, see [46, pp. 189–191].

13.78 Exercise. This exercise is included to illustrate the notion of dimension (13.53) for the reader who has not encountered this notion before. Prove (a)–(f) below using only 13.53 and the hints. All spaces are separable metric spaces.

(a) Dim(X) = 0 if X is a nonempty countable space, a Cantor set (use 7.11), the space J of irrational numbers in R^1, $J \times J$, or the set of all $(x,y) \in R^2$ with exactly one of x or y being irrational.

(b) Dim(R^1) = 1 and dim(R^n) $\leq n$. It is true, but hard to prove, that dim(R^n) = n for $n \geq 2$.

(c) If X is a nondegenerate continuum, then dim(X) $\geq$ 1. [Hint: Recall (f) of 5.17.].

(d) If dim(X) = 0, then X is totally disconnected (7.10). [Hint: First prove that any subspace of a 0-dimensional space is 0-dimensional.] See (e).

(e) If X is compact, then dim(X) = 0 if and only if X is totally disconnected. [Hint: Half is from (d). For the other half, use 5.2.] The converse of (d) is false [26, p. 22].

(f) If X is a nondegenerate rational continuum (10.57), then dim(X) = 1. [Hint: Use (a) and (c).] In particular, if X is a nondegenerate dendrite, dim(X) = 1 by 10.20.

We note that the collection $\mathfrak{C} = \{C\colon C$ is closed in X and dim(C) $\leq n - 1\}$ is an additive–hereditary system in the sense of 10.17 (see Theorem III.1 of [26, p. 26] and Theorem III.2 of [26, p. 30]). Thus, 10.18 gives that if X is a continuum, dim(X) $\leq n$ if and only if any two points of X are separated in X by a closed subset of X of dim $\leq n - 1$ (see A of [26, p. 36]).

13.79 Exercise. Recalling 13.52, show the necessity of assuming the range is a Peano continuum by giving an example of a universal map of the continuum in 1.5 onto the continuum in 1.6 which is not weakly confluent.

REFERENCES

1. R. H. Bing, Higher-dimensional hereditarily indecomposable continua, Trans. Amer. Math. Soc., 71(1951), 267–273.

2. Karol Borsuk, *Theory of Retracts*, Monografie Mat., Vol. 44, Polish Scientific Publishers, Warszawa, Poland, 1967.

3. Jeffrey A. Brooks, Weakly confluent images of the sinusoidal curve, Master's Thesis, West Virginia University, 1989; director: Sam B. Nadler, Jr.

4. J. J. Charatonik, Confluent mappings and unicoherence of continua, Fund. Math., 56(1964), 213–220.

5. J. J. Charatonik and W. J. Charatonik, Images of the Cantor fan, Topology and its Appls., 33(1989), 163–172.

6. Paul Civin, Two-to-one mappings of manifolds, Duke Math. J., 10(1943), 49–57.

7. H. Cook, Continua which admit only the identity mapping onto nondegenerate subcontinua, Fund. Math., 60(1967), 241–249.

8. H. Cook, Tree-likeness of dendroids and λ-dendroids, Fund. Math., 68(1970), 19–22.

9. C. H. Dowker, Mapping theorems for non-compact spaces, Amer. J. Math., 69(1947), 200–242.

10. C. A. Eberhart, J. B. Fugate, and G. R. Gordh, Jr., Branchpoint covering theorems for confluent and weakly confluent maps, Proc. Amer. Math. Soc., 55(1976), 409–415.

11. Samuel Eilenberg, Sur les espaces multicohérents, I., Fund. Math., 27(1936), 153–190.

12. Samuel Eilenberg, Sur les transformations continues d'espaces métriques compacts, Fund. Math., 22(1934), 292–296.

13. Samuel Eilenberg, Sur les Transformations d'espaces métriques en circonférence, Fund. Math., 24(1935), 160–176.

14. Samuel Eilenberg, Transformations continues en circonférence et la topologie du plan, Fund. Math., 26(1936), 61–112.

15. Samuel Eilenberg and Norman Steenrod, *Foundations of Algebraic Topology*, Princeton Univ. Press, Princeton, N.J., 1952.

16. Gary A. Feuerbacher, Weakly chainable circle-like continua, Fund. Math., 106(1980), 1–12.

17. E. E. Grace and Eldon J. Vought, Semi-confluent and weakly confluent images of tree-like and atriodic continua, Fund. Math., 101(1978), 151–158.

18. J. Grispolakis and E. D. Tymchatyn, Continua which admit only certain classes of onto mappings, Top. Proc., 3(1978), 347–362.

19. J. Grispolakis and E. D. Tymchatyn, Weakly confluent mappings and the covering property of hyperspaces, Proc. Amer. Math. Soc., 74(1979), 177–182.

20. O. H. Hamilton, Fixed point theorems for interior transformations, Bull. Amer. Math. Soc., 54(1948), 383–385.

21. O. G. Harrold, Jr., Exactly $(k,1)$ transformations on connected linear graphs, Amer. J. Math., 62(1940), 823–834.

22. O. G. Harrold, Jr., The non-existence of a certain type of continuous transformation, Duke Math. J., 5(1939), 789–793.

23. Jo Heath, Every exactly 2-to-1 function on the reals has an infinite set of discontinuities, Proc. Amer. Math. Soc., 98(1986), 369–373.

24. Jo W. Heath, k-to-1 functions on arcs for k even, Proc. Amer. Math. Soc., 101(1987), 387–391.

25. Sze-Tsen Hu, *Theory of Retracts*, Wayne State Univ. Press, Detroit, Mich., 1965.

26. Witold Hurewicz and Henry Wallman, *Dimension Theory*, Princeton Univ. Press, Princeton, N.J., 1948.

27. J. Krasinkiewicz and Sam B. Nadler, Jr., Whitney properties, Fund. Math., 98(1978), 165–180.

28. K. Kuratowski, *Topology*, Vol. II, Acad. Press, New York, N.Y., 1968.

29. A. Lelek, On confluent mappings, Colloq. Math., 15(1966), 223–233.

30. A. Lelek, Some problems concerning curves, Colloq. Math., 23(1971), 93–98.

31. A Lelek and David R. Read, Compositions of confluent mappings and some other classes of functions, Colloq. Math., 29(1974), 101–112.

32. T. Maćkowiak, Continuous mappings on continua, Dissertationes Math., 158(1979).

33. T. Maćkowiak and E. D. Tymchatyn, Continuous mappings on continua II, Dissertationes Math., 225(1984).

34. Venable Martin and J. H. Roberts, Two-to-one transformations on 2-manifolds, Trans. Amer. Math. Soc., 49(1941), 1–17.

35. Stefan Mazurkiewicz, Sur l'existence des continus indécomposables, Fund. Math., 25(1935), 327–328.

36. Louis F. McAuley (editor), *Monotone Mappings and Open Mappings*, SUNY at Binghamton, 1970.

37. T. Bruce McLean, Confluent images of tree-like curves are tree-like, Duke Math. J., 39(1972), 465–473.

38. Harlan C. Miller, On unicoherent continua, Trans. Amer. Math. Soc., 69(1950), 179–194.

39. J. Mioduszewski, On two-to-one continuous functions, Dissertationes Math., 24(1961).

40. Sam B. Nadler, Jr., Confluent images of the sinusoidal curve, Houston J. Math., 3(1977), 515–519.

41. Sam B. Nadler, Jr., *Hyperspaces of Sets* Monographs and Textbooks in Pure and Applied Math., vol. 49, Marcel Dekker, Inc., New York, N.Y., 1978.

42. Sam B. Nadler, Jr., Universal mappings and weakly confluent mappings, Fund. Math., 110(1980), 221–235.

43. Sam B. Nadler, Jr., and L. E. Ward, Jr., Concerning exactly $(n,1)$ images of continua, Proc. Amer. Math. Soc., 87(1983), 351–354.

44. J. H. Roberts, Two-to-one transformations, Duke Math. J., 6(1940), 256–262.

45. A. D. Wallace, Quasi-monotone transformations, Duke Math. J., 7(1940), 136–145.

46. Gordon Thomas Whyburn, *Analytic Topology*, Amer. Math. Soc. Colloq. Publ., Vol. 28, Amer. Math. Soc., Providence, R.I., 1942.

47. Gordon T. Whyburn, Characterizations of certain curves by continuous functions defined upon them, Amer. J. Math., 55(1933), 131–134.

48. G. T. Whyburn, Interior transformations on compact sets, Duke Math. J., 3(1937), 370–381.

Special Symbols

Index